3D and HD Broadband Video Networking

For a listing of recent titles in the
Artech House Telecommunications Library,
turn to the back of this book.

3D and HD Broadband Video Networking

Benny Bing

ARTECH
HOUSE

BOSTON | LONDON
artechhouse.com

Library of Congress Cataloging-in-Publication Data
A catalog record for this book is available from the U.S. Library of Congress.

British Library Cataloguing in Publication Data
A catalogue record for this book is available from the British Library.

ISBN-13: 978-1-60807-051-0

Cover design by Vicki Kane

685 Canton Street
Norwood, MA 02062

10 9 8 7 6 5 4 3 2 1

To my Mother.

Contents

Preface

Streaming live and on-demand digital video content over the Internet and in telecommunications and broadcast networks is prevalent. In addition to broadband service providers (e.g., cable, digital subscriber line, satellite, digital video broadcast), Web content providers, including video aggregators, have increasingly large volumes of video on their sites and are making them more discoverable, helping drive usage and ad revenue. A large number of online marketing initiatives now employ video to help promote products in a far more enriching, entertaining, and informative manner than typical 30-second TV slots allow. The last digital island, the TV, is finally joining the PC and mobile phone as an Internet connected device. Leading vendors such as Sony announced that 90% of their HDTVs will be broadband-enabled in the future. Panasonic and LG have both released Skype-enabled HDTVs with embedded HD webcams, and video codecs and processors. These developments will have a profound impact on the distribution and consumption of digital media.

Online video companies have raised nearly half a billion dollars in new capital for 2009. The vast potential of the market is only evident since the two years or so and this potential is matched by impressive statistics—nearly 250 billion online video views were reported for 2009. Investors recognize the low cost of deployment and pervasiveness of the service will dramatically change the way consumers access video entertainment and the way providers and advertisers compete. More significantly, the Internet can indeed deliver crystal-clear, high-quality video on a big screen, comparable to payTV service but without the inconvenience of appointment-based viewing. Even retail giants such as Best Buy, Sears, and Walmart are joining the online video ecosystem. To counter the online TV revolution, major cable, telephone, and satellite companies have also started to place premium content online (e.g., TV shows, sports, movies), just like many content owners and distributors (e.g., CBS, ESPN3, Starz, Netflix, Hulu).

This book addresses the key challenges facing cable, DSL, and wireless providers in delivering high quality next-generation video and discusses solutions to enhance customer satisfaction via improved quality of experience and service. It describes important techniques that can be exploited to enhance video transmission over a broad range of networks: bandwidth-constrained managed private networks (e.g., payTV networks), error-prone over-the-air broadcast networks (e.g., terrestrial wireless and satellite networks), as well as unmanaged online networks (i.e., the public Internet) that often lose packets due to network congestion. The applications are wide-ranging—in addition to payTV service, the techniques can be adapted to maximize the bandwidth utilization of diverse video transport platforms such as video-on-demand, mobile video, digital video broadcast, Internet-enabled TV, third-party and user-generated video streaming. This will, in turn, benefit service providers by enabling more channels or videos to be carried in their channel lineup and help attract new customers or give existing customers more value for their subscriptions.

While the Internet has been a fountain of information, a busy marketplace, a thriving social scene, it is becoming the epicenter of global entertainment and a clear migration towards various forms of IP video is emerging. Accompanying this trend are various challenges of transporting bandwidth-intensive video traffic. Internet traffic today contains a significant amount of compressed video traffic, dominated by popular sites like YouTube and Hulu. While a large volume of Internet traffic is still taken up by peer-to-peer (p2p) applications (according to the Ipoque Internet Study, http://www.ipoque.com/resources/internet-studies/internet-study-2008_2009), in some cases, more than 70%, p2p usage has been declining in favor of streaming media. This is due to the significant increase in online video portals serving billions of video streams, including HD video. Hulu, for example, currently streams over 1 billion videos in a single month whereas YouTube serves over 13 billion streams per month. With increasing Internet penetration and greater availability of high-quality videos online, including 3D HD videos, the proportion of streaming traffic will likely increase in the future.

As I was working diligently on this book during the final weeks of December 2009, I attempted to locate some popular charity songs released in the mid-1980s, which I have not heard for quite a while. Despite my best efforts, I could not find these songs from my satellite service. I was however, able to locate and stream the music videos from YouTube almost instantly and replay them whenever I want to. I had the bonus of viewing the videos for the first time, in addition to listening to the songs. Thus, while many innovative technologies have changed the last decade at an incredible pace with iPod, Skype, Flash, podcasts, Wi-Fi, iPhone, e-commerce, Facebook (bringing over 550 million users or nearly 10% of the world's population together), social media, broadband-HDTV, Blu-ray, and Wii, I couldn't be more excited about what online media content will bring to the next decade. Already, the *Avatar* movie started the ball rolling by creating a huge demand for 3D movies. 3D sports channels were next. Google introduced the Android-based Nexus One smartphone that works with multiple wireless carriers and plans to launch a tablet computer to rival the iPad. Sprint released the Evo 4G Android smartphone that allows HD video calling. Perhaps the Android Internet set-top and HDTV isn't too far behind. Looking forward, one thing is certain—I will continue with online and digital TV services. Looking back, one botch springs to my mind—bing.com beat me to using my last name for my startup company.

I was humbled to receive the 2010 National Association of Broadcasters (NAB) Technology Award. The award recognizes organizations that bring exhibits and demonstrations of significant merit to the NAB Show, presenting advanced research and development projects in communications technologies. It quickly brought my attention to the people who have supported my work in the past few years. Special mention should go to Vince Groff for his sharp technical and business insights and to Jeff Finkelstein for bridging the gap between research and practical reality. I also wish to acknowledge the excellent work performed by my team of graduate students. I am grateful for the support shown by the editorial staff at Artech House—it has been 10 years since the publication of my first book that was adopted by Cisco Systems and I sincerely hope this new book will be equally successful.

As an engineer, I always believe the work that I do should be of the highest utility. I hope this book fulfils this important objective and equips you with the technical expertise for bandwidth-efficient video networking. At the same time, I hope the thought-provoking exercises challenge you to think about how the associated technologies will evolve in the future and enable you to fully master the concepts presented in the text. There are supplementary materials to complement the book. An online tutorial is posted in the IEEE Educational Activities and the IEEE Communications Society Tutorial Now websites. The solutions to the exercises, an additional set of exercises, and a set of slides for each chapter can be made available to instructors who adopt the book as a class text. In addition, if you would like to see the following demos or learn about my current projects, feel free to drop me a note.

- 1080p HD streaming at 1.5 Mbps on an HDTV with a high-speed Internet connection
- 720p HD streaming at 700 Kbps on a laptop with a 4G wireless connection
- 480p SD streaming at 300 Kbps on an Android smartphone with a 3G wireless connection
- Bandwidth-efficient 3D HD video streaming
- Performance comparison of raw streaming, progressive streaming, and frame smoothed streaming
- HD video quality improvement via error resilience and error concealment
- Removal of error resilience at the receiver so that error concealment can occur on any regular decoder/player
- Touch-free and wireless human-TV interface enabled by hand gestures and a single webcam (supports all brands of webcams, including 3D webcams)
- Android Internet set-top and HDTV with personalized smartphone control
- Interactive gesture-based smartphone set-top
- Suspicious activity detection for video surveillance applications
- Human fall detection for telehealth applications
- Impact of information loss on decoded H.264 video quality

As you continue to read the remaining chapters, reflect upon these emerging trends as time- and place-shifted TV viewing becomes pervasive. You are also welcomed to send me your thoughts and feedback at bennybing@yahoo.com as we move into an exciting online entertainment decade.

- The shift to subscription services like Netflix, which added almost 3 million subscribers in 6 months, growing its subscriber base by 26% to 14 million subscribers

- The decision by Hollywood Video, the second largest operator of video rental stores, to close the last of its movie rental stores, which once totaled over 2,400
- YouTube's nascent move into movie rentals
- The launch of Google TV
- The delivery of 3D TV to the living room
- The rise of TV and movie viewership on smartphones
- The emergence of a fourth screen, the tablet PC
- The proliferation of online HD video delivery by CDNs
- The massive investments on online video advertising, which will help long-form content monetization

Chapter 1

Empowering High-Quality Digital Video Delivery

The Internet is poised to support increasing video traffic load. It has become the key distribution platform for delivering TV and has opened up new ways for discovering, sharing, and consuming media content in and out of the home. Online TV providers are trumping payTV providers with a dramatic increase in subscription and advertising revenue in recent years. Over 80% of Internet users now watch video while 20% of these users watch TV. The number of Americans watching online TV shows has doubled in the last 2 years. In 2009, some 40 million households worldwide watch online video regularly on their TV sets. The service is either free or cheaper than payTV, which is the main consideration for many consumers. Online TV makes truly global events possible, reaching millions of consumers all around the globe. The popularity of online TV is accompanied by several challenges facing service and content providers. For instance, how can the bandwidth crunch associated with online video transmission be addressed? The available bandwidth must also support other content-rich services such as peer-to-peer applications, multiplayer games with audio/video chat, streaming media, on-line collaboration, video-on-demand, and interactive TV. In this chapter, we evaluate some game-changing trends in broadband TV.

1.1 Wither the Set-Top Box and Digital Video Recorder?

Over 80 million U.S. households subscribed to payTV services of cable TV companies, telcos, and satellite TV providers at the end of 2008. However, the Internet is becoming more popular than digital video recorders (DVRs) for accessing on-demand TV content. With content providers and consumer electronics vendors teaming to enable online or over-the-top (OTT) delivery of TV services directly to the TV (thus enabling a host of new Internet-based video services, including Skype video calling on the living room TV), proprietary set-top boxes (STBs) are no longer the only mode of TV connection. This partnership yields a highly differentiated "product"; for instance, zero upfront costs to the consumer and the convenience of anytime, anyplace access to TV content. With the exception of live shows and sports programs, allowing consumers to watch TV content without a predetermined schedule is a definite plus point. Nevertheless,

Google, Intel, Sony, and Logitech have teamed up to develop an Android platform called Google TV (http://www.google.com/tv) to bring the Internet into the living room via a new generation of HDTVs and STBs [1]. The open and Web-based Google TV box will have a browser, bringing all of the Internet to the TV, including the full range of online video Web sites.

1.2 The Rise of HD and 3D Video

High-definition (HD) video is growing in importance and popularity, as evidenced by the intense competition between satellite, cable, and telcos to offer the highest number of HD channels. Google Trends also reveals a higher number of search results for HD over standard definition (SD), as shown in Figure 1.1. YouTube started offering HD-quality videos in December 2008. Emerging wireless home network standards are geared toward HD (e.g., IEEE 802.15 Task Group 3c, IEEE 802.11ac/ad, and IEEE 802.11aa). With the transition from analog to digital TV, HDTVs now come equipped with in-built digital tuners and hardware video decoders, facilitating over-the-air HD (and SD) video reception. While the improved video quality for HD over SD is a desirable asset, service providers may need to upgrade their networks to support higher capacity or new video codecs in order to transport HD video. Besides the obvious need for higher capacity, another key requirement in servicing HD video is the need to dimension the buffer size at the client device. A small buffer size may lead to high packet losses whereas an overprovisioned buffer size will increase channel change latency, which is undesirable. A specific challenge associated with OTT HD services is the need to manage the bandwidth properly in order to ensure that users do not oversubscribe and compromise the service quality of other users. Scalability is another key issue. For instance, CNN Live served 1.3 million concurrent live streams in the moments leading up to President Barack Obama's inaugural address on January 20, 2009, and served a record-breaking 26.9 million live streams during the President's speech. This shatters the previous record of 5.3 million live streams set on 2008 Election Day.

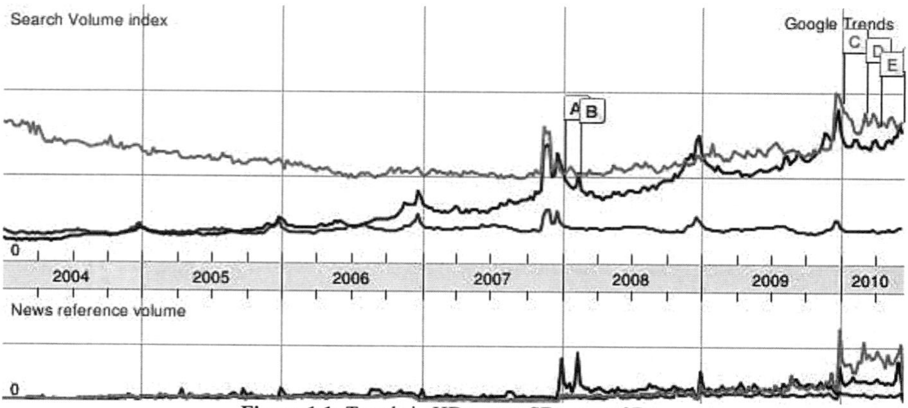

Figure 1.1: Trends in HD versus SD versus 3D.

Next-generation HD may emerge in the form of three-dimensional (3D) HDTV that brings actors and sports stars to life, popping them off the screen and unlocking details otherwise bound to 2D. 3D displays project image pairs to the viewer. Stereoscopy is the most widely used method, which captures stereo pairs in a two-view setup, with cameras mounted side by side, separated by the same distance as between the pupils of a pair of human eyes. 3D HD could be further fueled by autostereoscopic 3D displays that remove the need to wear polarized glasses, thereby making it far more convenient to watch these movies. The *Avatar* 3D movie grossed a record breaking $615 million after 11 days since its debut, reached $1 billion by the first week of 2010, and over $2 billion by March 2010, far surpassing the cost for producing the movie ($300 million) and the hit movie *Titanic* ($600 million). Besides *Avatar*, over 170 3D movies were released in 2009 (http://all3dtv.com/categories/movies), and 4 of the top 10 box office successes of 2009 were 3D. Like broadband-enabled HDTVs, the emergence of 3D HDTVs may become a new revolution. The global demand for such TVs are expected to exceed 100 million in the next 3 years. 3D video camcorders, laptops, video games, as well as HDTVs that convert video content from 2D to 3D, are also available (http://www.cnn.com/2010/TECH/01/06/toshiba.3d.tv). ESPN and Sky plan to launch 3D channels in 2010, broadcasting live sports in 3D to the living room, including the 2010 World Cup. Other types of 3D content include live concerts, natural landscapes, and commercials. Online 3D videos are now available in a number of Web sites (e.g., http://www.3dmovies.com) as well as YouTube. However, even with improved compression, streaming 3D HD movies remains a tough challenge and may require peak rates as high as 300 Mbps. Other technical challenges include light-field stereoscopic 3D image acquisition, 2D to 3D conversion, computer-generated 3D virtual space, human perception relating to stereopsis, 3D tracking, and smooth 3D video image creation for continuous viewing as the viewer moves.

New ultra-HD technology is also in the horizon, with 33 million pixels comprising 7,680 (4 × 1,920) horizontal and 4,320 (4 × 1,080) vertical pixels per frame, dwarfing current 1080p resolution by 16 times. Note, however, that this is still a far cry from the processing power of a single human eye–126 million pixels. The experimental digital video format was developed by NHK Science and Technology Research Laboratories (http://www.nhk.or.jp/strl). The frame rate is 60 Hz and the bandwidth is 600 MHz, giving a bit rate that ranges from 500 Mbps to 6.6 Gbps. Several years ago, NHK demonstrated a live satellite relay over IP for display over a 450-inch (11.4m) screen. Video was compressed from 24 Gbps to a rate of about 100 Mbps whereas 22.2 channels of surround sound audio was compressed from 28 Mbps to roughly 7 Mbps.

1.3 Video Content Distribution

Many incumbent cable, telco, and satellite payTV providers share the common trait of employing closed "walled gardens" systems that offer selected (and often repeated) video content and mostly appointment-based viewing. This is in contrast to the open Internet model where users are able to access any content they choose and watch videos anytime and anywhere. The emergence of OTT devices and service providers offers more video choices to the consumer by seamlessly integrating live TV with stored video, on-demand movies, and online Internet video, providing replacement or supplementary TV services. Here, the network may not be owned by the video provider. Some of these providers (e.g., Apple TV, Hulu, Netflix) may complement and supplement TV content provided by existing payTV providers while others (e.g., Sezmi) are meant to compete directly. Sezmi (www.sezmi.com) provides three types of TV sources: broadcast stations, cable channels, and Internet content. The first two are captured over the air via a powerful antenna in the Sezmi 1 Terabyte DVR/STB combo. With online videos becoming popular, satellite TV providers stand to lose out more since high-speed Internet and voice over IP services are typically unavailable. Many cable providers, for example, are experiencing high growth in voice and Internet subscription, even as they see a decline in video subscribers. However, as will be described shortly, satellite TV providers are countering the rise of the online TV revolution by placing their content online and are evaluating ways to make online TV more accessible to their subscribers.

Many payTV providers allow the same video content to be broadcast on the same channel at different times or on different channels. In online TV, duplicate broadcasts of the same video content is avoided. For example, if a movie or TV show is available on Crackle, someone who searches for the video on the Hulu Web site will be referred to Crackle. Clearly, online video portals provide users with more control in choosing the desired content and in discovering new content (via popularity ranking, feed of video recommendations from experts and friends). In doing so, only videos requested will be streamed, thus making more efficient use of bandwidth. Tables 1.1 and 1.2 show some popular online video portals. Most of these services require Adobe's Flash player and the H.264 coding standard. While Hulu (www.hulu.com) serves over 1 billion streams per month, YouTube (www.youtube.com) boasts over 13 billion video views per month. YouTube has the largest library of both user-generated [2] and increasingly premium video [3]. YouTube is a video-sharing Web site where users can upload, view, and share video clips. The site displays a wide variety of user-generated video content as well as movie clips, product demonstrations, and commercials. Unregistered users can watch the videos, while registered users can upload an unlimited number of videos. YouTube is evolving from a consumer destination site to a full-fledged video distribution platform. To this end, YouTube recently published an expanded set of application programming interfaces (APIs) to allow third parties (e.g., TiVo devices, STBs, mobile handsets, Web sites) to gain easier access to YouTube's content. Even though online video entertainment has a very short history, competition is intense and some companies are already facing profitability problems. More significantly, payTV companies not embracing the online video revolution are facing increasing pressure to provide more value for their existing services.

Table 1.1: Online Video Portals Requiring Consoles or Broadband TVs

	Console	TV Episode	Movie Rental	Movie Purchase	Comments
Netflix	$99	Starting $4.99/month			14 million users, over 17,000 movies/episodes.
Xbox	$199	$2	$4 (SD), $6 (HD)		Video game console. Partners ESPN.
PlayStation	$299	$1.99–$2.99	$2.99	$9.99	Video game console.
Apple TV	$229	$1.99	$3.99	$14.99	Online video streaming and retail. Sold 200 million TV programs, over 32,000 movies.
Vudu	$299	$1.99	$3.99	$19.99	2,000 1080p HD movies, 16,000 movies. Acquired by Walmart.
Blockbuster	$99	$1.99	$3.99	$9.99	Supported by Yahoo!/Intel Widget Channel.
Roxio CinemaNow	Optional	N/A	$2.99–$3.99	$9.99–$19.99	
Amazon	Optional	$1.99	$2.99	$14.99	Online video streaming and retail, over 50,000 titles. Supported by TiVo, Sony's Bravia, Xbox 360, Windows Media Center, and Roku.

Table 1.2: Online Video Portals with Direct Viewing

	Episodes and Movies	Comments
Metacafe	Free short-form online video entertainment.	TV clips, movie trailers, music videos, sports clips, video games.
Blinkx	Free short-form online video entertainment.	Powerful video search engine.
Crackle	Free, ad-supported episodes and movies.	Supported by Sony Entertainment.
Adobe TV	Free, need to install Adobe Media Player.	Blip.tv, CBS, Comedy Central, Epicurious, KQED, MTV, MyToons.com, Nickelodeon, and more.
Hulu	Free, ad-supported episodes and movies.	Partners Sling Media and Disney. Owned by NBC Universal and News Corp. Began HD videos in August 2008. Over 1 billion views per month.
Joost	Free browser-based streaming.	WiFi video streaming via iPhone.
ESPN3	Depends on subscription to participating high-speed ISP.	Many live games. Video sharing available on Facebook.
Fancast	Free, ad-supported episodes and movies.	Supported by Comcast. Xfinity service requires paid subscription.
TV.com	Free, ad-supported episodes and movies.	Owned by CBS.
CBS	Free, ad-supported episodes and movies.	Owned by CBS.
Sling	Free, ad-supported episodes and movies.	Sling box needed for live TV.
Starz Play	Free, ad-supported episodes and movies.	Similar movie download model as Amazon (pay-per-view). Free previews. Over 2,500 titles from Starz, Encore, Movieplex. Partners Verizon.
Veoh	Mostly free, subscription needed for premium channels.	Open-platform supports user-generated video. Requires Veoh player.

A further distinction relates to the business model: free download (e.g., Hulu) versus paid subscription (e.g., Netflix). Hulu streams mainly TV shows and selected movies, which does not overlap paid premium video services such as Netflix. While the free download model is popular, one has to put up with commercials (although there are fewer ads than payTV, for now). Interestingly, a free business model is also adopted by Lala (www.lala.com), an ad-free music streaming service. Lala allows users to select any of the 8 million songs of their choice (an important difference from Web radio stations). Unlike iTunes, which only provides paid music downloads, Lala fills a huge void with unlimited streaming of free premium music. These developments point to the fact that the Internet is rapidly becoming an ecosystem for entertainment.

A key advantage of OTT providers is the possibility of making video content available to a global audience. Netflix, for example, is expanding its reach beyond the boundaries of North America to an international audience (over 300 million broadband subscribers worldwide versus 80 million subscribers in the United States [4]). However, major cable, telco, and content providers have recently launched TV Everywhere (TVE) [5] online services. Currently, TVE is a bundled service, which means subscription to both TVE and payTV services are needed. If the TVE concept becomes unbundled (i.e., customers subscribe only to TVE), this revolutionary approach will enable payTV providers to attract new customers outside their franchise areas. For instance, some condominium residences subscribe to bundled service with a contract payTV provider and do not allow third-party providers of the tenant's choice. If this approach is pursued aggressively, it may actually turn cable and telcos into the biggest OTT players themselves. However, the biggest drawbacks are the need to sacrifice traditional video service (where significant infrastructure investments have been made on enhancements such as switched digital video and IPTV) and the increased competition (e.g., cable operators may compete more intensely against each other).

Not to be outdone, satellite TV providers are also putting their content online to reach customers who may have satellite dish restrictions. On November 10, 2008, British Sky Broadcasting (BSkyB), the dominant payTV operator in the United Kingdom with over 9 million subscribers, announced they will offer online TV to any U.K. resident with a broadband Internet connection. Satellite TV channels are broadcast over the Internet and no satellite receiver is needed. This is facilitated by the in-built Sky player in Windows 7, which brings Internet, broadcast, and recorded content together. The Windows Media Center also connects to CBS Audience Network, the full Zune video podcast library, MSNBC, and Netflix. Satellite TV provider Dish Network recently announced a similar initiative for its online international TV service. Customers need not subscribe to Dish in order to access the new service. Thus, potential customers need not change from their existing payTV provider (e.g., cable). An Internet STB is shipped to the customer when the online service is established. The use of a STB prevents the customer from being restricted to on-demand shows from one or two networks, as in the case of many online video portals. With Sling box functionality now incorporated into the STB (http://www.slingbox.com), Dish customers will have instant online access to all their TV programs and DVR content via any broadband connection–be it stationary or mobile, at home or remote, wireless or wired.

Internet-enabled HDTVs with embedded hardware video decoders are now gaining traction, superseding many legacy set-tops, and making online video viewing even more convenient and pervasive. These new broadband HDTVs use either a wired or wireless Ethernet connection to access content onscreen from

Yahoo, Flickr, YouTube, and more. The widget-based interface and scrollbar run along the bottom of the screen for easy browsing of the latest programs and content. Active OTT households are projected to grow from 40 million in 2009 to 170 million in 2014. A basic high-speed Internet connection is now indispensable whereas payTV services are optional. For instance, major cable providers have reported losing video customers for successive quarters since the start of 2008 although the loss is compensated by a stronger growth in high-speed Internet and voice customers. In contrast, Netflix boasted strong growth in both revenue and subscribers during economic downturn and over half of their customers use the Watch Instantly streaming service. With video streaming becoming prevalent, DVDs have become less popular. Walt Disney reported a 32% drop in quarterly net income in December 2008 primarily due to a huge decline in DVD sales.

Once thought to be a problem with online TV, live sports shows are now being broadcast by Web portals such as ESPN3, Fox Sports, and Univision. Among the major sporting events broadcast online include Summer and Winter Olympics, March Madness, MLB.tv, and Sunday Night Football. The 2010 World Cup witnessed an unprecedented level of online and mobile video streaming and sharing, which includes 3D videos, Dolby 5.1 surround sound, and ads. For ESPN3, key highlights and goals are marked on the online player controls, which can be selected for review at any time during the match. The video is made available right after each live broadcast has ended, making it convenient for busy professionals and local fans who cannot watch the match at the scheduled time.

The strong demand for online video entertainment is not only well-sought after by consumers, but also by video portal owners and advertisers who wish to outdo the competition. As online video portals become mainstream, eventually we may see a clear migration from managed payTV services (operating over private networks) to cheaper online video services (operating over unmanaged networks such as the public Internet). This is an interesting development since the IPTV service launched by the telcos is meant to compete with payTV services provided by cable and satellite. As it turns out, the IPTV service has not been a cost-effective competitor, and all payTV service providers now have to contend with the competition posed by OTT providers. This trend is not new as we have also witnessed, not too long ago, peer-to-peer voice and voice-over-IP services taking over the reins from the venerable landline phone service (operating over a managed network specially designed for voice communications).

1.4 Content Quality versus Video Quality

Among the key factors that will determine the success of broadband video services are content, pricing, convenience, and video quality (VQ), in the order listed. A 500-channel payTV channel lineup seems limited when compared to the vast amount of free videos that are available on the Internet. The movie theater currently provides the best VQ but does not enjoy the biggest audience. The key motivation for watching movies in the theaters is perhaps the release of new content, rather than the larger screen and better VQ. More people are watching online videos even if the VQ is not the best at times. This is primarily driven by cost (free or low fee), which is highly appealing to nonworking adults (e.g., students, retirees), and convenience (can watch whatever you want anytime and anywhere). Many online TV providers allow users to rate the video content, including the commercials, but VQ is not rated. Unlike payTV services, which require a proprietary STB to access the content, online Internet TV can be accessed via a variety of devices, including broadband-enabled HDTVs and multimedia computers. Nevertheless, the ability to assess VQ as perceived by the user is key to improving the coding efficiency of codecs, which will ultimately allow better quality video to be transported at lower bandwidth.

The utility of measuring VQ in a point-to-multipoint access network is less apparent. There are three reasons to support this observation. First, VQ assessment performed on an access network should be done without decoding and comparing video content at thousands of customer premise devices (e.g., STBs) spread across multiple locations. Unfortunately, current VQ meters all require video decoding and therefore may not be able to detect video artifacts in a responsive manner or prevent them. Second, error concealment (EC) can be implemented to conceal any unavoidable packet losses or errors associated with video transmission, thus removing the need for constant monitoring of VQ. The technology not only conceals visual artifacts and improves VQ, it also takes up less overheads than bit-level forward error correction (which is usually powerless against burst packet losses arising from router congestion or buffer overflow). The real value of EC lies in the ability to improve VQ for larger screen devices such as laptops and TVs. It is also invaluable in enhancing high frame rate HDTV (120 or 240 Hz) for supporting full-motion sports videos. For instance, EC is used in some new 240-Hz HDTVs to remove motion blur and image judder that can plague some displays when displaying fast movement onscreen. Motion in every frame is analyzed and then adjusted so that nothing goes by in a blur. The frame rate is quadrupled from 60 to 240 Hz without repeating the same image to make more frames. Instead, interpolation (a form of EC) is employed to insert new frames between existing frames to create smoother transitions. Third, the migration from constant bit rate (CBR) to variable bit rate (VBR) systems implies that video frames are now encoded with constant quality. VQ is less of an issue for VBR encoding because the VQ level for every encoded video frame is the same and this level can be set to a very high level. It makes sense to adopt VBR encoding in online video streaming since the Internet will not guarantee a constant bandwidth anyway and hence, maintaining a CBR encoded output does not represent a significant advantage. On the other hand, the CBR approach makes more sense over a managed payTV network with fixed-bandwidth channels (e.g., 6-MHz channels). However, many payTV service providers are now migrating to VBR video systems to avoid:

- Complex and expensive VQ optimization associated with CBR encoding
- Expensive and time-consuming evaluation/monitoring of measured VQ levels
- Less efficient coding associated with CBR
- Less efficient bandwidth utilization with bonded (aggregated) channels

Besides VBR encoding, migrating from legacy MPEG-2 codecs to new H.264 or VC-1 codecs is equally important in improving VQ. The majority of VQ measurement devices are mostly passive monitoring equipment designed for the older MPEG-2 standard. This precludes many coding and network enhancements in H.264, including a far more superior coding efficiency, and powerful error resilience and video transport features. This is one of the reasons why video image breakups are less frequent when watching H.264-coded online movies compared to payTV or broadcast TV service. Unfortunately, some payTV service providers are unable to migrate to H.264 receivers in the near term due to the high number of MPEG-2 STBs at the customer premise, some of which are owned by the consumers. However, the superior coding efficiency and VQ available with H.264 will motivate these providers to migrate to the platform sooner rather than later, just as they are countering the emergence of OTT providers with subscription-based online video portals.

1.5 Multiscreen Video

The grand challenge facing many service providers today is effective bandwidth management for supporting high-quality video delivery. While bandwidth in metro and core networks supporting the Internet are mostly overprovisioned, routers and switches connecting these networks are susceptible to network congestion and as a result may discard packets randomly. These losses can be detrimental to compressed video transport and can render forward error correction remedies ineffective in defending against such losses. On the other hand, the access network (e.g., cable, DSL, wireless) poses a bandwidth crunch, which service providers must address carefully in order to enable pervasive HD connectivity.

Consumers today wish to watch movies and TV shows whatever they choose, whenever they desire, and on whatever device is currently available, from a TV to a laptop to a tablet to a smartphone. Service providers are therefore pushing content beyond the TV to any video-enabled device, including PCs and smartphones (Figure 1.2). A multiscreen bundled service provides a single offering of one price, one point of customer contact, and one integrated electronic program guide. This means subscribers do not sign contracts with different providers, receiving different packages of content, and paying different fees. An optimized end-to-end video transport platform is needed to support the concept. Bandwidth consumption should be low so that packet losses are minimized and bandwidth bottlenecks overcome, thereby ensuring instantaneous connections and high-quality, smooth video playback. Overall, the platform should stream videos at a low rate and at reduced cost, and utilize condensed storage requirements.

Highly efficient encoders such as H.264 introduce a high degree of bit rate variability in the compressed video stream. Many video transport mechanisms (e.g., switched digital video or SDV) convert the video streams from variable to constant bit rate. This presets the total bit rate of each video stream to a fixed value, which is usually set lower than the desired value for optimal VQ. This implies streams are seldom allowed to peak to the maximum rate and so video information run the risk of being dropped due to buffer overflows. Thus, the use of fixed bit rate allocation for H.264 video transmission may cause variable VQ during playback. With DOCSIS 3.0 channel bonding (multichannel) capabilities, underutilization of the channel bandwidth is another acute problem that needs to be addressed if the current CBR approach is adopted. As more aggressive channel bonding becomes available in the future (up to 24 channels or more), there will be a pressing need to optimize the bandwidth usage in the CBR approach or to migrate to the VBR approach involving the aggregation of VBR video streams.

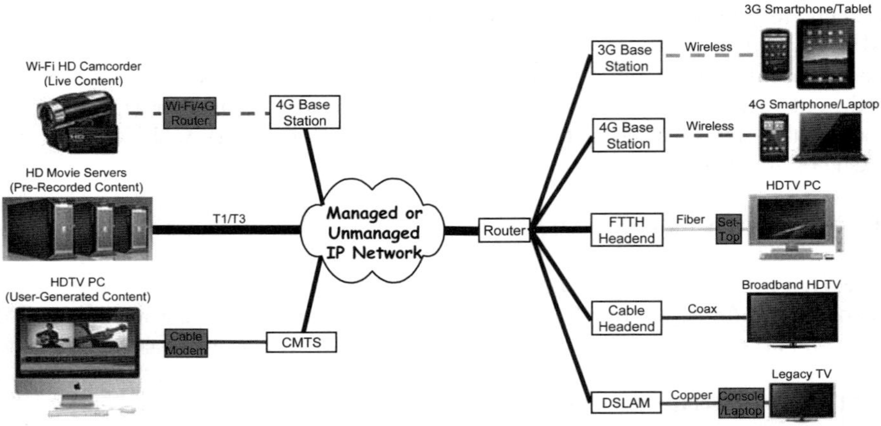

Figure 1.2: End-to-end network management for delivering high-quality multimode video connections.

Efficient transport of compressed videos over the Internet is fundamental to the success of online video portals. Many online portals employ the highly efficient H.264 codec, and yet these systems sometimes encounter stalling in video playback, degrading end-user quality of experience, especially with HD videos. The reason behind this failure is because the codec is only part of the equation—managing packet losses and the high peak-to-average rate associated with VBR videos are equally important. The peak bandwidth demands of compressed HD videos can easily exceed the bandwidth limits of the access network.

Freeze frames and choppy online video playback are primarily caused by packet losses when video data is transported across the Internet. Such instances are more pronounced with HD videos and may happen whenever the coded video encounters a scene change, requiring an entirely new and different frame to be encoded, which in turn imposes a high amount of data to be carried by the network. Even if the network is able to cope with the higher-than-normal traffic load and carries the video data without any losses, the player may become overwhelmed with a sudden burst of video data, resulting in buffer overflow and subsequent losses. More often than not, the higher load will stress the network, leading to a higher number of losses, thereby preventing the receiver from buffering sufficient video data before it is displayed. This results in data starvation, causing frames to be skipped and the player to pause.

In addition, the ability to proactively monitor link quality and identify video artifacts and packet losses without decoding the video content are important, together with reactive measures to correct artifacts and minimize or eliminate future losses. To counter possible bandwidth hogs associated with OTT users, intelligent policy-based resource allocation should be implemented to effectively manage oversubscribed users, including peer-to-peer users, in a noninvasive and nondiscriminatory manner. This also creates the need for a dynamic bandwidth allocation algorithm to optimize the bandwidth utilization. If the algorithm can be made predictive (see Chapters 5 and 6), this will also reduce the buffering resources at the STB and allow the admission control mechanism at the video headend to allocate bandwidth resources ahead of time.

1.6 Mobile Video

One-way broadcast TV was the original TV delivery method. Broadcast TV standards have evolved from analog to digital to mobile standards. Mobile digital TV is delivered via the same infrastructure as digital over-the-air broadcasts to home TVs but with special enhancements for viewing on mobile devices. With this feature, timely information, including emergency messages and video, can be broadcast to mobile users.

The entry of two-way 4G broadband wireless networks with mobile Internet enables on-the-go video streaming on laptops and smartphones, providing a much wider network coverage than the traditional home Internet service. WiMax and LTE are promising 4G wireless broadband solutions that offer ubiquity and immediate access for both fixed and mobile users with quality of service support. The technologies are especially attractive for bridging the last mile and the digital divide, enabling voice, video, and Internet connectivity both inside and outside the residential premise. Unlike wired access, a large portion of the deployment costs is incurred only when a subscriber signs up for service. Given that these networks are cheaper and easier to deploy and maintain, the entry barriers to new competitors are low, which may lead to more service providers deploying such networks in the future. In addition, when LTE networks become available, users will be able to hop from one wireless carrier to another and back again, all while keeping the same device.

Many U.S. cable operators are targeting 4G wireless networks to provide converged quadruple-play services (i.e., voice, video, data, and mobile access) for their customers, to reach remote or sparsely populated areas not economically serviceable by wireline access networks, and to compete with cellular operators for wireless revenue. For instance, Comcast and Time Warner have invested heavily in a joint-partnership with Sprint and Clearwire to roll out WiMax service in the United States. The WiMax standard leverages on many features from the DOCSIS cable standard (developed by CableLabs) and this creates a positive incentive for cable operators to deploy WiMax networks. Although terrestrial WiMax access offers considerable advantages over satellite and digital TV broadcast, including two-way connection capabilities, new technical and deployment challenges must be overcome (e.g., fixed-mobile convergence, bandwidth challenges). In Atlanta, which currently has the biggest WiMax footprint in the United States, one can capture a live event on video and stream it instantly via WiMax to another location with Internet connectivity and even broadcast it on YouTube Live (http://www.youtube.com/user/Live) so that the event can be watched on any Internet-enabled HDTV or PC/laptop. Clearwire's WiMax network will expand from 27 markets in 2009 to all major U.S. metro areas (80 markets) by end-2010, covering 120 million points of presence. There are currently 555 WiMax deployments in 147 countries worldwide.

The increased bandwidth in 4G wireless networks allows mobile video applications to be supported. Mobile video is an exciting extension of the broadband video experience although mobile TV (over TV bands) has suffered a slow start primarily due to the lack of new wireless infrastructure supporting the service. However, the emergence of sleek smartphones has changed the mobile video landscape. These devices make watching full-length videos on a 3.5-inch screen more enjoyable. The concept has now broadened to larger screen portable devices such as tablet computers and laptops. With the availability of high-speed wireless Internet access and improved battery power consumption, these devices are rapidly adopted by the enterprise. For instance, the iPhone design extends battery life through the use of a hardware video decoder and supports the high-speed packet access (HSPA) cellular standard [6]. HSPA supports 7.2 Mbps download and 2 Mbps upload speeds. The industry-leading Apple App Store has generated developers more than $1 billion in revenues while serving up over 5 billion application downloads since its inception two years ago. There are currently 225,000 Apple apps versus 35,000 Android apps. The Android platform is an open platform that supports efficient embedded application development. The Nexus One Android phone from Google is a competitor to the iPhone for the mobile video market share. Unlike the iPhone, the Android phone supports Adobe's Flash player, making it possible to watch online videos from popular Flash-based video portals such as YouTube and Hulu. In addition, the device is carrier-agnostic or "unlocked," which means users can switch from a T-Mobile or AT&T network to a Verizon or Sprint network without purchasing a new device. It can also be used in other parts of the world (e.g., Vodafone network) with a compatible SIM card. The new phone supports HSPA, 802.11n, Bluetooth 2.1, GPS, 100,000:1 contrast ratio (higher than some HDTVs), and has a microphone and an in-built 480p SD video camcorder [7]. Sprint has released an Android smartphone called Evo 4G [8], the world's first 4G smartphone that works on the broadband WiMax network. The phone has a 4.3-inch touchscreen display and comes equipped with two HD cameras: 8 Megapixels on the back and 1.3 Megapixels on the same side as the screen to facilitate video calls. The device supports personal hotspot capability whereby the device can connect up to eight Wi-Fi-enabled devices and an FM radio. Apple has released the larger screen (9.7 inch) iPad tablet computer that supports 720p video and HSPA connectivity. The early adoption of iPad is strong with 1 million units sold in its first 4 weeks of availability. Apple has also released a slimmer version of the iPhone (9.3 mm) that comes equipped with two cameras (for video chat, video shooting and editing) and a high-resolution screen (960 × 480) with 326 pixels per inch (pixel density that is indistinguishable to the human eye, which can process 300 pixels per inch). Google is planning to launch a tablet computer soon. These developments may drive an explosion of mobile video and social TV applications. Table 1.3 summarizes the typical online video streaming rates.

Table 1.3: Video Streaming Rates

Network	Video Resolution	Screen Size	Streaming Rate
Home	1080p	50" HDTV	> 3.5 Mbps
4G Wireless	720p	15" Laptop or 10" Tablet	> 2.5 Mbps
3G Wireless	480p	3.5" Smartphone	> 1.5 Mbps

1.7 Streaming Protocols

Streaming protocols determine how well compressed video can be transported across the network. In addition, they simplify digital rights management (DRM) since a small amount of the video content is typically cached at the client device for a short period of time, reducing the risk of someone reusing the data. This makes it more acceptable to stream video directly to a PC and not via a proprietary STB. Enhanced security can be achieved via secure protocols such as HTTPS. Streaming also reduces memory requirements at the client device. By providing content storage and streaming on high-end servers, "cloud" entertainment may become the bedrock of future cloud computing applications. While streaming is the most popular mode of watching video, there are instances where downloads may be more appropriate (e.g., downloading movies to watch on a plane).

Loss-free protocols namely TCP and HTTP (a Web delivery standard) are popular choices in online video streaming whereas protocols designed for real-time multimedia traffic such as RTP over UDP are popular for managed payTV networks. Highly compressed audio/video is sensitive to information loss, in particular, random information loss that is common in public Internet transport. Thus, loss-free protocols are more desirable and practical for such networks, although the end-to-end delay may have to be calibrated properly, especially for live video service. In addition, many content delivery networks (CDNs) have massive deployments of HTTP servers. By not using proprietary video streaming protocols such as RTSP and RTMP, they can avoid additional capital expenditure.

Adaptive bit rate (ABR) or adaptive streaming is commonly employed together with HTTP. The technology is a combination of scalable video coding (which allows the server to send a video encoded at a rate suitable for the connection) and progressive streaming or download (which requires the compressed video to be segmented into smaller fragments of 2 to 10 seconds). ABR streaming was pioneered by Move Networks [9]. Microsoft and Adobe have also developed their own ABR streaming approaches. Recently, Apple has released a new operating system that supports ABR streaming [10] and a compatible iPhone media player. ABR streaming dynamically adapts the VQ to the link quality, channel capacity, and computer resources, thereby reducing instances of player stalling. For example, when network bandwidth decreases (based on feedback from the receiving client), video content is delivered at a lower rate. This process increases the latency as the server and the player switch to a lower quality video and make necessary adjustments. Depending on the severity and variability of the network congestion, the latency can sometimes exceed several seconds. Thus, although ABR streaming obviates users having to make choices like "high-quality" or "low-quality" video streaming, the main drawback of this approach is a noticeable drop in VQ as well as reinforced artifacts whenever there is a sudden degradation or improvement in network bandwidth or link quality. If such instances occur frequently, it could compromise end-user quality of experience. As you will discover in the later chapters of this book, the alternative is to maintain the VQ suitable for the subscribed network bandwidth by making efficient use of the available bandwidth and employing error concealment, thereby ensuring smooth playback without compromising VQ at all times.

1.8 The User-TV Interface

The user-interface for the TV has been revolutionized by the Internet and the World Wide Web, and as a result, the thin-client model has become more and more popular. Netbooks are now the fastest growing category of portable computing devices, requiring only a browser on the client while the "heavy-lifting" is done by servers in the network. Even the Evo 4G smartphone comes equipped with a HDMI output, allowing online videos to be displayed on the big screen. These devices allow new user-interface logic to be customized to the user's preference at any time merely by linking in additional Web pages or by downloading "apps" without a STB firmware reload or reboot.

1.9 Conclusions

Video is driving the growth in Internet traffic with the proliferation of video and peer-to-peer applications and the continued, or even accelerated growth in demand for video downloads. Equally as exciting is the demand for being able to enjoy these applications while on the go. Thus, OTT video content suppliers will play an important role in broadband wireless access networks. Online video raises several interesting propositions when compared to payTV video-on-demand and premium video services. For example, which service provides the better value, Netflix or HBO? When comparing online video streaming with digital video recording, which is more convenient? Existing cable and satellite TV programs are now pushed online. Conversely, Internet-only content and bonus material are pushed directly to the TV. The migration from payTV to online video is gaining momentum—both consumers and advertisers are moving in the same direction. As a result, managed payTV networks may become underutilized in the same way as the public switched telephone network. Managed networks are limited in reach, whereas the Internet is global. This trend may gradually erode the business model for payTV operators. Open access (with open STBs and open platforms such as Android) and virtual operator services are emerging. An interesting observation is that online video is taking the lead on the latest video technologies, ranging from H.264 VBR encoding to MP4 encapsulation to progressive ABR streaming and it appears payTV providers are also starting to embrace these technologies.

Networked video delivery must strive to provide a high-quality TV experience. Improving the quality of Internet video delivery requires managing the packet losses and video artifacts. Packet losses due to congestion can be as high as 20% or more and the one-way end-to-end latency can range anywhere from 10 to 500 ms. Feedback control on the Internet may not be responsive, which is a drawback since accessing digital movies should be as simple and fast as flipping a channel on a remote. Thus, Internet video transport can be more challenging than sending video over terrestrial wireless networks. These networks typically deal with channel errors on a single link and delay is not significant since bit error correction and packet retransmission can be performed on a local link. Bandwidth conservation methods are needed to reduce video packet losses over the network and at the receiver. Bandwidth management challenges remain, including end-to-end QoS-guarantees, seamless connectivity, Net neutrality compliance, and effective policy/traffic management.

References

[1] N. Bilton, "Google and Partners Seek TV Foothold," *The New York Times*, March 17, 2010.

[2] Network World's "The 10 best IT videos of 2009," http://newsletters.networkworld.com/t/4401735/258733820/88528/0.

[3] YouTube TV Shows, http://www.youtube.com/shows.

[4] United States of America Internet Usage and Broadband Usage Report, http://www.internetworldstats.com/am/us.htm.

[5] E. Buskirk, "Cable departs from Hulu model with TV Everywhere," June 26, 2009, http://www.cnn.com/2009/TECH/biztech/06/26/wired.tv.everywhere/index.html.

[6] iPhone Specifications, http://www.apple.com/iphone/specs.html.

[7] Nexus One Specifications, http://www.google.com/phone/static/en_US-nexusone_tech_specs.html.

[8] Evo 4G, http://now.sprint.com/evo.

[9] Move Networks Adaptive Streaming, http://www.movenetworks.com/move-media-services/move-adaptive-streaming.

[10] HTTP Live Streaming, http://tools.ietf.org/html/draft-pantos-http-live-streaming-01.

Exercises

1.1. Many online TV, cable, and satellite providers prefer to use the STB to offer their online services to subscribers. Subscribers need to sign up for the service to get the box or have to buy from a third-party vendor. Describe the pros and cons of using the STB to provide online TV services. Contrast the approach taken by the FCC. FCC approved plug-and-play standards for equipment that allows digital cable to plug directly into TV sets without the need for an STB (see http://www.fcc.gov/cgb/consumerfacts/plugandplaytv.html).

1.2. What benefits do satellite providers stand to gain by offering their subscribers online channels and TV content? Why do some channels broadcast their premium content for free on digital TV (in HD too) and on the Internet when the same content is available on payTV channels? Justify if the following statement is true. If mobile digital TV remains an “after-market” phenomenon, requiring dongles and other appendages to access the service on mobile devices, consumers may not use the service but instead rely on 4G/3G wireless service for mobile video content that are available from free online TV websites as well as OTT service providers (Netflix, for example, is available on the iPad and the iPhone 4).

1.3. Broadcast TV and 4G wireless networks typically employ licensed radio spectrum whereas Wi-Fi uses unlicensed spectrum. Explain the key advantage of using licensed spectrum in relation to users’ ability to access the wireless service. What is the main technical limitation of digital broadcast standards such as ATSC compared to 802.16 and LTE? With smartphones becoming an expensive investment for personal use, describe the pros and cons of business models that employ locked and unlocked carriers.

1.4. The ABR adaptive streaming technology is normally used in conjunction with loss-free, two-way protocols such as HTTP, which implies all packets sent by the server will eventually be received, even though some of these packets may be retransmitted. In addition, progressive streaming or download using segmented video chunks is used to reduce the traffic load on the network and to provide smoother playback and connection scalability. Why is there a need to reduce the video quality and bit rate when network congestion occurs? Is ABR suited for networks with high bandwidth fluctuation, such as the public Internet? Is ABR more suited for video delivery to small-screen devices or large-screen HDTVs? Is ABR needed if videos are efficiently encoded? Why do content delivery network providers tend to focus on the streaming protocols instead of the encoding?

1.5. The lack of Flash support in iPhones, iPods, and iPads forced some content providers to offer video on HTML5 (http://www.youtube.com/html5). HTML5 is the proposed new standard for HTML that aims to reduce the need for proprietary Rich Internet Application (RIA) plug-ins such as Adobe Flash and Microsoft Silverlight. Google may gradually replace the use of the H.264 Adobe Flash Player on YouTube with a mixture of HTML5 and VP8, a new open-source codec. Analyze the implications of the emergence of HTML5 and VP8 for video playback.

1.6. Unlike DVDs, which play on any DVD device, consumers are not comfortable with limited usage of online video content. For example, videos that are compatible with Flash may not be compatible with the Windows Media Player or the Apple Quicktime. Leading entertainment and consumer-electronics companies have therefore formed a consortium, the Digital Entertainment Content Ecosystem (DECE, http://www.decellc.com), to develop technical specifications that content distributors and manufacturers can follow to ensure that consumers are not locked to a specific platform. The idea is to let consumers know that content and devices carrying a special logo will play nicely with one another.

Notably absent from DECE is Apple. Explain how the DECE standard will affect video rental and video purchase.

1.7. Retail giants such as Best Buy, Sears, and Walmart are joining the digital media ecosystem. Best Buy, for example, is providing their high-dollar customers with free online video rentals from CinemaNow. Explain how this development will impact incumbent payTV services.

1.8. Many consumers will watch the best available content on the best available screen at the best available video quality and price. Rank these four criteria in the order of importance and justify your answer.

1.9. The iPad has been adapted by Comcast to be the new personalized "remote," allowing users to view channels, browse the full channel lineup, and invite friends to watch movies. The iPad application connects the iPad to the STB using the Enhanced TV Binary Exchange Format (EBIF), the cable industry's specification for delivering one-to-one interactive applications to STBs. Explain why the iPad works better as a user and social TV interface compared to the laptop or netbook. Will video sharing become more pervasive via Facebook or Twitter on this iPad application and if so, what are the business implications for payTV service? Will the use of iPad by a cable operator rival the use of the iPhone by a cellular provider? With Netflix going mobile by targeting the iPad (as well as the iPhone and Android phones), how will this development impact cable operators?

1.10. Although multiscreen video connectivity has become widespread, matching the nature of the video content to the screen size is important for the user experience. For example, watching *Avatar* in the movie theater is a much better experience than watching the same movie on a smartphone. However, watching a 30-minute cartoon or comedy works well for the smartphone. Moreover, mobile users seldom watch full-length movies on 4-inch displays but will watch trailers, music videos, news conferences, and sports highlights. Thus, short videos are perfect for informing and entertaining mobile users. Analyze the role of visual effects, storyline, and video content in enhancing user experience for multicreens.

1.11. One reason why music piracy is more prevalent than video piracy is because compressed audio requires small storage space and bandwidth requirements. Will the use of next-generation video codecs with vastly improved coding efficiency create DRM problems in video distribution?

1.12. The front camera on the iPhone 4 is meant for video calling but does not work on 3G cellular networks for now, only limited-range Wi-Fi networks. However, the Evo 4G smartphone allows video calls over a 4G cellular network. Analyze the capabilities of 3G and 4G networks in supporting high-quality Skype video calling on a smartphone.

1.13. Comment on the validity of the following statement: TV moved from a wireless to wired (i.e., cable) environment while computer networks and Internet access are moving in the opposite direction.

Chapter 2

The Access and Home Networks

The majority of broadband services are supported by a wide variety of broadcast or switched access networks. Satellite, terrestrial wireless (e.g., digital video broadcast, 4G wireless), and traditional cable networks employ broadcast (shared) transmission. On the contrary, telcos operate a point-to-point packet-switched access network (e.g., DSL, fiber) with dedicated connections. This fundamental infrastructure difference may provide telcos with an advantage but is offset by the lower data rates available in the copper wires in DSL networks. Although fiber access networks may increase the rates, the increase is not substantial (upstream/downstream rates range from 5/15 to 20/50 Mbps). Moreover, the cost per Mbps is significantly higher than the cable and DSL options, resulting in a low take-up rate (about 30% based on the fiber-to-the-home council report), and the service is only available in a limited number of neighborhoods in the United States. This chapter focuses on cable access networks and their evolution to a switched digital video (SDV) architecture to support high-speed information and entertainment delivery. In addition to the SDV architecture (that only sends active video channels to the set-top devices), encoding and link transport issues as well as video delivery methods involving broadcast, multicast, and peer-to-peer networks will also be covered. We will then examine home entertainment networks, broadband convergence, and next-generation network initiatives.

2.1 Introduction

Broadband is an economic driver for the 21st century. Globally, over 300 million households subscribe to broadband Internet and this is expected to increase to 525 million in 2011. Broadband Internet can bring significant economic/social benefits, including improved education and public-safety, and enhanced healthcare through telemedicine and electronic medical records. It can also bring efficiencies by ushering smart grids, smart homes, and smart transportation. The FCC task force estimates the total cost of broadband deployments in the United States between $20 billion and $350 billion. It assumes services provided 100 Mbps or faster. The actual broadband speeds may lag behind advertised speeds by at least 50%, possibly more during busy hours. For example, the peak usage hours (e.g., 7 to 10 p.m.) create network congestion and speed degradation. In addition, about

1% of users drive 20% of traffic while 20% of users drive up to 80% of traffic. With smartphone sales to make up the majority of wireless device sales by 2011, much more wireless spectrum will be needed for mobile broadband.

The National Science Foundation report on Residential Broadband [1] concluded that the core network is largely overbuilt and that access network represents a significant bottleneck. This is compounded by the fact that there is a diverging set of options and enabling technologies (e.g., DSL, cable, fiber, satellite, and terrestrial wireless), each with differing deployment cost and time, service range, and performance. Since the core network capacity and Internet traffic are both growing exponentially, whereas the access capacity is only growing linearly, bottlenecks may soon appear at the network edges. The capacity of home networks is also growing quickly. An example is the migration from 54 to 600 Mbps Wi-Fi connected networks. An emerging concern is that while considerable efforts have been made to improve the so-called last mile capacity, and this has been successful to some extent with improved speeds for cable, DSL and wireless networks, the capacity bottleneck now appears to be in the second last mile.

The most widely deployed solutions today are DSL and cable networks, with cable taking the lead with over 60 million U.S. subscribers (subs) reported in 2009 [2]. The annual cable industry revenue is $86.3 billion, with advertising revenue contributing $26.6 billion and capital expenditures in last 10 years standing at $127 billion. In contrast, Verizon's FiOS fiber-optic TV service experienced a downward trend in TV customer additions in 2009, specifically 300,000, 299,000, 191,000, 153,000 additions in each quarter of 2009 although it has 2.9 million subs in total (out of 9.2 million broadband subs). AT&T added 248,000 U-verse TV subs in the fourth quarter of 2009, up slightly from the 240,000 in the previous quarter. AT&T ended 2009 with 2.1 million U-verse TV subs.

The wireless access option can be considered the easiest and quickest to deploy. For instance, the digital video broadcasting (DVB) and the Advanced Television Systems Committee (ATSC) standards allow free standard definition (SD) and 720p/1080i high-definition (HD) digital TV (DTV) channels to be broadcast over-the-air to an HDTV, PC, or laptop, without the need for a proprietary customer premise equipment (CPE) such as a set-top box (STB). The number of DTV channels in the United States is increasing steadily since the demise of the analog channels in June 2009. The ATSC (www.atsc.org) is an international, nonprofit organization developing voluntary standards for DTV. Over 195 ATSC members represent the broadcast, broadcast equipment, motion picture, consumer electronics, computer, cable, satellite, and semiconductor industries. ATSC channels are currently broadcast in over 200 cities by over 1,500 TV stations.

ATSC is partnering the Open Mobile Video Coalition (OMVC) to launch mobile DTV, which will allow viewers to watch their favorite programs on portable and mobile devices. OMVC (http://www.openmobilevideo.com) is an alliance of U.S. commercial and public broadcasters formed to accelerate the development and rollout of mobile DTV products and services. OMVC includes more than 800 TV stations. Consumer devices, transmission technologies, and data services include netbooks, cellphones, accessory receivers for Wi-Fi phones and laptops, in-vehicle devices, DVD players, and USB receivers for laptops that are all equipped with mobile DTV reception capability. Electronic service guide (ESG) suppliers can also take advantage of mobile DTV's two-way capability, which also allows upload of personalized information, content consumption, and user behavior data. More importantly, digital TV broadcasters can offer user interactivity and more choices to viewers. ATSC is optimized for a fixed reception (using highly directional fixed antennas) and uses 8-VSB modulation [3]. It is not robust enough against Doppler shift and multipath RF interference caused by mobile environments. To address these issues, additional channel coding mechanisms are introduced in ATSC—Mobile/Handheld (ATSC-M/H), which is a U.S. standard for mobile DTV. It is an extension of the terrestrial ATSC DTV standard and is therefore compatible to the standard but does not cause interference in legacy reception. This is achieved by flexibly dividing the bandwidth into a variable-size mobile DTV part and a legacy DTV part. ATSC-M/H employs 3 layers namely presentation layer (audio/video coding, close captioning), management layer (transport, streaming and nonreal-time file transfer, ESG), and physical layer (RF transmission and forward error correction or FEC). To date, 45 U.S. broadcast TV stations are sending test mobile DTV signals using the ATSC-M/H standard, allowing mobile DTV service to feature programs similar to fixed DTV. The A/153, ratified in late 2009, standardizes the characteristics of the emitted M/H signal and the functionality within the signal.

Many network operators have come to realize that the Internet Protocol (IP) platform is a scalable solution capable of delivering integrated voice, video, and data services to the end user. While any broadband service provider may use MPEG over IP transport technology, the cable industry was the first to adopt it in live deployments. IP technology enabled digital cable networks to evolve towards switched networks. IP also provides a natural platform for interactive TV services such as video-on-demand (VoD), gaming, and television-commerce. However, IP-based video can stress access network infrastructures, especially as the number of on-demand services and HD customers grows over a large geographic area.

Broadcast TV, near-VoD (time-shifted programming), pay-per-view, and so forth, are services controlled by the service provider. In broadcast TV, live channels are broadcast continuously. On-demand services refer to prerecorded or DVD-type video services (stored by provider), where the user has full control on how the video is played (e.g., pause, fast forward, slow-motion, rewind, and other trick modes). For instance, VoD allows subscribers to view selected movies or TV channels on demand and may be the best form of video service since subscribers watch TV on their own time and at their convenience. Channels can also be customized to the subscriber's preference without paying for idle bundled channels. However, the latency for accessing VoD services tends to be higher than broadcast services. In addition, more bandwidth is required since VoD streams are normally unicast. As such, VoD load and utilization need to be monitored and appropriately sized. Bandwidth conservation becomes more critical for such services. Online TV is essentially Internet VoD service, although it can also support live video. It is inherently more bandwidth-efficient than broadcast services since users actually watch the videos. It is cheaper to deploy and does not require expensive bandwidth reclamation equipment such as SDV equipment.

The most expensive component of access deployment is related to the home. Thus, service providers have started to look at the implications of home networking and end-user devices to serve the digital home of the future. Specifically, they are addressing new technologies and services surrounding the most costly, time-consuming, and customer-impacting component of the IP video implementation. Thus, the success of IP video may be more dependent on enabling the consumer with in-home network capabilities to access service content.

2.2 IPTV over DSL

Internet Protocol Television (IPTV) is a method of distributing TV content over a private (managed) IP network and is a term largely used to refer to a telco TV service. This should be distinguished from Internet or online TV (e.g., YouTube, Hulu) that distributes TV content over the public Internet (an unmanaged network). IPTV content distribution can be multicast (supporting live TV) or unicast (supporting interactive services, targeted advertisements, and VoD). Currently, there are 33 million IPTV subscribers worldwide.

IPTV over DSL employs switched digital video with dedicated DSL connections in the access network to guarantee quality of service (QoS). In the unicast approach, high channel change latency and bandwidth consumption on the access link can result in a significant bottleneck in homes or businesses with simultaneous users. Channel change latency can be improved with complex buffer management and video playback solutions but this may inevitably lead to increased network overheads and set-top complexity. The multicast approach is the core technology driving IPTV deployments with switched video. Every channel maps to a multicast address and flipping to a new channel results in joining the multicast group corresponding to the desired channel.

2.3 Broadband Cable Networks

Broadband cable technology provides high-speed Internet access, voice services, and HD video transmission to homes and businesses, as well as schools, hospitals, and other community centers of activity. Driven by new technologies such as bandwidth aggregation (via DOCSIS 3.0), flexible and efficient edge quadrature amplitude modulation (QAM) bandwidth allocation and sharing on the downstream, and switched digital video (SDV) transmission, broadband cable can compete with fiber access in bandwidth and service provisioning with only a small fraction of fiber deployment and maintenance costs.

The hybrid-fiber coax (HFC) network is a vast improvement over the older, one-way cable TV (CATV) distribution plant and is capable of supporting a wide variety of services for the access network. CATV was originally designed to extend broadcast analog TV signals to homes obstructed from line-of-sight reception, which resulted in viewers not being able to pick up a good signal. CATV overcame terrain issues by erecting mountaintop sites and tall towers in strategic locations. The HFC network has been extensively deployed in over 92% of all U.S. homes. An HFC network uses a bidirectional shared-coaxial cable medium in a tree-branch architecture combined with fiber to increase the bandwidth capabilities of the access network. The combination of the shared medium and tree-branch topology allows a high degree of aggregation of user's traffic, which is not possible in dedicated telephone connections using traditional twisted pair local loops between each home and the central office.

2.3.1 DOCSIS Standard

The Data over Cable Service Interface Specification (DOCSIS) standard is a de facto HFC standard developed by CableLabs. DOCSIS enables high-speed Internet-based services, including packet telephony, video conferencing, and multimedia services. The DOCSIS specification allows for 100 miles of HFC (or 800-μs propagation delay) between the headend and a subscriber's home. The upstream and downstream are divided into two separate chunks of bandwidth ranging from 5 to 42 MHz and 54 to 864 MHz or 1 GHz, respectively, which are further subdivided into 6-MHz channels. The downstream bandwidth varies from 54 to 750 MHz in older systems with legacy analog channel bandwidth ranging from 54 to 550 MHz and digital channels taking up the rest. The upstream and downstream data is transferred between the cable modem termination system (CMTS) at the headend and the cable modems (CMs) at the subscriber premises. The CMTS can manipulate their internal queues to give certain traffic classes preferential treatment. For instance, data traffic going to specific network addresses or to specific service-based addressing (associated with service IDs) can be given priority or have internal buffers or bandwidth reserved for them. Although such prioritization is not as functional as latency and error rate QoS guarantees, it can be usable for service level agreements when combined with restricted entry (e.g., attempted connections at high priority are rejected if they increase the traffic load to the point when the entire network will slow down).

Unlike DSL, where each subscriber has a dedicated connection to the head-end (DSLAM), the cable upstream path to the CMTS is shared by all subscribers on a given cable segment. If that upstream is saturated, downstream speeds may be compromised for all subscribers on that segment since many applications (e.g., applications based on TCP and HTTP) require both upstream and downstream to operate. Although a shared medium slows down during congestion, the upstream and downstream are separate channels. The contention nature of the upstream minislots allows other users to request upstream transmission opportunities, even when there are a few heavy users on the upstream.

DOCSIS 1.0, 1.1, and 2.0 are earlier standards developed by CableLabs. DOCSIS 1.0 and 1.1 provide asymmetrical data rates on the upstream and downstream. DOCSIS 1.1 enhances DOCSIS 1.0 with QoS features. DOCSIS 2.0 enhances the upstream modulation of DOCSIS 1.0/1.1 to provide rates that are comparable to the downstream rates. The data rates for these standards are shown in Tables 2.1, 2.2, and 2.3. Clearly, the upstream bandwidth and data rates are more scalable. The upstream channels can be aggregated, for example, four upstream channels per downstream channel per cable segment, to support 1,000 subscribers (a recommended maximum). DOCSIS uses the raised cosine low-pass filter to remove high-frequency harmonics for both upstream and downstream transmission. As an example, the occupied downstream bandwidth for 256-QAM becomes 5.36 × (1 + 0.25) or 6.7 MHz. The data rate is 8 bits/symbol × 5.361 Msymbols/sec or 42.88 Mbps. Similarly, for 64-QAM on the downstream, the QAM modulator must run at exactly 5.057 Msymbols/sec, giving rise to a rate of 6 bits/symbol × 5.057 Msymbols/sec or 30.34 Mbps and a bandwidth of 5.057 × 1.25 or 6.3 MHz. The upstream rates can be computed in the same manner. For EuroDOCSIS (European version of DOCSIS), the channel bandwidth is 8 MHz and the symbol rate is fixed at 6.952 Msymbols/sec, which gives a maximum downstream data rate of 55.62 Mbps for 256-QAM and 41.71 Mbps for 64-QAM. The upstream rates are the same as the North American version of DOCSIS. Demodulation of the QAM signals is specified in terms of channel impairments (e.g., adjacent channel interference, incoming phase noise, echoes, power line hum) and the input signal to noise ratio.

Table 2.1: DOCSIS 1.0/1.1 Upstream Data Rates

Symbol Rate (Msymbols/sec)	Bandwidth (MHz)	QPSK Data Rate (Mbps)	16-QAM Data Rate (Mbps)
0.16	0.2	0.32	0.64
0.32	0.4	0.64	1.28
0.64	0.8	1.28	2.56
1.28	1.6	2.56	5.12
2.56	3.2	5.12	10.24

Table 2.2: DOCSIS 2.0/3.0 Upstream Data Rates

Symbol Rate (Msymbols/sec)	QPSK (Mbps)	8-QAM (Mbps)	16-QAM (Mbps)	32-QAM (Mbps)	64-QAM (Mbps)	128-QAM (Mbps)
0.16	0.32	0.48	0.64	0.8	0.96	1.12
0.32	0.64	0.96	1.28	1.6	1.92	2.24
0.64	1.28	1.92	2.56	3.2	3.84	4.48
1.28	2.56	3.84	5.12	6.4	7.68	8.96
2.56	5.12	7.68	10.24	12.8	15.36	17.92
5.12	10.24	15.36	20.48	25.6	30.72	35.84

Support for 128-QAM is optional. Channel bonding possible with DOCSIS 3.0.

Table 2.3: DOCSIS 1.0/1.1/2.0/3.0 Downstream Data Rates

Modulation	Symbol Rate (Msymbols/sec)	Bandwidth (MHz)	Raw Data Rate (Mbps)
64-QAM	5.057	6.3	30.34
256-QAM	5.361	6.7	42.88

Channel bonding possible with DOCSIS 3.0.

DOCSIS 3.0 is the latest cable specification [4]. DOCSIS 3.0 allows cable operators to add more HD channels and increase broadband Internet speeds. Two key features of DOCSIS 3.0 are bidirectional channel bonding (i.e., aggregating channels to increase bandwidth) and IP multicast forwarding support. With the introduction of channel bonding, the potential scope of multicast applications in a cable network is much greater than with earlier DOCSIS implementations.

DOCSIS 3.0 has enabled PCMM (see Chapter 10) to leverage multicast by introducing differentiated QoS per individual multicast group. PCMM can potentially use bonded as well as nonbonded channels for multicast traffic. Channel bonding also allows more efficient bandwidth utilization when a higher number of variable bit rate (VBR) video streams are aggregated.

The specification provides bonding of multiple channels in both upstream and downstream directions. The most basic implementation bonds four channels, providing 160 Mbps on the downstream (256-QAM) and 120 Mbps on the upstream (64-QAM). As many as 24 channels or more can be bonded, giving an aggregated data rate of over 1 Gbps on the downstream. The higher downstream bandwidth supports IP video downloads and streaming, as well as traditional client/server Internet applications involving Web browsing and FTP. The slightly asymmetrical upstream and downstream rates cater for p2p, voice over IP, online gaming applications as well as business services. In addition, DOCSIS 3.0 supports IPv6 and increased edge QAM flexibility for downstream VoD support. The session and resource manager (SRM) can allocate bandwidth to a digital stream in a static or dynamic fashion using entire QAM carriers or a subset of QAM carriers. Each digital stream may contain MPEG transport stream packets or IP packets. Unlike previous versions of DOCSIS (i.e., 1.0, 1.1, and 2.0), DOCSIS 3.0 supports a decoupled modular-CMTS (M-CMTS) architecture where all components must cooperate to provide the required level of QoS (Figure 2.1). For instance, the downstream edge QAM cannot delay, drop, or reorder packets received from forwarder (also known as staging processor).

The Gigabit-Ethernet architecture supports oversubscription by sharing the Ethernet payload over more QAMs, provided not all QAMs will exceed their peak usage rate simultaneously. Ethernet switching enables any unicast or multicast stream from the headend (which may be encrypted) to be directed to any remote QAM, and thus enable a more efficient utilization of the capacity. Multiple service operators (MSOs) can now start a deployment with a low number of streams but have the flexibility to direct any stream to any serving area in the network.

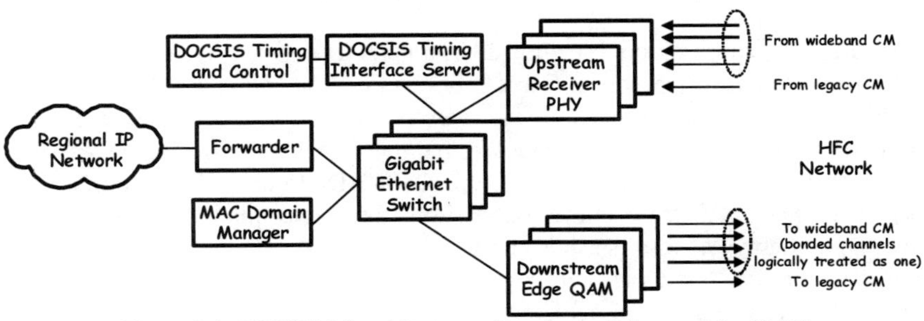

Figure 2.1: DOCSIS 3.0 architecture with a decoupled or modular CMTS.

2.3.2 Switched Cable Services

Switched cable deployment was first used for VoD applications, where movies are sent to individual consumers on request. In early 2003, many leading U.S. cable

MSOs began deploying IP/Ethernet-based VoD architectures (Figure 2.2). After the initial success of VoD and subscription VoD (SVoD), cable MSOs began considering extending the IP/Ethernet narrowcast infrastructure to carry additional video content from their broadcast lineup. Switched digital video (SDV) is a new video delivery architecture that allows operators to expand their offerings to a virtually unlimited number of broadcast programs while releasing precious bandwidth for the revenue-generating services. SDV improves bandwidth utilization and allows unused bandwidth to be reclaimed when inactive channels are not accessed by the users. In addition, significant cost benefits can be achieved from bandwidth sharing and optimization through high-density video processing. SDV allows each node or region to operate with the level of programming complexity once reserved for the main distribution center. It moves complex processing of the channel lineup far closer to the subscriber (e.g., at network edge), placing heavy demands on edge video processing equipment. The combination of a high-bandwidth pipe to the home and a switched network capability places cable operators at an excellent competitive position compared to other broadband service providers.

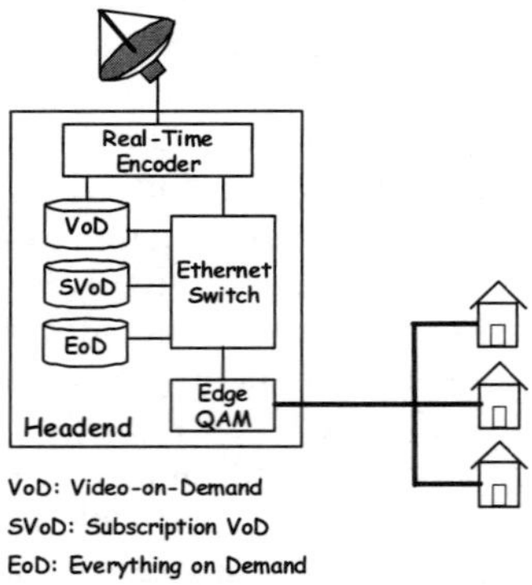

Figure 2.2: Video-on-demand (VoD) architecture.

In the past, cable plants transmit all channels in a given tier to all STBs that reside at the customer's premise. Cable pipes can be 750 MHz, 860 MHz or 1 GHz but the typical available capacity is between 500 to 750 MHz, enough to simultaneously carry 83 to 125 broadcast channels. Each 6-MHz channel can carry data, up to four multicast streams, a 720p/1080i HD stream, or a combination of these services. SDV frees more bandwidth for cable plants by dynamically activating a channel as a subscriber views it (Figure 2.3). The narrowcast approach broadcast popular channels continuously whereas less popular channels are dynamically activated as subscribers view them. This potentially allows for faster channel change compared to the unicast approach. For example, if a channel is selected by one viewer, and another viewer selects the same channel, the newcomer simply joins the existing stream, rather than having a second stream created, which consumes additional bandwidth and increases channel change latency. The narrowcast approach also allows multicast video services to be incorporated easily. When the last viewer in a service group or node leaves a particular channel, it is no longer sent to that node, freeing up the bandwidth. The procedures for channel change are illustrated in Figure 2.4. This signaling process should complete within a 350-ms window so as to keep the "total" channel change time under 500 ms (taking into account the startup delay introduced by the STB before it can start decoding the received video).

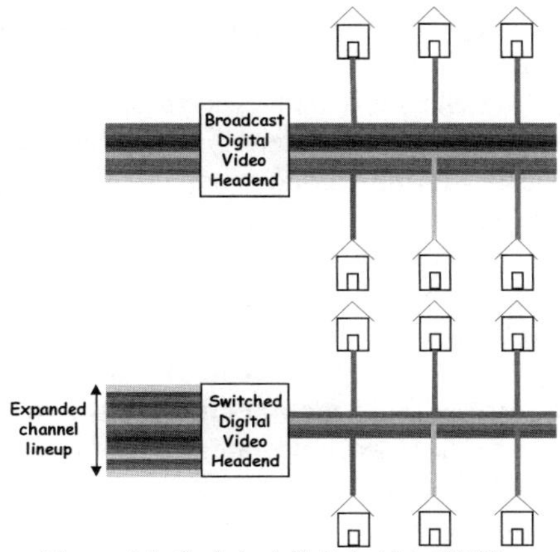

Figure 2.3: Switched digital video (SDV).

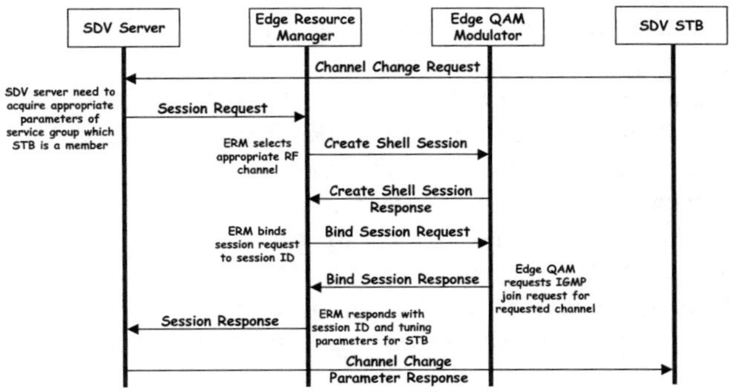

Figure 2.4: Channel change signaling in SDV.

Each program targeted for SDV is typically encoded at constant bit rate (CBR) and then encapsulated into UDP/IP/Ethernet packets. Next, the content must be injected into the IP network as a single program transport stream (SPTS) over a dedicated IP multicast group. This means that video streams must be converted from a multiprogram transport stream (MPTS) to SPTS. Digital program insertion may be performed to insert local ads into the source content. The edge QAM treats the switched channels as standard IP multicast services throughout the network. The channels will be dropped until requested by a subscriber. When a request is made, the service will be forwarded only to the required portion of the network.

Since SDV utilizes 50% to 75% less bandwidth compared to traditional broadcast architectures, migrating to SDV allows the operator to reclaim much-needed bandwidth and expand the video channel lineup and other services (e.g., high-speed data and voice services). After sending the basic analog channels, the cable operator can still offer 1,000 or more digital channels (including HD channels) without having to send them all simultaneously or sacrifice existing channels. SDV can also deliver video in a similar manner as IPTV—one channel at a time. In addition, SDV provides an efficient mechanism for migration to H.264 and the new DOCSIS Set-top Gateway (DSG). A DSG employs an extension of the DOCSIS standard that describes how out-of-band information (e.g., data program guides or channel lineups) is delivered to a cable STB. A DSG can enter into a downstream-only mode when the upstream becomes impaired, avoiding the need to scan for new downstream channels and periodically attempting to reacquire the upstream channel.

The main drawback with current SDV installations is that video streams are converted from VBR to CBR. This presets the total bit rate of each video stream to a fixed value (typically 3.75 Mbps for SD video). This bit rate is usually set lower than the desired value for optimal video quality (VQ), which implies streams are seldom allowed to peak to the maximum rate. In addition, highly efficient codecs such as the H.264 encoder introduces higher bit rate variability in the compressed video stream, resulting in higher peak-to-average ratios and drastic short-term rate changes, which can lead to undesirable losses within the STBs' buffers. Thus, the use of fixed CBR allocation for the transmission of compressed video may lead to severe underutilization of the channel bandwidth or degradation in the quality of video playback.

A key advantage of the VBR approach is the economies of scale with regards to bandwidth utilization when multiple VBR streams are stacked together. In the CBR approach, with 3.75 Mbps per SD stream, a 38.88 Mbps 256-QAM channel can accommodate 10 SD streams. However, more VBR SD streams can be accommodated over the same 256-QAM channel, improving the bandwidth efficiency by more than 40%. The improvement is possible with the use of bonded channels. Although some SDV vendors advocate the use of VBR even in an SDV deployment, some operators have shied away from this approach due to:

- An increase in the required network computing resources of routers, switches, edge QAMs, and other components;
- A reduction in the relative throughput of these devices while increasing the chances of queuing overflows or underflows;
- An increase in the network communication overhead for both the data plane (video delivery) and control plane (control and signaling).

Some solutions to these problems include smoothing each individual VBR video stream to CBR before transporting the multiplexed video streams over one or more bonded channels and statistical multiplexing of the VBR streams (see Chapter 9). Both solutions can achieve considerable bandwidth gains, allowing service providers to increase the number of channels in their lineup.

2.4 Transport and Streaming Protocols

RTP, UDP, and TCP are Internet transport protocols that complement the IP network protocol. IP provides only best-effort packet delivery and thus, makes no guarantee that a given packet will be delivered to the destination, nor does it guarantee its latency if delivered. Thus, there is a need for the transport protocol, which conditions the network protocol to the needs of the application (e.g., with QoS provisioning).

2.4.1 Real-Time Transport Protocol (RTP)

RTP is designed to meet the needs of real-time multimedia streaming (e.g., voice and video) over UDP/IP. RTP ensures that IP packets arrive on time and in the same sending order. To this end, RTP provides a timestamp and sequence number to IP packets containing media information, which can be used by the receiver to synchronize playback and manage buffers to minimize delay jitter. It also supplies information to allow a receiver to resynchronize media for lipsyncing or having text appear at the correct time in relation to an image or word. This can be configured for low-latency applications such as interactive conversations and streaming media. The bitstream can be encrypted for improved privacy against eavesdropping. Like UDP, RTP does not provide reliable transmission, thus out-of-sequence packets can be an issue. RTP can be enhanced for better monitoring, streaming capabilities, and video codec support, using the following RFCs:

- RFC 3550: RTP: A Transport Protocol for Real-Time Applications
- RFC 3551: RTP Profile for Audio/Video Conferences with Minimal Control
- RFC 3640: RTP Payload Format for MPEG-4 Streams
- RFC 3984: RTP Payload Format for H.264 Video
- RFC 4425: RTP Payload Format for Video Codec 1 (VC-1)

2.4.2 Real-Time Transport Control Protocol (RTCP)

This is a companion protocol to RTP. RTCP gathers performance statistics on a media connection such as bytes sent, packets sent, lost packets, delay jitter, and round-trip delay. The application can use this information to monitor the quality of its connections and make adjustments as required, such as changing from a low-efficiency to high-efficiency codec or add an adaptive jitter buffer. The combination of RTP/RTCP is normally used for synchronizing multiple RTP streams. H.460.19 allows several RTP or RTCP sessions to be multiplexed over the same network socket and facilitates firewall implementation.

2.4.3 Real-Time Transport Streaming Protocol (RTSP)

RTSP (defined by IETF RFC 2326) is a client-server protocol for multimedia remote control over IP. This is an IP application-level protocol for controlling delivery of multimedia content, similar to SIP or H.323. It enables the client device to support live or stored Web content streaming and is designed to take advantage of lower-level protocols to provide a complete streaming service over the Internet. Examples of these complementary protocols include RTP for streaming and RSVP for QoS assurance. RTSP is mostly used for video application signaling (just like SIP is mostly used for VoIP signaling). It provides TV- or DVR-like remote control functionality for audio/video stream services such as content navigation (e.g., pause, reverse, fast forward), and absolute positioning and programs for later operations. It provides a means for choosing delivery protocols such as UDP, multicast UDP, and TCP as well as delivery mechanisms based on RTP. It is highly beneficial for both large audience multicasting and real-time multimedia-on-demand unicasting.

2.4.4 Real-Time Messaging Protocol (RTMP)

RTMP was developed by Adobe. The protocol provides multiplexing and packetization services for the higher-level multimedia stream protocol, ensuring timestamp-ordered end-to-end delivery of all messages when used in conjunction with TCP (Figure 2.5). In addition to the play command, RTMP incorporates a play2 command that can switch to a different bit rate stream without changing the timeline of the content played (Figure 2.6).

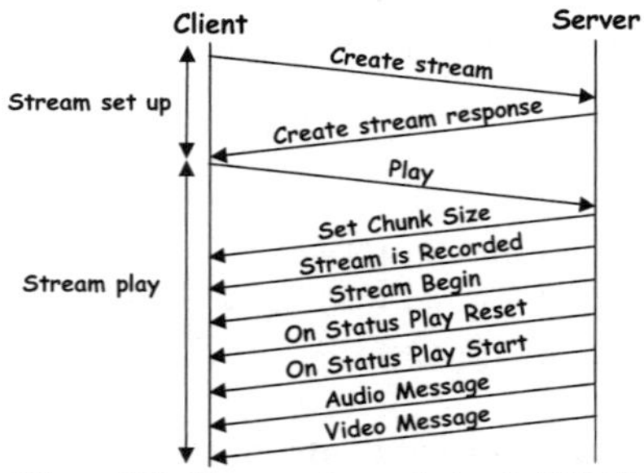

Figure 2.5: Stream setup and play in RTMP.

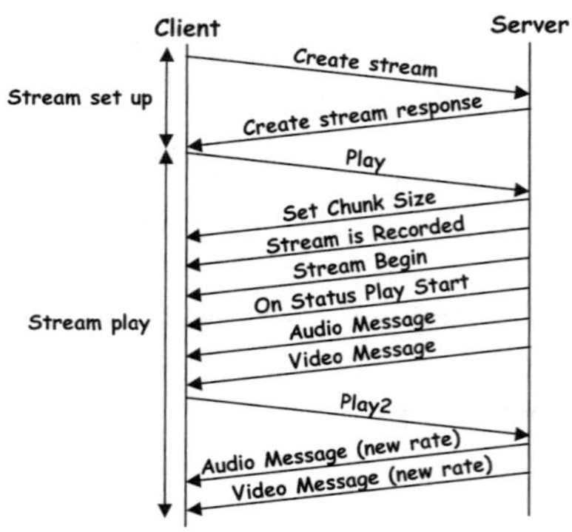

Figure 2.6: Switching streams in RTMP.

2.4.5 TCP and HTTP

Loss-free protocols that ensure ordered delivery of packets are desirable for compressed video transport because compressed video is highly sensitive to information loss. Unlike UDP, which does not add much to IP other than specifying the source and destination ports of the process, and a checksum to detect corrupted packets, TCP is a connection-oriented transport protocol that guarantees delivery of packets. The reliability is achieved by first establishing a session and then retransmitting corrupted or lost packets. HTTP is largely based on TCP. HTTP allows standard caching (which can scale to a high number of connections) and eliminates any issues with firewalls since it uses the same ports as Web browsing.

TCP incorporates congestion and flow control using a sliding window mechanism to minimize lost packets. Thus, the end-to-end throughput is highly dependent on how fast the window rotates. This constitutes a throughput bottleneck in high-speed networks in metropolitan and wide areas (i.e., windows usually close). Therefore, the end-to-end throughput is often specified in terms of network response time and window size. Four congestion control algorithms are defined in standard TCP namely slow start, congestion avoidance, fast retransmit, and fast recovery. Congestion is signaled by packet losses that are detected via timeouts and duplicate acknowledgments, and the missing packets are subsequently retransmitted. A congestion window is maintained by the sender to control the amount of data sent to network before an acknowledgment is received. The ideal value of the window is the bandwidth-delay product (BDP) or the product of the available bandwidth and the round trip time (RTT). Another state variable, the slow start threshold (ssthresh), is used to determine whether the slow start or congestion avoidance algorithm is used to control data transmission. During congestion avoidance, TCP probes the available bandwidth by increasing the congestion window slowly, approximately one segment per RTT, and reduces the congestion window dramatically when network congestion is detected.

Despite the widespread deployment of the protocol, standard TCP exhibits poor performance in broadband networks. Standard TCP assumes that network congestion is a direct result of packet loss and employs the additive increase multiplicative decrease (AIMD) algorithm to solve the problem. However, AIMD is too conservative when applied to broadband networks where the BDP and the RTT are both very large. A long RTT reduces the throughput because the TCP congestion window is increased in the time unit of the RTT. Thus, a very small probability of packet loss will be sufficient to prevent standard TCP from utilizing the available bandwidth efficiently. To operate in broadband networks, TCP needs to be modified from its original low-speed network design and to more aggressive congestion control algorithms. Some of these methods are described below.

Besides the simple solution of using multiple TCP connections in parallel, better broadband transport protocols can be designed by either modifying the standard TCP protocol or redesigning new congestion control schemes based on

UDP. The authors in [5] compare the performance of two implementations of standard TCP (Tahoe and Reno) as well as two modifications of TCP Reno (New-Reno and selective ACK or SACK). TCP Tahoe operates with slow-start, congestion avoidance, and fast retransmission. TCP Reno operates with a fast recovery enhancement. TCP New-Reno eliminates TCP Reno's wait for a retransmission timeout when multiple packets are lost within a window interval. In SACK, the receiver reports to the sender, a noncontiguous set of data segments that have been received and queued, thereby enabling the case of multiple packet losses to be handled more efficiently.

Other efforts to modify standard TCP include HighSpeed TCP [6], Scalable TCP [7], and Explicit Control Protocol (XCP) [8]. HighSpeed TCP behaves in a similar manner to traditional TCP for small BDP but updates the congestion window aggressively for high BDP. In Scalable TCP, a more aggressive multiplicative increase multiplicative decrease (MIMD) congestion control algorithm is used instead of AIMD. Based on the observation that packet loss is a poor indicator of congestion (since this occurs after congestion has occurred), XCP extends the idea of Explicit Congestion Notification and adds a congestion header field. Because the congestion state (per data flow) is carried in header of the packets, intermediate routers can help senders choose appropriate congestion windows. However, the requirement of modifying all intermediate routers in the path may not be realistic. Rate-based congestion control algorithms built on the top of UDP, including SABUL, Tsunami, Hurricane, and RUNAT, are reported in [6]. These solutions are implemented at the application layer above UDP and therefore avoid modifications to operating system kernels and routers.

Based on the assumption that packet losses are indications of network congestion, the AIMD congestion control of standard TCP reaches the steady state, which reflects the protocol's efficiency in terms of throughput and link utilization. Any packet loss is considered to be the result of congestion and the congestion window size is reduced dramatically as a precaution to avoid congestion. However, this assumption does not hold when the end-to-end path includes wireless links. Wireless links may experience transmission errors and higher latency due to signal fading, shadowing, handoff, and other radio impairments. This may cause TCP to falsely assume the presence of packet losses, which in turn causes the radio link to be underutilized. Extensive research has been conducted to overcome these limitations. Suggested solutions can be categorized under end-to-end solutions (which require modifications at the client and/or server), link layer solutions, or proxy-based solutions (which require some changes in the network without modifying the end nodes).

2.4.6 TCP Operation in an Access Network

Unlike UDP, TCP-based protocols such as FTP and HTTP require acknowledgments (ACKs), which imply upstream and downstream links are needed. In a DOCSIS cable access network, a downstream TCP flow will have to receive ACKs on the upstream. Methods to enhance TCP throughput include the use of cumulative or concatenated ACKs. With cumulative ACKs, several TCP packets can be acknowledged using a single ACK. In concatenated ACKs, more ACKs are sent in a single upstream burst. Both methods will speed up the upstream ACKs significantly and increase the downstream TCP window.

In order to transmit a packet upstream, a CM must first request bandwidth. The CMTS grants the bandwidth and the CM then waits for its scheduled time before it can transmit. This cycle is referred to as the request-grant cycle (RGC) or the turnaround time. For a video download application, the rate of TCP packets that can be received by a CM on the downstream is inversely proportional to the RGC duration plus the time to transmit the TCP ACK. The RGC duration is typically between 3 to 5 ms, which comprises the MAP duration (1.5 to 3 ms range) and the MAP ahead time that includes the propagation delay [9]. If each 40-byte upstream TCP ACK takes a further 1.5 ms to send, the maximum packets per second (pps) for the upstream ranges from 154 to 222 pps or equivalently, 49 to 71 Kbps. Assuming maximum-length (1,500-byte) TCP packets on the downstream, this gives a maximum downstream data rate of between 1.848 and 2.664 Mbps. Thus, even if the upstream slows down to dial-up rates, the downstream can still achieve the maximum rate. For a 64-QAM (30.34 Mbps) upstream DOCSIS channel and a 256-QAM (42.88 Mbps) downstream DOCSIS channel, the maximum bandwidth utilization on the upstream and downstream is 0.23% and 6.21%, respectively. This means that up to 16 simultaneous TCP connections (running at maximum speed) can be accommodated on a single 256-QAM downstream channel. Since the TCP window size is typically 2 packets to 1 ACK, the downstream rate and bandwidth utilization can be doubled. In addition, if the upstream channel becomes saturated (i.e., approaches 100% bandwidth utilization), then as many as 6.21/0.23 or 27 bonded channels can be supported on the downstream. In practice, if the upstream saturates, the downstream rate will drop to about the same upstream speed (30.34 Mbps), a slowdown of about 30% (30.34 Mbps versus 42.88 Mbps). This slowdown is more dramatic for DOCSIS 1.1 and 1.0 (66%) since the maximum upstream speed is limited to about 10.24 Mbps (Table 2.1).

2.4.7 UDP Operation in an Access Network

UDP requires only a unidirectional link because the protocol does not require acknowledgments. Assuming the same RGC duration (3 to 5 ms) and a video download application, the maximum data rate for the downstream can be higher for UDP and ranges from 2.4 to 4 Mbps (corresponding to 200 to 333 pps).

2.4.8 Optimizing TCP Operation on the Internet

One of the goals is to counter the high network latency, which can be as high as 500 ms (one-way) on the Internet. If TCP window scaling is used, the network throughput may improve 100% to 500% on WAN links, although the improvement is less in LAN environments. The receive TCP window specifies the number of bytes a sender may transmit without receiving an acknowledgment (equivalent to the amount of bytes in the receiver's memory buffer). Reducing the receive window size effectively causes an ACK to be sent to the sender for data received in a shorter period of time. This reduces the probability that the sender will timeout while waiting for an ACK. However, it will also increase the amount of backlog traffic at sender, thereby lowering the throughput. In general, larger receive windows will improve performance over high-delay, high-bandwidth networks. For high efficiency, the receive window should be an even multiple of the TCP maximum segment size (MSS). The default setting of 64 Kbytes is fine for most LANs but too low for Internet connections. The value should be set to 256 Kbytes for T1 lines or lower and 2 to 4 Mbytes for T3, OC-3 or even faster connections. The optimal buffer size = 2 × bandwidth × delay. There are several enhancements for high-loss environments.

- RFC 2582: The New Reno Modification to TCP's Fast Recovery Algorithm.
- RFC 2883: An Extension to the SACK-based TCP (defined in RFC 2018) reduces the number of retransmissions to improve overall throughput.
- RFC 3517: A Conservative SACK-based Loss Recovery Algorithm for TCP performs loss recovery when duplicate ACKs are received.
- RFC 4138: Forward RTO-Recovery (F-RTO): An Algorithm for Detecting Spurious Retransmission Timeouts with TCP and the Stream Control Transmission Protocol (SCTP) prevents unnecessary retransmission of TCP segments when there is a sudden or temporary increase in the RTT.

2.4.9 Optimizing Transport Protocols for Video Streaming

A proper delivery rate at the receiver prevents buffer underflow or overflow and maintains smooth video playback. It is useful when streaming with TCP or UDP (without using RTP) because timing synchronization is absent in these transport protocols. This requires the number of frames in the video file or segment to be determined. The average frame size or segment can be computed using the file size or segment divided by the number of frames. Suppose the size of a compressed video file of 1,732 frames of duration 72.2s is 16.6 Mbytes. The average frame size is 16.6 Mbytes/1,732 frames or 9.58 Kbytes. The video frame rate is 1,732/72.2 or 24 frames/sec (24 Hz). Thus, the desired average receiving rate becomes 9.58 × 8 × 24 or 1.84 Mbps.

2.5 Link Quality Measurement

Continual performance and quality monitoring of the transmission link is needed to quickly identify, locate, and fix problems before they impact the video and sound (e.g., audio, voice) quality during playback. It is also important to separate errors or losses associated with the network transport from errors associated with the limitations of the video codec. Link quality measurements (e.g., delay jitter and packet loss rate) can be collected from the customer premise device. Many emerging wireless deployments allow such information to be collected by the base station so that transmission parameters can be adjusted. This may be a more responsive alternative to monitoring VQ at the STB since it does not involve the decoding of the received content to compare with the actual video content.

A link quality indicator can be used to track the video transport performance by sampling the video packets at predefined intervals. The indicator may employ network-level measurements to identify and measure delay jitter and packet loss. A lightweight and scalable indicator serves as a proactive and preventive measure to forecast, with high probability, impairments that may result in unacceptable video delivery or on conditions that result in unacceptable network margins before VQ is impacted. This is in contrast to the use of error concealment (covered in Chapter 8), which is a reactive measure taken *after* the video is degraded. The link quality indicator is also useful for dimensioning the appropriate buffer size at the client device to prevent losses and for determining the amount of retransmission, in particular, for higher priority video packets or partitions. For the latter case, the relative importance of the video information can be extracted from the H.264 NAL unit header without examining the video content.

There are two key parameters employed by the link quality indicator:

- Delay jitter rate (DJR): This indicates the amount of buffering required at the receiving side to deal with delay variation in the packet interarrival times. It is defined as the ratio of the buffer size and the average time between request and consumption of video data. This can be measured in timestamp units (expressed as a 32-bit unsigned integer) such as the RTP timestamp.
- Video packet loss rate (VPLR): This indicates the number of out-of-order/lost packets over a specified duration.

The impact of the client buffer size and the video delivery rate on DJR and VPLR can be evaluated further. We employ an H.264-encoded *Foreman* video (300 frames) with a sampling interval of 1 second for computing DJR and VPLR. The results are shown in Table 2.4. The results for the *Terminator-2* HD video are shown in Table 2.5. A few key observations can be made:

- The DJR improves (decreases in value) as the delivery rate increases. This is expected since the amount of time that the video needs to be buffered is less for a high delivery rate.
- The buffer size does not influence the DJR as much as the delivery rate since the values are nearly the same. This is expected for the following reason: as long as sufficient buffer is available for storing at least one video packet, more buffer space is not necessary.
- Even when the buffer size is small (e.g., 20 KB) and cannot hold most of the packets (i.e., buffer size is less than the average packet size), the DJR does not suffer. However, the VPLR becomes affected. VPLR indicates the proportion of lost packets (i.e., packets that are delivered late compared to their playback/delivery time). It is seen that if the buffer size is small and as the delivery rate increases, more packets are expected for delivery. Due to the small buffer size, these packets are not available. Hence, the VPLR suffers. For a delivery rate of 8 Mbps, the loss rate is nearly 4.8%.

- It is also seen that when there is sufficient buffering space available at the client (receiver), then there is no loss (VPLR = 0).

Table 2.4: DJR and VPLR Performance for an H.264 *Foreman* CIF Video

	Rate of Delivery (Mbps)	DJR (bytes/sec)	VPLR (lost packets/total packets)
Client Buffer Size = 20 Kbyte	9.6	15.39	0.048
	8.0	24.27	0.048
	6.4	37.59	0.045
	4.8	59.79	0.021
	3.2	104.18	0.003
	1.6	237.39	0
Client Buffer Size = 50 Kbyte (average packet size = 59.2 KB)	9.6	15.22	0
	8.0	24.07	0
	6.4	37.39	0
	4.8	59.49	0
	3.2	103.68	0
	1.6	236.38	0
Client Buffer Size = 100 Kbyte or more	9.6	15.22	0
	8.0	24.07	0
	6.4	37.34	0
	4.8	59.45	0
	3.2	103.68	0
	1.6	236.38	0

Table 2.5: DJR and VPLR Performance for an H.264 *Terminator-2* HD Video

	Rate of Delivery (Mbps)	DJR (bytes/sec)	VPLR (lost packets/total packets)
Client Buffer Size = 100 Kbyte	9.6	60.49	0.117
	8.0	78.33	0.091
	6.4	105.12	0.0145
	4.8	149.81	0
	3.2	239.22	0
	1.6	507.48	0
Client Buffer Size = 120 Kbyte (near to average packet size)	9.6	60.47	0.113
	8.0	78.28	0.085
	6.4	104.35	0
	4.8	148.26	0
	3.2	237.67	0
	1.6	502.56	0
Client Buffer Size = 150 Kbyte	9.6	60.47	0.104
	8.0	78.28	0.081
	6.4	104.35	0
	4.8	148.26	0
	3.2	237.67	0
	1.6	502.56	0
Client Buffer Size = 200 Kbyte or more	9.6	60.45	0
	8.0	78.21	0
	6.4	104.19	0
	4.8	147.98	0
	3.2	236.26	0
	1.6	501.78	0

To summarize, moderate-sized buffers (greater than the average size of the packets) with an appropriate delivery rate reduce the DJR to a reasonable value while maintaining losslessness of transmission (VPLR = 0). A small buffer size at the client device may lead to high packet losses whereas an overprovisioned buffer size will increase channel change latency, which is undesirable.

2.6 MPEG Video Encapsulation

Delivering a video program over IP requires several layers of encapsulation. Typically this involves combining a few elementary streams such as a video stream, one or more audio streams, and optional data streams (subtitles, close captions). The MPEG-4 standard specifies compression of audio-visual information into an audio or video elementary stream. The MPEG-4 Part 14 or MP4 file format is a multimedia container format specified as a part of MPEG-4. In MPEG-4 encapsulation, these streams take the form of audio-visual objects that may be arranged into an audio-visual scene by means of a scene description. Information on the type of MPEG-4 stream carried in the payload is conveyed by the Multipurpose Internet Mail Extension (MIME) format parameters. Each MPEG-4 elementary stream consists of a sequence of access units (AUs), which may include compressed audio and video frames. The AU is the smallest data entity that contains timing information. For audio, the AU may represent an audio frame and for video, a video frame. All AUs are byte-aligned and padded zero bits are added for frames that are not byte-aligned. AUs within an RTP packet can be interleaved to improve error resiliency. The sender can dynamically adjust the interleaving pattern based on the AU size, error rates, and other factors but need to inform the receiver about buffer resources needed for deinterleaving. The RTP receiver does not need to know the interleaving pattern used; it only needs to extract the index information carried in each AU and insert the AU into the appropriate sequence in the decoding queue. For example, if an RTP packet contains 3 AUs, if the AUs are numbered 0, 1, 2, 3, 4, ..., and if an interleaving group length of 9 is chosen, then RTP packet(i) contains the following AUs:

RTP packet(0): AU(0), AU(3), AU(6)
RTP packet(1): AU(1), AU(4), AU(7)
RTP packet(2): AU(2), AU(5), AU(8)
RTP packet(3): AU(9), AU(12), AU(15)
RTP packet(4): AU(10), AU(13), AU(16)
RTP packet(5): AU(11), AU(14), AU(17)

The RTP timestamp must carry the sampling instant of the first AU in the RTP packet. When multiple AUs are carried within an RTP packet, the timestamps of subsequent AUs can be calculated if the frame period of each AU is known. For audio and video, this is possible if the frame rate is constant. MPEG-4 defines two types of timestamps:

- Composition timestamp (CTS)
- Decoding timestamp (DTS)

The CTS represents the sampling instant of an AU, and hence is equivalent to the RTP timestamp. The DTS may be used in MPEG-4 video streams that use bi-directional coding (i.e., when frames are predicted in both forward and backward directions using either a reference frame in the past or a reference frame in the future). Since the DTS cannot be carried in the RTP header, it may be carried in the RTP payload for each AU. To efficiently code the timestamps, each CTS or DTS is coded as a difference. For the CTS, the offset from the RTP timestamps is provided, and for the DTS, the offset from the CTS is provided.

Because service providers have more experience with MPEG-2, some deployments have employed MPEG-2 transport stream (TS) to carry MPEG-4 video. Like UDP, MPEG-2 TS is unidirectional and works well in networks with low losses. An MPEG-2 TS consists of a series of 188-byte TS packets (TSPs), where each TSP typically requires 4 bytes of header for video (i.e., 184-byte payload) or 7 bytes of header for audio. The start of the TSP contains a synchronization byte of 0x47. This is followed by the payload unit start indicator (1 bit), adaptation field control (1 bit), packet identifier (10 bits), continuity counter (4 bits), and payload byte offset (8 bits). The payload of each TSP may contain program information and video or audio streams that are collectively called packetized elementary streams (PESs). The audio and video PESs are sequentially separated into access units and each TSP contains information from one elementary stream only. The PES header starts from the first byte of the TSP payload. The length of the PES header is variable—8 bytes for video without B-frames and audio, 13 bytes for video with B-frames. An 8-byte PES header contains the start code prefix (3 bytes, 0x000001), the stream identifier (1 byte), the PES length (2 bytes, which gives a maximum PES length of 65,536 bytes), and the timestamp for synchronization purposes (2 bytes). The length of the PES payload is also variable since the length of the audio and video streams are variable. If the audio payload is encoded using advanced audio coding (AAC), a 12-bit sync word of the 7-byte audio data transport stream (ADTS) header is first located. The frame length is then obtained from the ADTS header and the audio data corresponding to that length is encapsulated with the audio stream identifier in the PES. The audio frame number is computed from the beginning of the stream and is included in the 2-byte timestamp. Each AAC frame contains 1024 samples, as defined by the standard. The time for specific audio frame playback can therefore be computed by multiplying 1024 with the current frame number and the sampling duration. If the video payload is encoded using H.264, the network adaptation layer flag and the picture and sequence parameter sets are first identified (see Chapter 3). The parameter sets are encapsulated in a separate PES with a frame number of zero. The video data corresponding to the frame length is then encapsulated as the video payload together with the stream identifier. Just like the AAC audio stream, the frame number for the video payload is computed from the start of the stream and encapsulated in the timestamp.

The TSPs are grouped together and can be encapsulated within a UDP datagram, forming an IP packet. Alternatively, the TSPs can be wrapped within an RTP packet before UDP encapsulation. While a TS stream contains multiplexed data, RTP does not provide any multiplexing capability. RTP relies on underlying encapsulation such as UDP to provide multiplexing over an IP network. Each RTP stream must then carry timing information to allow the receiver to synchronize the streams. Figure 2.7 shows an RTP packet containing several MPEG-2 TSPs within its payload, all encapsulated using UDP in an IP packet. Seven 188-byte transport packets typically constitute the maximum RTP payload. Each encapsulation adds additional header overheads. For general delivery over the unmanaged Internet without QoS guarantees, streaming protocols such as RTP may be used, but even then, packets may be lost in delivery, resulting in artifacts in the media presentation. If the network has sufficient QoS (e.g., private managed network), it is possible to deliver media packets without the RTP overhead. The packets are instead inserted directly into UDP packets. Figure 2.8 shows how MPEG-2 TSPs can be encapsulated within a UDP/IP packet.

The overhead to carry an MPEG TS is 46 bytes over UDP/IP/Ethernet and 58 bytes over RTP/UDP/IP/Ethernet. Together with the TSP header, this equates to a video overhead of about 5.7% and 6.7%. For audio, a 2-byte header for each TSP and a 2-byte header length for each IP packet are needed, giving a total overhead of 111 bytes and 123 bytes respectively. The number of TSPs per IP packet may vary. Encapsulating more TSPs per IP packet improves the network utilization but this increases jitter across the network. However, if jumbo packets supporting up to 47 TSPs per IP packet are used, network utilization can be increased to 99%. In practical deployments, the TS overheads are closer to 9% (including FEC overheads). As an example, a DOCSIS 3.0 CMTS serving four 256-QAM outputs on the downstream provides a data rate of 42.88 Mbps each. The actual TS rate of 256-QAM is about 38.88 Mbps, which equates to about 4 Mbps of overheads.

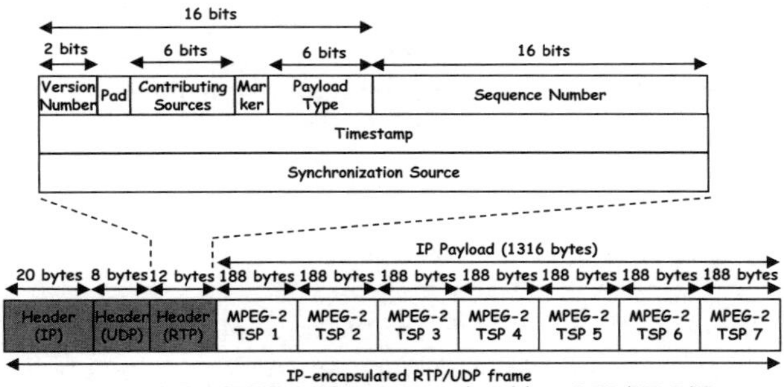

Figure 2.7: MPEG-2 TSPs encapsulated into RTP/UDP/IP.

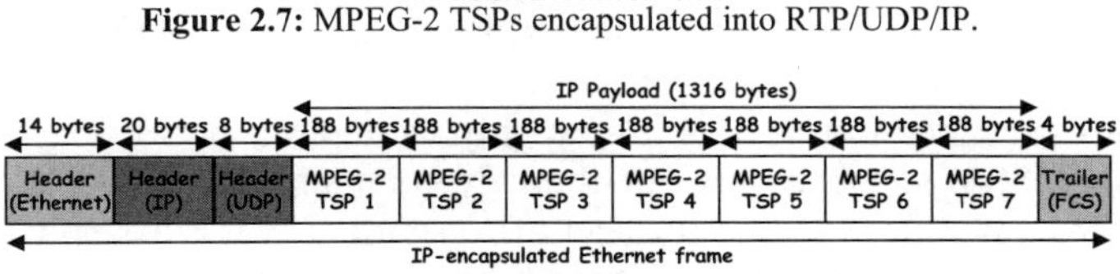

Figure 2.8: MPEG-2 TSPs encapsulated into UDP/IP/Ethernet.

2.7 IP Multicast

The DOCSIS 3.0 standard supports IP multicast, which can provide cost-effective, efficient, and high-quality content delivery, including new services like IP video.

This is a natural evolution from IP data and voice cable services. With IP multicast, considerable bandwidth savings can be achieved by packaging TV channels and video programs into different multicast groups since, given a service group, there is a high probability that subscribers will be viewing the same channels, especially popular channels. This creates the need to evaluate subscriber and channel profiles to determine the most effective channel packaging solutions without compromising QoS for isochronous services like video (i.e., low delays, low jitter, consistent bit rate). In addition, the DVB standards for secure pay-per-view (PPW) video multicast can also be adopted. Some benefits of multicast video transmission for a DOCSIS 3.0 network are listed in Table 2.6.

Table 2.6: Impact of Multicast Video on a DOCSIS 3.0 Network

Factor	**Influence on Performance**
Number of customer movie requests	Initial increase in request blocking can be reduced with multicast because requests are grouped
Buffering at set-top	More required for multicast compared to unicast
On-demand nature	Near on-demand, instead of true on-demand for unicast
System complexity	Higher compared to unicast
Total number of channels	Multicast increases capacity further because each channel can satisfy multiple requests

2.7.1 Mechanisms

IP multicasting conserves network bandwidth by reducing the amount of unnecessary network traffic in one-to-many or many-to-many communications. It has a wide range of potential applications and can provide cost-effective and high-quality content delivery. For example, the one-to-many model can be used for IP video while the many-to-many model can be used for video conferencing, online gaming, and so forth. Three mechanisms are required, namely

- Group addressing mechanism;
- Host joining/leaving mechanism;
- Multicast-enabled routing protocols.

Two popular IP multicast models are the any source multicast (ASM) model and the source specific multicast (SSM) model. ASM supports both one-to-many and many-to-many communication models, where the receiver joins an available multicast group. The use of network level source discovery in ASM simplifies applications at the expense of a highly complex network deployment. Alternatively, SSM supports only one-to-many communication models, where the receiver joins a multicast source (instead of a group). The use of application level source discovery in SSM reduces network complexity, thereby lowering the cost of operation. SSM also provides resistance against denial-of-service (DoS) attacks.

Multiple multicast trees allow for fine-grained traffic engineering while a smaller number of trees carrying multiple programs reduces signaling and other overheads. Two extremes are described next. Each TV channel can be allocated its own IP multicast tree, potentially allowing for bandwidth-efficient delivery of only requested channels to different branches of the IP multicast tree. This typically results in higher management overhead and increased channel change latency. Alternatively, the entire set of broadcast channels can be packaged into a single IP multicast tree.

2.7.2 Internet Group Management Protocol (IGMP)

The IGMP manages multicast data flows. With multicast, a single source sends data to multiple destinations at the same time. Each broadcast TV channel would have a unique IP multicast group. Using IGMP protocol, clients can receive the broadcast packets and enable the routing of the broadcast stream to the user device via the network. While multicast saves on network bandwidth, there is no reliability mechanism, and packet losses are possible. In addition, retransmissions due to an error-prone link or overloaded router can make multicast as inefficient as repeated unicast. Thus, there is a trade-off in many multicast applications when using TCP or UDP. Many multicast systems employ UDP exclusively.

Protocols used in IP multicast can be divided into group management protocols and multicast routing protocols. Group management protocols include:

- IGMP v1 (RFC 1112): explicit-join/leave by timeout;
- IGMP v2 (RFC 2236): adds an explicit-leave message;
- IGMP v3 (RFC 3376): optimizes support for SSM model by including source filtered multicast (SFM).

An example of the exchange of IGMP signaling messages in multicast IPTV is shown in Figure 2.9. Here, the IGMP Join time is network-dependent and the system may glitch if latency Join is less than the latency Leave. Reducing the Leave and Join signaling delays minimizes the channel change latency. The SDV system in cable systems uses the QAM modulator to issue join and termination (leave) requests for the IP multicast streams.

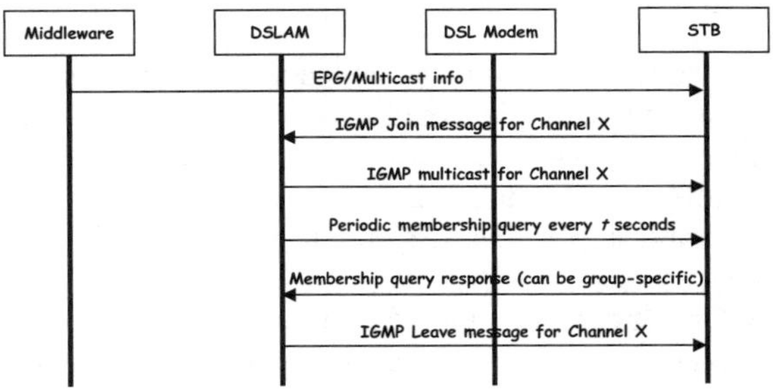

Figure 2.9: Channel change signaling in multicast IPTV.

2.7.3 Multicast Routing Protocols

These include:

- DVMRP (dense-mode/implicit-join/source-based tree): The first of the multicast routing protocols. Uses the flood-and-prune or implicit-join method to deliver traffic everywhere and then determines where the uninterested receivers are. Source-based distribution trees are used. Limitations make this method unattractive for large-scale Internet deployment.
- MOSPF (dense-mode/explicit-join/source-based tree): Adapts the Open Shortest Path First (OSPF) protocol for multicast operation. Employs an explicit-join message so that routers do not have to flood the entire domain with multicast traffic from every source. Source-based distribution trees are used.
- PIM-DM (dense-mode/implicit-join/source-based tree): Allows a router to use any unicast routing protocol and performs reverse path forwarding (RPF) checks using the unicast routing table. The use of an implicit-join message implies routers have to use the flood-and-prune method to deliver traffic everywhere and then determine where the uninterested receivers are. Source-based distribution trees are used, just like all dense-mode protocols.
- PIM-SM (sparse-mode/explicit-join/source-based tree): Similar to PIM-DM but uses an explicit-join message, so that routers can determine where the interested receivers are and send join messages upstream to their neighbors, building trees from receivers to the reverse path. Typically does not use source-based distribution trees.
- CBT (sparse-mode/explicit-join/shared-tree): Shares all characteristics of PIM-SM but tends to be more efficient at finding sources than PIM-SM.

2.7.4 Challenges for Multicast Access Networks

The key challenges for deploying IP multicast in DOCSIS networks are

- SSM and IGMP v3 support throughout the network;
- Enhancement of DOCSIS to provide QoS support for IP multicast services;
- Support for layer 2 virtual private network (VPN) service.

The use of multicast streams poses a challenge for implementing QoS. Unlike unicast, which involves a single transmitter and receiver, multicast involves multiple receivers, each with a potentially different service level agreement (SLA), communicating with the same transmitter. Furthermore, the dynamic nature of multicast group membership makes it difficult to predict in advance the network resources consumed by multicast streams.

Although IP multicast allows efficient delivery of streaming video to thousands of receivers by replicating packets throughout the network, problems arise when the node is located far away from the multicast publishing points. Streaming video that uses interframe compression require a reference frame. Out-of-order video packets or missing reference frames may cause the video to freeze. To deal with this problem, one can reproduce the multicast closer to the user. An alternative is to employ peer-to-peer (p2p) multicast streaming [10] for video sharing (in the same way people share general data files) and video streaming.

Multicast IPTV over DSL poses new challenges. The bandwidth bottleneck in the DSLAM backhaul, router, video headend is solved but the access link bottleneck remains even though DSL does not require explicit bandwidth reservation. Specifically, multichannel viewing in a single household can result in bandwidth bottlenecks, due to the dedicated connections in DSL networks and the limited bandwidth of the copper wires that degrades with distance. As an example, Figure 2.10 shows the bandwidth bottleneck at premises C and D when accessing multiple channels on a multicast DSL network.

2.7.5 Peer-to-Peer Multicast

P2p traffic presently accounts for a significant portion of Internet backbone traffic and overcomes the current lack of IP multicasting support by major Internet service providers. Like wireless for the last mile, it is a disruptive but promising technology. Although p2p transmission may consume high bandwidth (compared to unicast or even multicast transmission), it is scalable to many users and long distances, and is particularly well suited for transatlantic communications. Napster (www.napster.com) is the first p2p file-sharing platform. It was invented by Shawn Fanning in 1998 while he was a college student in Northeastern University in Boston. Another popular p2p platform, BitTorrent (www.bittorrent.com), currently has over 160 million installed clients worldwide. Widely used p2p video streaming platforms include PPLive (www.pplive.com/en), Sopcast (www.sopcast.com), Veetle (www.veetle.com), TvAnts (tvants.en.softonic.com), and UUSee (www.uusee.com).

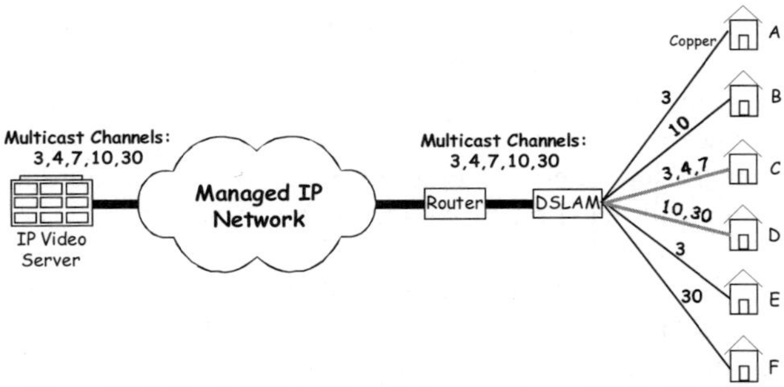

Figure 2.10: Bandwidth bottleneck for accessing multiple channels.

P2p voice applications such as Skype (www.skype.com) are challenging traditional VoIP. Skype captured a significant portion of the international voice calling with 2 million users in the first 3 months since it was launched, and 1 million simultaneous subscribers 1 year later. Skype allows users of the same application to make unlimited free voice calls over the Internet and works with all firewalls, network address translators (NATs), and routers. With Skype, no VoIP media gateways are required since the p2p application effectively allows a voice packet to be downloaded quickly by the recipient from multiple sources. Skype is able to intelligently route encrypted calls through the most effective path possible. Skype even keeps multiple connection paths open and dynamically chooses the one that is best suited at the time. This has the noticeable effect of reducing latency and increasing call quality throughout the network. The success of p2p applications like Skype is also boosted by an FCC ruling in February 2004 that voice communications flowing entirely over the Internet (i.e., voice services that do not interconnect the telephone system) are not subject to traditional government regulations and taxes that applied to public-switched telephone networks. Skype partnered with Boingo to bring voice over Wi-Fi service to 18,000 Wi-Fi hotspots in the United States. Skype video calling has increased substantially in recent years, with an average of 34% of Skype-to-Skype calls now including video. The new Skype 4.2 beta processes 720p HD video at 30 Hz. LG and Panasonic have developed HDTVs with built-in webcams and Skype software, allowing users to communicate with their friends and family from the comfort of the living room and on a big screen. These Skype-enabled HDTVs include voicemail, landline, mobile calls and free Skype-to-Skype call features as well as 720p HD webcams and microphones. In addition, the on-board video encoding and processing power may help out older computers.

In p2p multicast, a distribution tree is established between the source and the receivers. In a VoD environment, the source sends the video to a number of clients (peers) that, in turn, send to other peers until all peers receive the video. The peer nodes need to contribute upload bandwidth, which relieves the bandwidth bottleneck at the video source [11]. This application-layer multicast approach presents a number of technical challenges. First, the distribution tree cannot grow indefinitely, since nodes closer to the leaves may experience high delays. This may not be a real issue for VoD service. However, for other applications, such as live video, delays beyond a certain threshold cannot be tolerated. Another challenge is inherently related to the dynamic nature of p2p communications: nodes in an overlay may join/leave whenever they want, without any notice. A third problem concerns the complexity of establishing the distribution tree. Indeed, nodes are extremely heterogeneous in terms of receiving, transmitting, and storage capacities. For a small network, managing such a structure is feasible. For larger dynamic topologies (millions of receivers), it is likely that the overlay always operates in a suboptimum configuration. Solutions proposed to overcome such limitations include advanced algorithms for correctly placing nodes on the overlay.

The spread of worm viruses can be very rapid with p2p systems. Although the exponential data dissemination capability in these systems can be exploited to halt the spreading, the dynamic participation of peer nodes can reduce the effectiveness. In addition, the port-hopping capabilities in p2p systems imply that port controls (e.g., port 80 for http, port 20 for ftp) can be bypassed. This makes it difficult to manage p2p traffic.

Proper dimensioning of the upstream bandwidth is critical for supporting p2p applications since they will require continuous use of upstream bandwidth and upstream congestion will have an undesirable impact on the downstream. Suppose each CM is capped to an upstream rate of 0.96 Mbps (which can support up to four concurrent p2p video streams). The 30.34 Mbps upstream capacity for 64-QAM can handle roughly 30 (30.34 Mbps/0.96 Mbps) simultaneously active modems before saturation. Assuming only 50% online subscribers and 20% of them actually send data at the same time, this gives 5 × 2 × 30 or 300 modems and allows upstream economies of scale to oversubscribe. In general, the number of streams can be computed from homes passed/node or cable segment × % of cable penetration × % digital subscribers × % expected peak usage. For example, 1000 × 60% × 50% × 20% gives 60 streams. This means that two aggregated 64-QAM upstream channels are required.

2.7.6 Robust Peer-to-Peer Video Streaming

P2p video streaming allows the video sender to deliver a video stream directly to a limited number of peers, and those peers will redirect the video content to other peers. Although each peer keeps a control connection with the sender, the video traffic is delivered in a distributed manner. Figure 2.11 illustrates the benefits of real-time (live) p2p video streaming over unicast streaming using a wireless setup. A wireless base station (e.g., in a cellular network) or a wireless relay (e.g., in a mesh network) assures that all wireless peers are valid subscribers and hence, ensures that the video content is protected against unauthorized distribution. Even with a star topology, p2p streaming relieves the traffic bottleneck at the video source. Note that in a multihop mesh topology, the bottleneck at the video source and base station are removed. In Figure 2.11(a), the video source employs four uplink time slots to send the video stream to four other subscribers. In Figure 2.11(b), the video source utilizes two uplink time slots, with another two slots employed by other wireless peers. By comparing the figures, it is clear that the additional delay in relaying the video information by a wireless peer may be offset by a lower number of streams transmitted by the video sender. Note that noninteractive applications like video streaming are more sensitive to delay jitter and packet loss than a fixed delay. In this case, the network overheads are reduced since all peers are controlled by the base station and no routing protocol is required. Such an architecture may address the needs of mobile workers (e.g., first responders and transportation operators) and fixed outdoor structures (e.g., kiosks and traffic lights) that require IP video surveillance capabilities.

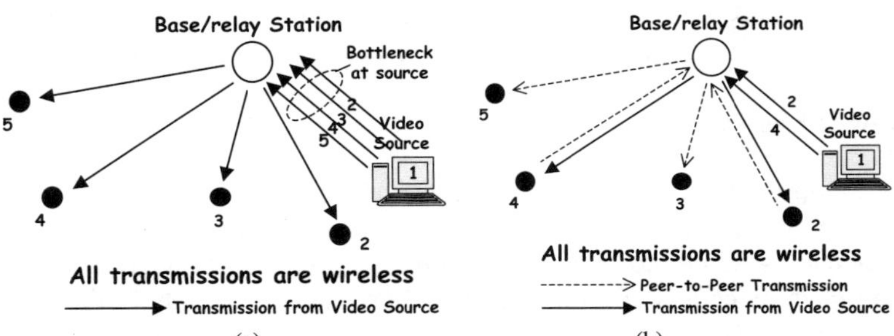

Figure 2.11: (a) Unicast video streaming, and (b) peer-to-peer video streaming.

2.8 Quality of Experience (QoE) versus Quality of Service (QoS)

Many payTV service providers have started to focus on video content programming and end-user QoE to complement network-based QoS monitoring in order to meet rising customer expectations. QoE augments end-to-end QoS by providing the connection to user perception. QoE has an economic dimension. Satisfactory QoE at a reasonable price helps a provider to stay in the market. Thus, the relationship between QoE, QoS, SLA, and pricing deserves attention. Among the items associated with QoE include:

- Measurements and models of user behavior;
- Perceptual QoS (PQoS) and quality measures;
- Interaction between subjective and objective QoE and QoS parameters;
- QoE and QoS feedback facilities and management approaches;
- Resource allocation, fairness, reliability, and dependability issues;
- Cost and pricing of QoE and QoS;
- Security-related QoE and QoS parameters.

2.8.1 Packet Losses versus Packet Errors

Packet losses can occur due to STB/server/router buffer overflow. Bit errors corrupt the packets, which may eventually reach the destination. Bit errors due to the channel generally occur as random or burst errors. The presence of these errors increases the bit error rate (BER) but affects the packet loss rate (PLR) differently. For example, burst errors may result in a lower PLR than random bit errors, although burst errors are more difficult to correct via FEC. The data rates allocated to the video stream also affects the maximum tolerable PLR. Suppose the bit rate allocated for MPEG-2 SD and H.264 HD video transmission is 3.75 Mbps and 6 Mbps, respectively. If one uncorrectable artifact can be tolerated in every 2 hours, then assuming the TS format is used, the PLR limits can be computed as:

$$\text{For MPEG-2 SD: PLR} \leq (7 \times 188 \times 8)/(3750000 \times 3600 \times 2) = 0.39 \times 10^{-6}$$
$$\text{For H.264 HD: PLR} \leq (7 \times 188 \times 8)/(6000000 \times 3600 \times 2) = 0.244 \times 10^{-6}$$

Thus, if the bit rate is doubled, one can tolerate nearly half of the loss rate. Packet loss can be limited for high-priority customers at the expense of lower-priority customers using unequal FEC, packet discard, and enforcement policies.

2.8.2 Codec Losses

MPEG employs lossy coding techniques. Hence, the coded VQ may vary considerably, even at the same coded rate. Assuming the same codec is used, some video content is harder to encode than others (e.g., old, noisy black and white films) so many encoders add prefiltering. Many cable headends further compress the received MPEG TS using statistical multiplexing. If the signal is not degraded to the point of uncorrectable errors (seen as blocking), the modulation method (e.g., QAM) is not very crucial. Some decoder chips in the STB can be fed errors that did not get corrected by the tuner/demodulator and they perform internal error concealment, such as repeating a frame or part of a frame. Some decoder chips can even accept a very corrupt transport stream that violates MPEG timing rules and more, and yet produce a watchable video.

Once the video is taken out via the baseband outputs, the factors that affect VQ are the STB digital-to-analog conversion back to analog and the filtering in the STB. Most STBs use very minimal filtering (too costly) and the digital-to-analog converter in the STB generates relatively clean video. A channel 3/4 remod is a common output for most audiovisual devices in North America. It is intended to be connected to a TV using a radio frequency (RF) signal. This channel option was provided because it was rare to have broadcast channels 3 and 4 used in the same service area. The choice allowed the user to select the unused channel in their area so that the connected device is able to provide video and audio on an RF feed to the television without excessive interference from a broadcast signal. The remod can be a weak link that affects VQ.

2.8.3 Network Coding and Fountain Codes

Network coding was first proposed about 8 years ago by a small group of researchers. Since then, it has attracted much interest and has the potential to improve the efficiency of information transmission in packet-based networks. It also reduces the delay and improves the robustness of the transmission. Network coding can be applied to video transmission in a lossy network. Segments of the original video content can be combined without additional overhead while previously downloaded segments allow recovering from future packet losses. This gives rise to implicit error correction, which is in contrast to explicit error correction that adds an overhead for every segment transmitted. In network-coded video transmission, multicast channels do not transmit original segments or independently encoded ones. Instead, original segments in the form of macro-blocks combine produce encoded segments with zero overhead. Another type of codes that can be applied to lossy networks is called Fountain codes (also known as rateless codes). These are erasure codes with the property that a sequence of encoding symbols can be generated from a given set of source symbols such that the original source symbols can be recovered from any subset of the encoding symbols of size equal to or only slightly larger than the number of source symbols. These codes do not exhibit a fixed code rate. They provide efficient encoding and decoding of large video files, which must be split into separately coded blocks.

2.8.4 Free Video Previews and Video Pausing

If a free preview window (e.g., 15 minutes) is offered before a decision to purchase is required, more streams may be needed since more people will watch a free preview than actually purchase. Video pausing is another customer impacting issue that requires careful consideration. For example, when a customer pauses the movie, how long is that stream reserved for that person? If the answer is 15 minutes then the customer loses the stream and risks the stream availability. Clearly, the customer cannot keep the stream indefinitely until completely viewing the movie because the customer may not return to viewing the movie for several hours. At some point the stream must be released for other viewers.

2.9 Home Entertainment Networks

For many service providers, home networking is an opportunity to increase the use of broadband data and video services. These companies are complementing traditional TV services with high-speed Internet access and advanced IP-based applications such as VoIP and IP video. A media gateway (also known as a service, residential, or home gateway) is a device that connects or bridges a home network to a broadband access connection. It enables communication and data transfer among networked devices in the home and across the Internet. A major benefit of a cable modem home gateway is the fact that the connection is always on, so the need to establish a connection is avoided.

The most popular home networking solutions today are 802.11 and IEEE 1394 (FireWire). Homeowners will not welcome new wiring installations that disrupt their premises. Therefore, alternative standards such as Multimedia over Coax (MoCA) and the Home Phoneline Networking Alliance (HPNA) allow the use of existing wiring to deliver home connectivity and maximize user acceptance.

Connecting diverse types of end-user devices, including TV sets, can be challenging. The Digital Living Network Alliance (DLNA) (http://www.dlna.org) is a collaboration of leading consumer electronics, PC, and mobile companies to create design guidelines for a new generation of products (e.g., TV, digital video recorder, PC, digital printer, portable music player) that can work together.

2.9.1 Display Technologies

LCD, plasma, and LED HDTVs are getting bigger, brighter, lighter, cheaper, and more energy-efficient. For a couple of years, plasma has been the choice for very large screens and LCD the choice for smaller sets. As LCD continues to grow, that distinction is diminishing. Light emitting diode (LED) TVs have a longer lifespan and consume less power. Using the LED as the primary light source (rather than traditional cold cathode fluorescent lamps) offers benefits such as ultrahigh contrast ratio and ultraslim depth (6.9-mm LED HDTVs are now available). The latest 3D HDTV models employ LED technology. Some LED models (e.g., Sharp HDTV) also come with "quad pixel" technology that add a fourth color (yellow) to the traditional red-green-blue (RGB) trio of primary colors, which will enable the display of more than a trillion different colors (24-bit RGB = 16.777 million colors, 30-bit RGBY = 1.074 trillion colors).

The laser TV is a projector TV technology that uses lasers instead of incandescent lamps to create light. Unlike plasma, LCD, and LED TVs that can only produce 40% of the range of colors visible to a human eye, laser TVs can easily double this production range because they use a narrower spectrum than LEDs to project the sharper light beams corresponding to the RGB colors. The intensity of the laser is maintained for the lifespan of the TV, consumes low power, and is lightweight. Organic LEDs (OLEDs) do not require a backlight, unlike all other display technologies, which implies they are very energy-efficient. They produce excellent image quality at 1,000,000:1 contrast ratio, resulting in wider viewing angles. Ultrathin screens measuring 3 mm in thickness are also possible. However, they are limited in size and are suited for smaller screen displays such as laptop and smartphone displays. Google's Nexus One Android phone and Apple's iPad employ OLED displays. Surface-conduction electron-emitter displays (SEDs) employ a matrix of tiny cathode ray tubes grouped in threes to form the RGB pixels. Flexible displays involving organic electro-luminescent and flexible crystal liquid displays are also emerging. Common 3D display technology for projecting stereoscopic image pairs to the viewer include:

- Anaglyphic 3D (with passive red-cyan glasses)
- Polarization 3D (with passive polarized glasses)
- Alternate-frame sequencing (with active shutter glasses)
- Autostereoscopic displays (without glasses)

Single-view displays project only one stereo pair at a time. Multiview displays either use head tracking to change the view depending of the viewing angle, or simultaneously project multiple independent views of a scene for multiple viewers (automultiscopic).

2.9.2 High-Speed Digital Interfaces

Newer STBs now come to digital interfaces, including high definition multimedia interface (HDMI), which reduces VQ problems significantly. HDMI is an audiovisual interface that carries uncompressed digital data. It is a digital alternative to analog interfaces, such as RF coaxial cable (described in Section 2.8.2), composite video, separate video (S-video), or video graphics array (VGA). While HDMI is primarily used for TV displays, HDMI converters to the Digital Visual Interface (DVI) for flat-panel computer displays are also available. HDMI version 1.3 provides a single-link bandwidth of 340 MHz or 10.2 Gbit/s, giving rise to a bit efficiency of 300 bits/Hz. The latest version, HDMI 1.4, supports 4K resolution and several stereoscopic 3D formats including field alternative (interlaced), frame packing (a full resolution top-bottom format), line alternative full, side-by-side half, side-by-side full, 2D + depth, and more. The displays use the HDMI signaling to identify the input format and then perform the necessary conversion to enable it to display the stereoscopic images.

2.9.3 Emerging Wireless Home Network Standards

IEEE 802.15 Task Group 3c (http://www.ieee802.org/15/pub/TG3c.html) was formed in March 2005. The task group (TG) aims to develop a millimeter-wave (60 GHz) wireless personal area network (WPAN) standard, providing data rates in excess of 2 Gbps for short-range indoor wireless networks. Several usage models define 60-GHz applications and environments, including single user and multiple user video streaming. Two multigigabit 802.11 TGs were formed in 2008:

- 802.11ac (under 6 GHz) was formed in September 2008;
- 802.11ad (60 GHz) was formed in December 2008.

These groups evolved from the Very High Throughput (VHT) study group established in 2007. Two frequency bands are considered, namely under 6 GHz and 60 GHz. The 60-GHz standard has significant overlap with 802.15 TG 3c. The new 802.11 standards, when ratified, will be backward-compatible to legacy 2.4- and 5-GHz devices, providing seamless handoff between 60-GHz and 2.4/5-GHz connections. Data rates in excess of 1 Gbps are expected and the maximum mandatory data rate for a single channel may exceed 500 Mbps. The 802.11 working group also established a new 802.11aa task group focused on video streaming [12].

The Wireless Gigabit Alliance (WiGig) (http://wirelessgigabitalliance.org) was established by more than 15 technology leaders within the consumer electronics, computer, semiconductor, and handheld industries to address faster wireless connectivity needs. It hopes to develop a 60-GHz band specification to achieve a data rate of up to 6 Gbps with a maximum throughput of over 5 Gbps. The low-power option will have a minimum throughput of 1 Gbps. Wireless HD is another 60-GHz standard promoted by consumer electronics and chipset vendors. Version 1.0 of the standard was completed in April 2008 and the standard converts the HDMI to wireless, enabling HD video streams to be transferred between audio/visual equipment without high-efficiency coding. It streams up to 4 Gbit/s at up to 30 feet. The first compliant equipment appeared in January 2009 with LG and Panasonic HDTVs.

Many 60-GHz standards are driven by 60-GHz CMOS RFIC technology, which lowers the cost of radio transceiver circuits. Tens of antenna arrays and "beam steering" devices are used with dynamic adjustment of the input voltage to each array or beam to adaptively control signal radiation angle. For example, the transceiver circuit developed by SiBEAM uses a ceramic package measuring about 20-mm square as antenna module. The surface of the module is covered with an array of about 36 antenna elements and voltage supplied to each element is adjusted to control radiation angle. Orthogonal frequency division multiplexing (OFDM) is employed to improve performance in nonline-of-sight use.

2.10 The Metro Network and Broadband Convergence

Metro Ethernet is growing in popularity, complementing Wi-Fi wireless Ethernet that is dominating the home network and public hotspots, as well as in access networks (via mesh Wi-Fi, Wi-Max, hybrid coax/Wi-Fi networks). Some industry experts predict that Ethernet plus an end-to-end tunneling technology such as provider backbone transport (PBT) will replace MPLS, and that optical transport network (OTN) will replace SONET/SDH.

2.10.1 Provider Backbone Bridges (PBB)

PBB is an alternative metro Ethernet technology that overcomes the service scaling limitations present in Ethernet networks based on IEEE 802.1ad QinQ implementations (or stacked virtual LANs). QinQ only allows for 4094 service instances per metro network due to the number of bytes of the VLAN tag in the Ethernet header, which may not be sufficient to accommodate the anticipated subscriber base for IP video, wireless, and other next-generation services. PBB provides greater scalability by a few orders of magnitude (up to millions of service instances per metro) by physically releasing more addressing space in the Ethernet packet. Additionally, it reduces the amount of state information the network is required to maintain and enhances security by clearly separating the customer and provider addressing space.

2.10.2 Provider Backbone Bridge Traffic Engineering (PBB-TE)

PBB-TE, as defined by IEEE 802.1Qay, adapts Ethernet technology to carrier class transport network. PBB-TE creates point-to-point Ethernet tunnels across a metro network to reserve bandwidth, support provisioned QoS metrics, and deliver 50-ms switchover to a provisioned protection path that matches benchmarks set by circuit-based SONET/SDH networks. Because the tunnel at the transport layer can be abstracted from the service layer, the technology is service and traffic agnostic.

PBB-TE is based on the layered VLAN tags as per IEEE 802.1Q, QinQ as per IEEE 802.1ad, and MAC-in-MAC encapsulation defined in IEEE 802.1ah PBB, but differs from PBB in disabling flooding/broadcasting, forwarding table creation, and spanning tree protocols, which are not needed in connection-oriented Ethernet. By using time-division multiplexing, the complexities associated with connectionless Ethernet LANs are avoided, which in turn simplifies operational administration and maintenance (OAM). Extensions to ensure path protection levels similar to unidirectional path switched ring (UPSR) protection in SONET/SDH are also available. For example, providers can leverage on Carrier Ethernet OAM standards (IEEE 802.1ag) that provide fault notifications in milliseconds, allowing carrier grade failover times to be achieved.

2.10.3 Carrier-Class Ethernet OAM Tools

The following tools bring standardized fault management and performance monitoring, giving service providers a comprehensive suite of tools to effectively manage their metro Ethernet networks.

- Fault management and performance monitoring (ITU-T Y.1731) defines performance monitoring measurements such as packet loss ratio, packet delay, and packet delay variation to assist with SLA assurance and capacity planning. For fault management the standard defines continuity checks, loopbacks, link trace, and alarm suppression for effective fault detection, verification, isolation, and notification in carrier networks.

- Connectivity fault management (IEEE 802.1ag) defines standardized continuity checks, loopbacks, and link trace for fault management capabilities in enterprise and carrier networks. This standard also partitions the network into eight hierarchical administrative domains.
- Link layer discovery (IEEE 802.1ab) defines discovery for all provider edge devices supporting a common service instance and/or discovery for all devices common to a single network domain.
- Ethernet in the first mile (IEEE 802.3 ah) defines mechanisms for monitoring and troubleshooting Ethernet access links. Specifically, it defines tools for discovery, remote failure indication, remote and local loopbacks, and status and performance monitoring.
- Ethernet protection switching (ITU G.8031) brings to Ethernet trunks, protection switching similar to SONET automatic protection switching/SDH multiplex section protection.

2.10.4 Next-Generation Network (NGN) Migration

There are key drivers behind the move by network operators to next-generation network (NGN) architectures, also known as broadband convergence networks (BCNs). Next-generation access networks promise the enabling of new services such as IP video, video-on-demand, and Web-based multimedia conferencing. This is in addition to cost-effective bandwidth-on-demand services where bandwidth pipes can be dynamically provisioned and released based on users' initiated demands. The access networks will offer telecommunications, broadcasting, and Internet access from a wide variety of devices. At the metro network level, there is a shift from closed ring (e.g., FOADM/ROADM) architectures to a switched optical layer (e.g., switched Ethernet) that integrates the packet-processing intelligence into the optical transport domain. How this migration is likely to unfold will depend on technologies that offer the best fit based on network requirements.

In NGNs, broad areas of convergence will take place, namely, convergence of packet and circuit-switched networks, convergence of wired and wireless networks, convergence of heterogeneous wireless networks, and convergence of telecommunications and video broadcasting. At the core of this convergence is the move towards an IP-based and Ethernet infrastructure that supports high-quality quadruple services. For efficient service provisioning, well-designed network operations and management functions with traffic engineering are essential. Key challenges include end-to-end QoS guarantees, seamless connectivity, implementation of capabilities within the network for state information management, resource allocation, and scheduling.

Along with network convergence, television and video have become the priorities for network operators around the world. Wireline operators are upgrading their network to support IP video, while wireless operators, faced with voice market saturation, are expanding towards mobile video offerings. To achieve seamless migration of these services, fixed mobile convergence for voice must be extended to video. In addition, video awareness must be built into the network. This is challenging since networks are typically unaware of the content of the video information carried. To ensure rich video experience or QoE while maintaining QoS calls for networks to handle several challenging tasks:

- Fast channel change;
- Error concealment;
- Admission control for oversubscription;
- Multicast and unicast video streams;
- Multirate SD and HD video streams;
- Bandwidth usage optimization.

These tasks are hard to enforce or implement since a typical subscriber today runs laptops, small to very large TVs, and smartphones, all with varied screen sizes and bandwidth requirements.

ITU-T has been operating a special expert group called Focus Group on Next-Generation Networks (FGNGN) to provide the architecture of NGN, and is currently working on the network management and operations issues in Study Groups 11 and 13. IETF has some working groups focused on network operations and management of IP/MPLS networks, but the operations and management for integrated networking with wired and wireless, telecommunication and broadcasting networks, have not been actively addressed yet.

2.11 Conclusions

A key challenge facing many broadband cable network operators is the need to optimize bandwidth resources for video transmission without compromising VQ playback. Many cable operators in the United States are migrating towards an SDV architecture that sends channels only to STBs in a service group that tune in to them, thereby conserving network bandwidth by not broadcasting signals to all STBs in all service groups all the time. SDV dynamically activates a channel as a subscriber views it using a narrowcast approach that allows for fast channel change. This is in contrast to the IPTV approach deployed over point-to-point DSL connections where channel change latency and bandwidth consumption may become significant bottlenecks in homes or businesses with simultaneous users in each premise. Attempts to improve the channel change latency with complex buffer management and video playback solutions inevitably lead to increased network overheads and set-top complexity. On the other hand, the broadcast approach employed by satellite networks suffers from poor bandwidth utilization when channels are not accessed by subscribers. This inability to reclaim the unused bandwidth of inactive channels restricts the addition of new services and channels. Using SDV, cable operators may potentially offer over 1,000 HD channels with ample room for future expansion of the channel line-up without sacrificing existing channels. Thus, from a technology and network architecture standpoint, broadband cable is poised to dominate over other wireline and satellite access options in the coming years. Nevertheless, further improvements in the existing coaxial-based access network can be made. Our ongoing research effort focuses on enabling OFDM to improve the efficiency of PHY layer transmission on the downstream and upstream paths, and the noise immunity on the upstream path. OFDM is widely deployed in Wi-Fi and 4G wireless networks.

References

[1] NSF Workshop Report, "Residential Broadband Revisited: Research Challenges in Residential Networks, Broadband Access, and Applications," January 20, 2004.

[2] National Cable and Telecommunications Association, http://www.ncta.com/Statistics.aspx.

[3] G. Davidson, et. al, "ATSC Video and Audio Coding," *Proceedings of IEEE*, Vol. 94, No. 1, January 2006, pp. 60–76.

[4] DOCSIS 3.0 Interface Specification, 2006, http://www.cablemodem.com/specifications/specifications30.html.

[5] K. Fall and S. Floyd, "Simulation-based Comparisons of Tahoe, Reno and SACK TCP," *ACM Computer Communication Review*, July 1996.

[6] A. Falk, T. Faber, J. Bannister, A. Chien, R. Grossman, and J. Leigh, "Transport Protocols for High Performance," *Communications of the ACM*, Vol. 46, No. 11, pp. 43–49, 2002.

[7] Q. Wu and N. Rao, "A Class of Reliable UDP-based Transport Protocols Based on Stochastic Approximation," *IEEE INFOCOM 2005*, March 2005.

[8] S. Floyd, "High-Speed TCP for Large Congestion Windows," RFC 3649, December 2003.

[9] T. Kelly, "Scalable TCP: Improving Performance in High Speed Wide Area Networks," *ACM Computer Communication Review*, 32(2), April 2003, http://www.deneholme.net/tom/scalable.

[10] D. Katabi, M. Handley, and C. Rohrs, "Internet Congestion Control for Future High-Bandwidth-Delay Product Environments," *ACM SIGCOMM*, Pittsburgh, PA, August 2002.

[11] "Advances in Peer-to-Peer Streaming Systems," *IEEE JSAC Special Issue*, Vol. 25, No. 9, December 2007.

[12] G. Venkatesan, "Multimedia Streaming over 802.11 Links," *IEEE Wireless Communications*, Industry Perspective article, April 2010.

Selected Bibliography

G. Dion, "SDV Troubleshooting: Gaining Visibility into the Complexities of SDV," http://www.cable360.net/ct/strategy/emergingtech/26402.html.

S. Krapp, "DOCSIS per sub Throughput Optimization," January 2007, http://www.cedmagazine.com/docsis-per-sub-throughput-optimization.aspx.

S. Palmer, *Television Disrupted: The Transition from Network to Networked TV*, Focal Press, 2006.

A. Sur, et al., "Extending the Service Bus for Successful and Sustainable IPTV Services," *IEEE Communications Magazine*, August 2008, pp. 96–103.
S. Vanhastel and R. Hernandez, "Enabling IPTV: What's Needed in the Access Network," *IEEE Communications Magazine*, August 2008, pp. 90 – 95.
J. Welch and J. Clark, "A Proposed Media Delivery Index (MDI)," RFC 4445, April 2006.

Exercises

2.1. Elaborate the key advantages and disadvantages of 4G wireless and OMVC. Highlight the key differences in the two-way connection capability.

2.2. Verify that the video overhead for encapsulating UDP/IP/Ethernet and RTP/UDP/IP/Ethernet TSPs ranges between 5.7% and 6.7%. Show that the total audio overhead ranges from 111 bytes and 123 bytes and compute the percentage overhead.

2.3. Routers on the Internet may drop packets, resulting in losses. Will losses occur on a local connection involving a home Wi-Fi router? How about on an access link? Justify your reasoning. Will FEC be effective in these cases?

2.4. Prove that a 24-bit RGB display gives 16.777 million colors and a 30-bit RGBY display gives 1.074 trillion colors. Will VQ improve significantly given that yellow is not a primary color and can be derived from a red/green combination? Will interlaced video work well on an SED display?

2.5. Since ATSC signals can be received by a digital TV via an antenna without an STB, are these signals protected by encryption? Assuming the same transmit power is used, explain why the ATSC signals for a specific channel give a better reception during the day than at night.

2.6. The ATSC-M/H standard employs a frame of duration 968 ms. Each frame can be subdivided into 5 subframes, each of duration 193.6 ms. Each subframe is further subdivided into 16 slots, each of duration 12.1 ms. Each slot can accommodate 156 TS packets. Calculate the maximum bit rate of the link. Note that this bit rate is similar to the IEEE 802.22 draft standard, which focuses on two-way communications using the white space of TV bands.

2.7. The 2-byte timestamp in the PES carries the audio or video frame number. Assuming an average frame size of 20 Kbytes for a compressed 1080p video, compute the minimum buffer size to accommodate all the video frames before the frame number rolls over. For a frame rate of 30 Hz, compute the duration of the video that needs to be kept in the buffer for proper playback. Note that for proper synchronization, a rollover takes place simultaneously for the audio and video frame numbers as soon as one of the streams crosses the maximum frame number.

2.8. The SDV approach can be adapted to online video streaming. The provider can deploy popular content as IP multicast on their networks and reserve unicast for only those services that are uniquely requested. With the multicast service, the service provider can select their preferred quality for while optimizing the total network bandwidth since only a single version is delivered to multiple subscribers concurrently. Explain whether the ABR approach discussed in Chapter 1 can be deployed with multicast services. What benefits will ABR bring to unicast services? How will trick modes affect the performance of ABR streaming?

2.9. Picture-in-Picture (PIP) is often used to watch one program while waiting for another to start or ads to finish. The technology requires two independent tuners or signal sources to display a main video program on the full TV screen and an inset TV window simultaneously. Sound is usually from the main program only. Two-tuner PIP TVs have in-built second tuners whereas single-tuner PIP TVs require external signal sources, which may be an external tuner, VCR, DVD player, or an STB with composite video outputs. Explain how SDV can be adapted for PIP systems. Note that ESPN3 provides PIP for its online sports portal where a highlight (e.g., shots on goal, goals) or commentary and a live match can be viewed side by side.

2.10. Suppose a cable headend is able to provide 500 channels. If 10,000 subscribers are serviced by the headend, will a SDV headend achieve more efficient bandwidth utilization than a non-SDV headend? Is the SDV concept useful for satellite broadcast services?

2.11. Given that the home, access, metro, and core networks are converging towards layer-2 Ethernet technology, are higher layer protocols such as RTP/UDP/IP and TCP/IP still relevant? Or will these protocols weigh on the end-to-end network with unnecessary overheads? Note that like TCP, Ethernet was originally designed to carry data traffic but has evolved to carry audio and video traffic. For example, the IEEE 802.1BA, the Audio Video Bridging (AVB) Task Group (http://www.ieee802.org/1/pages/802.1ba.html) has developed a suite of standards that enable audio and video streams over Ethernet LANs. Only 2 ms of buffering (maximum delay over 7 hops) is required to support the most time-sensitive traffic and network congestion will not cause dropping of stream packets. This standard, together with SMPTE 2022, are candidates to carry broadcast-quality uncompressed 3 Gbps video over IP.

2.12. Error resilience in video coding adds redundancy to cope with higher error or loss rates associated with network transmission. However, the additional bit overhead results in more network load, which can induce more losses and errors. How can error resilience methods be designed to work well with congestion-avoidance protocols such as TCP?

Chapter 3

Video Fundamentals

This chapter covers several important aspects of video compression, network transport, error resilience, video quality assessment, and an overview of emerging standards such as H.264, VC-1, and VP8. H.264 overcomes many limitations of motion estimation in MPEG-2 with improved interprediction via fine-grained motion estimation, multiple reference frames, unrestricted motion search, and motion vector prediction. H.264 also offers improved intracoding in the spatial domain using a context-sensitive deblocking filter with many techniques for mitigating errors, packet losses, and network variability including the use of scalable bitstreams (corresponding to different quantization levels) and slice coding (with no possibility of spatial error propagation from one slice to any other slice within the video frame). The improved coding efficiency and the new error resilient features make H.264 a perfect candidate for video streaming, broadcasting, and conferencing over a variety of fixed and mobile networks.

3.1 Display Resolution and Visual Quality

A list of common resolutions is shown in Table 3.1. A higher display resolution requires a higher density of picture elements (pixels or pels), which leads to better visual quality. However, the screen size of the display device plays an important role. For example, a 15-inch laptop monitor may be more suited for 720p than 1080p or QCIF videos. The number of bits representing each pixel and the quantization level impacts video quality (VQ). An HD video may suffer in VQ if a coarse quantization level is chosen. On the other hand, a CIF video coded with a fine quantization level can achieve good VQ on a smartphone. In general, we refer to videos coded with a fine quantization level as high-quality videos.

The number of pixels in a video frame is lagging behind digital image resolutions. For instance, the resolution of a 1080p video camcorder is nine times lower than an 18 megapixel camera that is commercially available. Unlike common video formats, which are usually specified in terms of vertical resolution (e.g., 720p, 1080p), digital cinema formats are usually specified in terms of horizontal resolution. These resolutions are often written as multiples of a base value of 1,024 pixels using a “K” or “Kilo” suffix. Thus, a 2K image (e.g., 2,048 × 1,536 pixels) will have 2,048 pixels wide while a 4K image (e.g., 4,096 × 3,072) will have 4,096 pixels wide. The vertical resolution varies with the pixel and display aspect ratios. The pixel aspect ratio is normally set to 1:1, corresponding to square pixels. This ratio may not be the same as the display aspect ratio, which is related to the pixel aspect ratio by the governing equation: display aspect ratio = width/height × pixel aspect ratio. For example: 720/576 × 64/45 = 16/9. The values for the pixel aspect ratio will not change when the video is cropped whereas the display aspect ratio will change.

Table 3.1: Common Display Resolutions

	Resolution (Pixels)
Digital cinema	2048 × 1080 (2K, 24/48 Hz), 4096 × 2160 (4K, 24 Hz)
High-definition 1080p	1920 × 1080 (16:9)
High-definition 720p	1280 × 720 (16:9)
Standard definition 480p	720 × 480 (NTSC), 720 × 576 (PAL)
Common intermediate format (CIF)	352 × 240 (NTSC), 352 × 288 (PAL)
Wide QVGA (used in PSP)	368 × 208
QVGA (used in iPod)	320 × 240
QCIF (quarter CIF)	176 × 120 (NTSC), 176 × 144 (PAL)

3.1.1 Serial Digital Interface (SDI)

SDI refers to a family of digital video interfaces standardized by the Society of Motion Picture and Television Engineers (SMPTE). These standards are used for the transmission of raw (uncompressed) and unencrypted digital video signals within TV facilities, and can also be used for packetized data. Due to the high bit rates, they are designed for operation over short distances (< 300m with coaxial cable). SD-SDI (SMPTE 259M) is a 10-bit serial digital interface operating at 143/177/270/360 Mbps. The current practice in recording studios for supporting 1080p video is to use a 2.97 Gbps dual-link HD-SDI (SMPTE 372) connection from the camcorder and input to the mixing and recording equipment. Dual-link connections require twice the number of cables as single-link connections, resulting in increased equipment cost and complexity. To this end, SMPTE is working to finalize a single-link 3G-SDI (SMPTE 424M) to replace HD-SDI. This standard transports a raw digital video stream using two HD-SDI (SMPTE 292M), each running at 1.485 Gbps, resulting in a single 2.97 Gbps video link interface.

3.2 Video Compression

Storage and streaming bandwidth dictate the need for compression. Raw videos require massive storage space. For example, an uncompressed 90-minute 480p SD video requires roughly 112 Gbyte of storage, which is more than 25 times the storage capacity of a standard DVD disk. Table 3.2 shows the various color formats and the corresponding uncoded efficiencies. The formats provide the luminance (brightness) sample, and the red chrominance (C_R) and blue chrominance (C_B) color samples, also known as YUV. Table 3.3 shows the coded bit rates and storage requirements. In general, coded HD video using MPEG-2 is very expensive to deliver, typically requiring 15 Mbps to support HD versus 3.75

Mbps for SD (4 times). Larger storage is also needed, typically 3 Gbyte/hour for HD versus 1 Gbyte/hour for SD (3 times). Thus, the increasing popularity of HD videos drives the adoption of more efficient standards such as H.264 and VC-1.

Table 3.2: Color YC_rC_b (YUV) Formats and Uncompressed Efficiencies

YUV 4:4:4	Typically 8 bits per Y, U, V plane No downsampling	24 bits/pixel 8 bits/sample
YUV 4:2:2	4Y pixels for every 2U and 2V 2:1 horizontal downsampling No vertical downsampling	16 bits/pixel 8 bits/luma sample 4 bits/chroma sample
YUV 4:2:0	2:1 horizontal downsampling 2:1 vertical downsampling	12 bits/pixel 8 bits/luma sample
YUV 4:1:1	4Y pixels for every 1U and 1V 4:1 horizontal downsampling No vertical downsampling	12 bits/pixel 8 bits/luma sample 2 bits/chroma sample

Table 3.3: Bit Rates for 30 Hz Videos Coded with an Efficiency of 0.25 Bit/Sample

Video Resolution (Progressive)	Color Format	Uncoded Bit Rate (Mbps)	Coded Bit Rate (Mbps)	Storage for 1-Hour Coded Video (Gbyte)
352 × 288 (CIF)	YUV 4:2:0	36.5	0.76	0.34
720 × 480 (SD)	YUV 4:2:2	165.9	2.59	1.17
1280 × 720 (HD)	YUV 4:4:4	663.6	6.91	3.11
1920 × 1080 (HD)	YUV 4:4:4	1493.0	15.55	7.00

In analog television, there are roughly 480 active lines with each line holding about 440 pixels. Thus, each frame has slightly more than 200,000 color pixels. Contrast 480p SD video, which has 720 × 480 or 345,600 pixels per frame. HD video may contain 720 or 1080 vertical progressive scan lines (720p or 1080p), 1080 interlaced lines (1080i), and is capable of displaying a 16:9 image and output digital audio. With 1080 active lines and 1920 color pixels in a line, each 1080p frame contains more than 2 million pixels (1920 × 1080 = 2,073,600), roughly six times more than 480p. With progressive scan, all active lines are displayed in each video frame whereas in interlaced scan, odd and even lines are displayed in successive frames (also called fields) at half the frame duration. Clearly, the disadvantage of interlacing is that the horizontal resolution is reduced by half, and the video is often filtered to avoid flicker and motion rendering artifacts. Interlacing relies on the phosphor persistence of a cathode ray tube (CRT) screen to blend the frames together. Since LCD/plasma screens are rapidly replacing CRT screens, interlaced content must be deinterlaced at playback time. However, a screen with poor deinterlaces can result in a jittery image. Thus, deinterlacing is sometimes performed before video coding. The interlaced video format is popular with broadcast and payTV services (480i for SD, 1080i for HD) whereas progressive format is widely adopted by online video portals.

The best VQ is obtained when the source video is captured uncompressed (e.g., via a HD/SD-SDI camcorder). For compressed videos, the VQ is largely dictated by the coding process, which can be lossy or lossless. When a compressed video is decoded, the decoded video will have the same size as the original raw video that is captured uncompressed. This is because each pixel in the raw and decoded video frames is represented by a fixed value, typically ranging from 0 to

255 (8 bits) for each color plane. Hence, if the video frames are not cropped, the frame sizes will remain the same. However, the VQ of the raw and decoded videos may not be the same (unless lossless coding is used). In many instances, lossless coding is not used since this usually takes the longest time to encode and decode. Table 3.4 shows an interesting comparison of the lossless data coding efficiency using WinRAR versus lossless H.264 video coding. Lossless coding normally uses a variable-length code (VLC), such as the Huffman code. Even with lossless coding, the H.264 coding efficiency is superior to data compression, especially for HD videos. With judicious selection of coding parameters, more savings can be achieved with lossy coding without sacrificing VQ (Table 3.5).

Table 3.4: Efficiency Comparison of Lossless Data and Video Compression

Video (Progressive Scan)	Coder	Compression Time (min)	File Size (Mbyte)	Efficiency
75 sec *Dell*, 720 × 480, 30 Hz, 1.18 Gbyte uncompressed	WinRAR	2.5	443.587	2.6575
	H.264	35.5	373.157	3.1591
72 sec *FCL*, 1280 × 720, 24 Hz, 2.39 Gbyte uncompressed	WinRAR	4.5	706.699	3.3880
	H.264	50	387.224	6.1833
125 sec *Terminator 2*, 1440 × 1080, 25 Hz, 6.92 Gbyte uncompressed	WinRAR	10	1,493.6	4.6356
	H.264	126	943.6	7.3376
72 sec *FCL*, 1920 × 1080, 24 Hz, 5.39 Gbyte uncompressed	WinRAR	10	1,311.0	4.1092
	H.264	111	772.1	6.9777
596 sec *BBB*, 1920 × 1080, 24 Hz, 44.53 Gbyte uncompressed	WinRAR	140	12,288.1	3.6235
	H.264	636	4,213.2	10.5682

Table 3.5: Efficiency Comparison of Lossless and Lossy H.264 Coding

Video (Progressive Scan)	Lossless File Size (Mbyte)	Lossy File Size with PSNR > 40 dB (Mbyte)	Relative Savings
75 sec *Dell*, 720 × 480, 30 Hz	373.157	19.260	19×
72 sec *FCL*, 1280 × 720, 24 Hz	387.224	8.637	45×
125 sec *Terminator* 2, 1440 × 1080, 25 Hz	943.595	32.298	29×
72 sec *FCL*, 1920 × 1080, 24 Hz	772.062	15.490	50×
596 sec *BBB*, 1920 × 1080, 24 Hz	4,213.150	101.954	41×

3.3 Video Containers

Movie files or videos are usually combined with audio information and encapsulated in containers such as AVI, ASF, MOV, RM, MP4, MPEG-PS (MPG), MPEG-TS, MXF, MKV. As can be seen from Table 3.6, the overhead for encapsulating a H.264 video in MP4 is insignificant, well below 0.01%, roughly 12 bytes per video frame and 4 to 6 bytes per audio frame. The MP4 container also incurs less overheads than the MPEG-2 transport stream (TS) container (Table 3.7), which has been widely used in many cable, satellite, and DTV systems. The difference in overheads grows as the video file size increases but the average percentage increase is in the region of 4%. A "stream" is a succession of data elements made available over time and the data elements in a stream are called frames. Each stream is coded by a codec that defines how the actual video data is encoded and decoded. Examples of codecs are DivX, MPEG-2, H.264, VC-1 (WMV-9), VP8. Packets are then read from the stream and each packet may contain complete frames or multiple frames. The general operations of a video codec are listed below and will be described in the following sections.

Encode: Predict (Spatial/Temporal) → Transform → Quantize → Scanning → Entropy Encode → Loop Filter
Decode: Entropy Decode → Predict → Dequantize → Inverse Transform → Loop Filter

Table 3.6: Overheads for MP4 Encapsulation (No Audio)

Video (Progressive)	H.264 File Size (Mbyte)	MP4 File Size (Mbyte)	Overheads (Kbyte)
75 sec *Dell*, 720 × 480	373.157	373.185	28
72 sec *FCL*, 1280 × 720	387.224	387.246	22
72 sec *FCL*, 1920 × 1080	772.062	772.085	23
125 sec *Terminator* 2, 1440 × 1080	943.595	943.632	37
596 sec *BBB*, 1920 × 1080	4,213.150	4,213.308	158

Table 3.7: Overheads for MP4 Versus MPEG-TS Encapsulation (with Audio)

Video (Progressive)	TS File Size (Mbyte)	MP4 File Size (Mbyte)	Difference (Mbyte)	% Difference
54 sec *The Dark Knight*, 1280 × 544	33.395	32.094	1.301	4.1
64 sec *The Hangover*, 1280 × 544	45.736	44.055	1.681	3.8
64 sec *Up Carl Goes Up*, 1280 × 720	46.705	44.994	1.711	3.8
73 sec *Fred Claus*, 1280 × 544	51.796	49.936	1.860	3.7
83 sec *Night at the Museum*, 1280 × 544	59.716	57.514	2.202	3.8
86 sec *Speed Racer*, 1280 × 544	62.907	60.595	2.312	3.8
93 sec *Oceans 13*, 1280 × 532	67.502	65.011	2.491	3.8
106 sec *300 HD*, 1280 × 544	54.579	52.479	2.100	4.0
125 sec *Terminator Salvation*, 1280 × 532	91.558	88.169	3.389	3.8
128 sec *Ugly Truth*, 1280 × 608	90.200	86.833	3.367	3.9
128 sec *Star Wars Clone Wars*, 1280 × 544	94.247	90.910	3.337	3.7
137 sec *The Astronaut Farmer*, 1280 × 544	94.140	90.663	3.477	3.8
139 sec *Serenity*, 1280 × 720	87.054	83.897	3.157	3.8
143 sec *10000 BC*, 1280 × 544	98.045	94.540	3.505	3.7
14[illegible] sec *Brothers Bloom*, 1280 × 544	100.209	96.505	3.704	3.8
146 sec *Children of Men*, 1280 × 688	55.050	52.523	2.527	4.8
150 sec *Transformers 2*, 1280 × 544	110.419	106.415	4.004	3.8
153 sec *Cars*, 1280 × 532	58.546	55.489	3.057	5.5
191 sec *Tetro*, 1280 × 544	141.744	136.442	5.302	3.9

3.3.1 Advanced Audio Coding (AAC)

AAC is a lossy (noninvertible) compression scheme for coding digital audio and is standardized as MPEG-4 Part 3 (MP4 audio). It is a successor to MP3 and AC3 (Dolby Digital), and generally achieves better sound quality at similar bit rates. MP4 high-efficiency AAC (HE-AAC) and AAC coders are also available (http://www.audiocoding.com). AAC supports three profiles and employs new techniques such as perceptual noise substitution, temporal noise shaping, backward adaptive linear prediction, and enhanced joint stereo coding. It supports a broad range of sampling rates (8 to 96 KHz), bit rates (16 to 576 Kbps), and audio channels (1 to 48). AAC is the standard audio format for Apple's iPhone, iPod, iPad, and iTunes; Sony's PlayStation 3; XM and Sirius satellite radio; Android smartphones; Nintendo's Wii, and has been adopted by the Digital Video Broadcasting (DVB) consortium (www.dvb.org) and 3rd Generation Partnership Project (3GPP, www.3gpp.org). The iPod Touch pocket computer encodes stereo audio at 160 Kbps using AAC with 48 KHz bandwidth. For stereo coding at 64 Kbps, AAC frames contain 200 bytes on the average, allowing 7 frames to be carried over a 1,500-byte Ethernet payload.

Two types of AAC are adopted for MPEG-4 transport. In low bit-rate AAC, the transport of one or more complete AAC frames of variable size is supported. The maximum size of an AAC frame in this mode is 63 bytes. For each AAC frame contained in the payload, the one byte AU-header provides:

- Size of each AAC frame in the payload;
- Index information for computing the sequence timing of each AAC frame.

High bit-rate AAC supports the transportation of variable size AAC frames. In one RTP packet, either one or more complete AAC frames are carried, or a fragment of an AAC frame is carried. The maximum size of an AAC frame in this mode is 8,191 bytes. For each AAC frame contained in the payload, the AU-header provides the same information as listed above for the low rate case. However, unlike the low rate case, each 2-byte AU-header is used to code the maximum size of an AAC frame (13 bits) and the AU-index (3 bits).

MPEG Surround or MPS (http://www.mpegsurround.com) is an emerging compression technique for multichannel audio signals. It is standardized as MPEG-D Part 1. The move towards MPS for audio parallels the trend towards multiview 3D video, with associated coding standards supporting spatial channels for audio and visual entertainment. When combined with HE-AAC, MPS can carry a 5- or 7-channel surround program at scalable data rates, usually 64 Kbps or less. These bass channels are sometimes accompanied by a low frequency effects (LFE) channel that requires only one-tenth of the bandwidth of the main audio channels (an LFE channel is sometimes known as a subwoofer channel), giving rise to 5.1 or 7.1 surround systems.

3.4 H.264, VC-1, and VP8 Standards

These are the most powerful standards that support efficient video delivery and storage, and achieve higher coding efficiencies than legacy standards. H.264 is standardized by the International Telecommunications Union (ITU) [1, 2] whereas VC-1 is standardized in SMPTE 421M [3] and was initially implemented by Microsoft as Windows Media Video (WMV) 9, mainly for supporting online video streaming. VC-1 is said to avoid the patent pool problem associated with H.264. VP8 is an open source video codec [4] formerly owned by Google but recently released as part of the WebM project [5]. It is currently royalty-free. Both H.264 and VC-1 are becoming widely adopted in video storage for consumer electronics (CE) devices (e.g., camcorders) as well as in narrowband and broadband network transport. Some examples include the DVD forum (http://www.dvdforum.org), the Blu-Ray Disc Association (http://us.blu-raydisc.com) for HD video recording and playback, satellite TV, and the European Broadcasting Union (http://www.ebu.ch).

3.5 H.264 Architecture

H.264 is the latest digital video coding standard developed by the Joint Video Team (JVT), a partnership between the ITU-T (SG16 Q.6) Video Coding Experts Group (VCEG) and the ISO/IEC (JTC 1/SC 29/WG 11) Moving Picture Experts Group (MPEG). The standard is also known as Advanced Video Coding (AVC) or MPEG-4 Part 10. H.264/AVC is widely adopted in CE equipment such as camcorders, surveillance cameras, smartphones. H.264 is the recommended codec for all DVB and 3GPP video services and is adaptable to different applications, client devices, and networks. This is necessary when transmitting interactive media over heterogeneous networks. For instance, the VQ may be prioritized over coding efficiency when bandwidth is abundant and is useful for applications when client device is not capable of displaying full resolution or full quality video. Many online video portals employ Adobe's Flash Player, which uses H.264. This jumpstarts OTT video services, allowing such services to achieve VQ comparable to that delivered over managed networks and legacy MPEG-2 set-top boxes (STBs). H.264 STBs have become indispensable for enabling IPTV over DSL.

Like prior MPEG standards, H.264 defines the syntax and semantics of the bitstream (i.e., the coded video representation) and the processing that the decoder requires. It does not define how encoding and other video preprocessing functions are performed, including coding optimization, thus enabling vendors to differentiate their encoders in terms of cost, coding efficiency, error resilience, or hardware requirements. Nevertheless, a noncommercial H.264 reference codec is available in [6]. Test sequences and bitstreams are available from [7]. Similarly, the VC-1 reference software can be purchased from [3]. Information on the decoder is available in [8].

H.264 defines three different types of frames:

- Spatially predicted intracoded I frames (without referring to other frames and hence can be decoded independently);
- Unidirectionally predicted P frames (predicted from previous I or P reference frames);
- Bidirectionally predicted B frames (predicted from previous and future reference frames).

Each video sequence comprises a group of pictures (GOP), which typically comprises an I frame, and one or more P and B frames. Typically, the first frame in a GOP is an intracoded I frame, which is often used as a point of resynchronization or re-entry to support trick modes (e.g., pause, fast forward, rewind, and so forth) and error recovery. For all other frames in the GOP (i.e., P and B frames), interpredictive or bipredictive intercoding is used.

Each color frame contains both luminance (luma) and chrominance (chroma) samples that make up the YC_BC_R components. Each frame comprises tiled macroblocks (MBs) that are separately coded spatially or temporally. Since the human eye is less sensitive to color (hue and saturation) than brightness, both chroma samples are normally subsampled by a factor of two in the horizontal and vertical directions (e.g., 4:2:0 color format). Thus, an 4:2:0 MB comprises 16 × 16 luma samples and two smaller blocks of 8 × 8 color samples. Each block can be compared with the corresponding block in the previous frame using a similarity metric called sum of absolute differences (SAD):

$$SAD(n) = \sum_{i=0}^{N-1} \sum_{j=0}^{M-1} \left| F_n(i,j) - F_{n-1}(i,j) \right| \qquad (3.1)$$

where F_n is the nth frame of size $N \times M$, and i and j denote the pixel coordinates. The encoder can use the SAD values stored in a predefined number of previous frames as a reference to make the best decision. SAD can also be used in temporal error concealment (EC) to identify the best replacement MB and generally ensures spatial continuity and produces visually good results.

Intracoded frames enable error resilience and random access to the bitstream but achieve moderate coding efficiency. These frames employ intraprediction from decoded samples within the same frame to minimize spatial correlation. The key objective is to remove predictable low frequency components, which can be perfectly reconstructed by the decoder. This is achieved using directional linear interpolation of adjacent edge pixels from neighboring MBs that are decoded before the current MB. Because of the relatively small areas, significant spatial correlation exists with each directional mode. If there is no directional pattern, the DC predictor is employed. Eight mode directions for intraprediction of 4 × 4 luma blocks are shown in Figure 3.1. In this case, a 16 × 16 MB is broken down into sixteen 4 × 4 blocks. In addition to these predictors, a DC luma predictor is supported. Here, the pixels of the current 4 × 4 block are predicted from the weighted mean of all pixels from the left and top decoded blocks. Another luma intraprediction method is useful if the luma component is slowly changing, such as nearly uniform image areas. Because a 16 × 16 luma MB is chosen in this case, there are fewer predictor directions, namely vertical, horizontal, DC, and plane. The plane intraprediction mode employs a linear function between the neighboring pixels to the top and left decoded MBs to predict the current pixel. For 4:2:0 chroma intraprediction, the vertical, horizontal, DC, and plane mode directions are used and prediction is performed on 8 × 8 blocks.

Unlike intrapredicted frames, interpredicted frames employ prediction from previously decoded reference frames and motion compensation to reduce temporal correlation among the frames. The prediction error is transformed, quantized, and entropy coded, and transmitted together with the prediction mode information. The primary task of the transform is to reduce the spatial redundancy of the prediction error. At the decoder, the scaled and quantized transform coefficients are inverse transformed and added to the predicted signal. In general, intercoding is more efficient using interprediction, which involves translating the reference array of pixels from a previously decoded reference frame to the current array. Bipredictive frames may also be used as references for coding other frames.

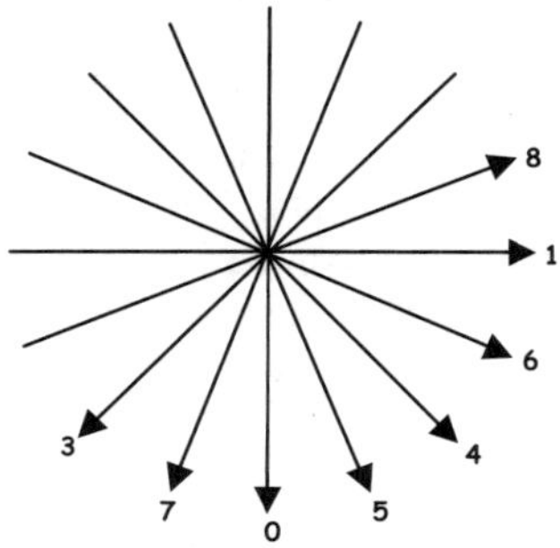

Figure 3.1: Intra (spatial) predictor directions for 4 × 4 blocks (direction 2 used by DC predictor).

3.5.1 Video Coding and Network Abstraction Layers

The H.264 video hierarchy is shown in Figure 3.2. The architecture comprises two conceptually different layers that allow the adaptation of the bitstream to different transport networks:

- Video coding layer (VCL);
- Network abstraction layer (NAL).

The VCL defines the core video coding engine that performs basic functions such as motion compensated prediction, transform coding of coefficients, and entropy coding. It is transport unaware and its highest data structure is the video frame, which comprises an independently decodable integer set of MBs typically arranged in raster scan order. The NAL is an interface between the VCL and the transport network. It is therefore responsible for encapsulation of the VCL data into transport entities, namely transport protocols (e.g., RTP/UDP, HTTP) and container formats (e.g., MPEG-4, MPEG-2 TS). The NAL operates on NAL units (NALUs), which are basic video fragments (of variable length) that enable packetized information to be transmitted across the network. Each unit comprises a 1-byte header and a bit string that constitutes the MBs of a video frame. Non-VCL information such as sequence and picture parameter sets, access unit delimiters, supplemental enhancement information (SEI) or video usability information (VUI), may also be added. NALUs are separated by a 4-byte flag (0x00000001). The RTP payload supports three NAL modes:

- Single mode: Each NALU transported in a single RTP packet;
- Noninterleaved mode: several NALUs of the *same* frame are packetized into a single RTP packet;
- Interleaved mode: several NALUs from *different* frames are packetized into a single RTP packet, not necessarily in their decoding order.

3.5.2 VCL and NAL Packetization

The packetization of the VCL and NALUs in the 3GPP framework is shown in Figure 3.3. The framework allows various wireless video applications in Table 3.8 to be supported. The elementary unit processed by H.264 is the NALU, which can be easily encapsulated into different transport protocols and file formats. Clearly, optimizing the size of the NALUs to be carried over RTP must be addressed for efficient delivery of the video segments. The bytestream format is defined in Annex B of the H.264 standard. RTP can be used to stream the H.264 video using the RTP payload format of RFC 3984. We briefly explain the format of the NALU header and the types of NALUs defined in RFC 3984 (Figure 3.4).

Sequence, consisting of
Frames or Pictures, consisting of } **NAL**
Slices and slice groups, consisting of
Macroblocks, consisting of
Blocks, consisting of
Pixels, consisting of
Samples } **VCL**

Figure 3.2: H.264 video hierarchy.

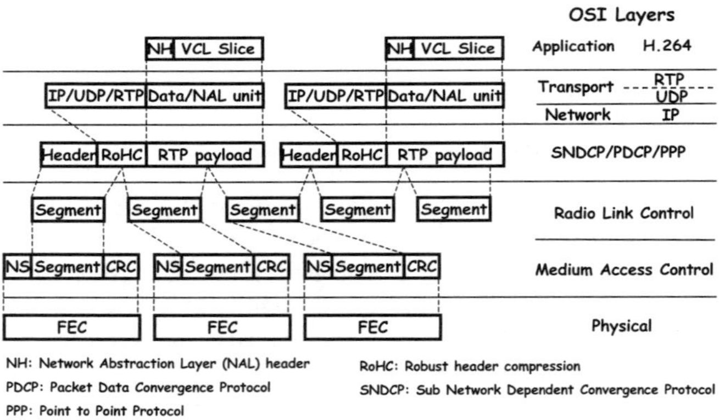

Figure 3.3: VCL and NAL encapsulation in IP packets.

Table 3.8: 3GPP Video Applications

Application	3GPP Service	Maximum Delay	Encoder Buffering	Transport Feedback	CSI	Encoding
Download and play	MMS	NA	None	Yes	NA	Offline
On-demand, pre-encoded streaming	PSS	1s	Yes	Yes	Partly	Offline
Live streaming	PSS	200 ms	Yes	Partly	Partly	Online
Multicast	MBMS	1s	Limited	Limited	Limited	Both
Broadcast	MBMS	2s	None	None	None	Both
Conferencing	PSC	250 ms	Limited	None	Limited	Online
Telephony	PSC	200 ms	Yes	Limited	Partly	Online

CSI: Customized Applications for Mobile Network Enhanced Logic (CAMEL) Subscription Information
MBMS: Multimedia Broadcast/Multicast Service
MMS: Multimedia Messaging Service
PSC: Primary Synchronization Code
PSS: Packet Switched Stream

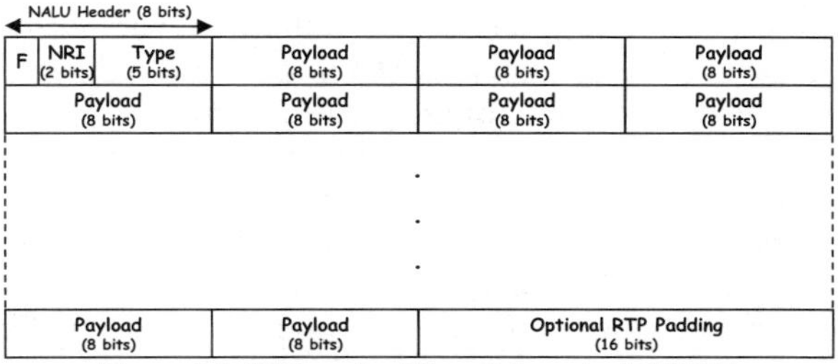

Figure 3.4: RTP payload format for a single NALU.

The first bit of the NALU header is a 0. The next two bits (i.e., the NAL reference indicator or NRI field) indicate the relative priority of the NALUs to be transported. The highest transport priority is 11, followed by 10, 01, and finally the lowest 00. A value of 00 indicates that the content of the NALU is nonreferenced, which means the NALU can be discarded without affecting the decoding of other NALUs. Values greater than 00 indicate that the decoding of the NALU is required to maintain the integrity of the reference frames. As for the Type field in the NALU header, the H.264 standard only specifies values from 1 to 23, which is the single NALU mode. The contents of the NALU types are shown in Table 3.9. Types 1 to 5 contain VCL information. Types greater than 5 are non-VCL NALUs. RFC 3984 defines type values 24 to 29. Of particular interest is type 24 (single-time aggregation packet or STAP-A) and type 28 (fragmentation unit or FU-A). "A" denotes the noninterleaved mode where several NALUs of the same frame are packetized into a single RTP packet. In most cases, one video frame is coded as one NALU, which is usually larger than the RTP maximum transmission unit (MTU). This implies that the FU-A fragmentation mode is expected to be widely used when the Annex B bytestream format is encapsulated as RTP packets. Each fragment will contain the original timestamp of the NALU. STAP-A can be used for out-of-band control information which is usually small, such as the sequence parameter set (SPS) and picture parameter set (PPS). The fields in the SPS provide sequence-level header information such as profile and level of codec, decoding order, number of reference frames, and aspect ratio and color format. The PPS provides picture-level header information on the entropy coding method, data partitioning, MB ordering, weighted prediction, initial quantization parameter (QP) values, and whether interpredicted MBs can be used for intraprediction. Although the contents of the SPS and PPS are not expected to change, they can be sent at regular intervals to improve error resilience.

3.5.3 An RFC 3984 H.264 Transport Framework

A possible H.264 transport framework is shown in Figure 3.5. The H.264 streamer at the video source first parses the Annex B bytestream to convert the NALUs to the RFC 3984 payload format and then encapsulates them as RTP packets. The sender buffer may keep a number of these RTP packets after they are transmitted for possible retransmission before releasing them. At the receiver, there are two buffers. The receive buffer stores the RTP packets, makes the retransmission decision (when a packet is lost), and converts the RTP packets from RFC 3984 NALUs to NALUs of Annex B format. NALUs of the same timestamp (indicating they belong to the same video frame) are then forwarded to the playback buffer. The H.264 decoder starts to decode when there is a minimum number of frames stored in the playback buffer. To maintain continuity in the playback process, the playback buffer keeps a minimum number of video frames. It will request the next frame (comprising a group of NALUs with the same timestamp) from the receive buffer, even if it is not completely received, to fill the playback buffer.

Both the sender and receive buffers can be maintained in manner similar to sliding windows. The receiver may send two types of acknowledgment: ACK and NACK, which are conveyed by RTCP. An ACK is a positive acknowledgment indicating no retransmission is required for any RTP packet with a sequence number (SN) lower than the current one. NACK indicates a request for an RTP packet of a specific SN. The receiver may send an ACK for a group packets (block ACK) to reduce the implementation complexity. When an ACK is received, the sender buffer will delete the packets with SN less than or equal to the SN in the ACK. The sender buffer may not keep a fixed window size. If video streaming is performed, the sender buffer must be large enough to store all packets that have been sent but not acknowledged. Note that selective video information dropping (Section 3.8.2) and EC (Section 3.6.3) can be performed at the Annex B NALU level whereas link (channel) quality measurement (Section 2.5) can be performed at the RTP (RFC 3984 NALU) level.

The receiver may request the retransmission of a lost NALU. The decision may depend on the priority of the lost NALU, the total number of retransmission requests, the link quality, and the available time for retransmission. The priority can be set based on the NRI field in the NALU header. As pointed out previously, a higher NRI value corresponds to higher priority. In order to decipher the NALU header of the lost RTP packet, a simple but efficient solution is to embed the NRI value in the RTP header of the neighboring packets. Assuming the NRI values of the Nth and $(N-1)$th RTP packets are NRI_1 and NRI_2 respectively, there are a total of 4^2 or 16 possible combinations. Since the values from 96 to 125 of the Type field in the RTP header are left for general purpose, the values from 96 to 111 to represent the NRI values. In this way, we can obtain the NRI value of the lost NALU of the next packet or even retain the complete NALU header from the neighboring packets if the lost NALU is of FU-A type.

The total number of retransmission requests depends on the link quality. With high loss rate, requesting for more retransmissions may be desirable since both the NACK and the retransmitted packets may become corrupted. In some situations, burst errors can corrupt many RTP packets at the same time. For these cases, the corrupted RTP packets may be discarded at the receiver (especially packets containing lower-priority NALUs), the bandwidth is conserved for new RTP packets (to be transmitted by the streamer), and EC can be activated at the receiver to correct the errors due to the discarded packets. Since the receive buffer cancels the retransmission requests for NALU passed up to the playout buffer, the redundant request can be eliminated by not requesting for a retransmission of a lost NALU if the retransmission is estimated to exceed one round trip time (RTT).

Table 3.9: NALU Types (includes RFC 3984 Types)

Type	Contents
0	Unspecified
1 - 4	Coded slice data of a non-IDR picture
5	Coded slice data of an IDR picture
6	Supplemental enhancement information
7	Sequence parameter set
8	Picture parameter set
9	Access unit delimiter
10	End of sequence
11	End of stream
12	Filler data
13 - 23	Reserved
24	Single-time aggregation packet (STAP-A)
25	Single-time aggregation packet (STAP-B)
26	Multi-time aggregation packet (MTAP16)
27	Multi-time aggregation packet (MTAP24)
28	Fragmentation unit (FU-A)
29	Fragmentation unit (FU-B)
30 - 31	Unspecified

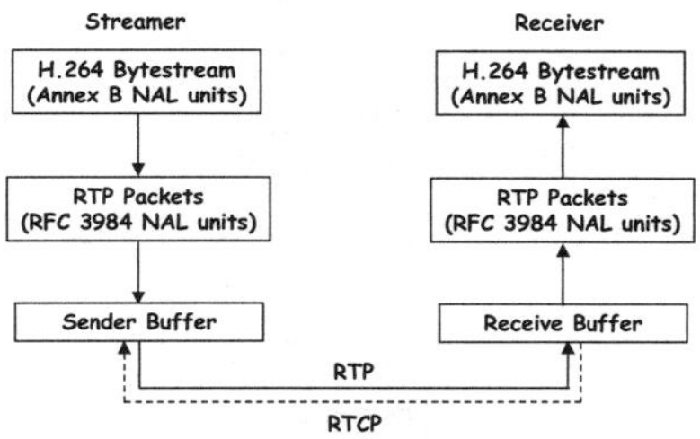

Figure 3.5: Transmission of NALUs in RFC 3984 framework.

3.6 Fundamental H.264 and VC-1 Benefits

H.264 and VC-1 offer high coding efficiency, lower storage requirements, and reduces visual artifacts. The bit rate and resolution of the content can be adapted to networking environment and display device. For the same content, H.264 typically achieves a two fold improvement in coding efficiency over MPEG-2 (designed for broadcast applications) and hence provides higher quality video using the same bandwidth or the same quality video using lower bandwidth. A higher coding efficiency can be achieved for HD than SD videos, allowing a large amount of content to be stored on a single optical disc.

3.6.1 Spatial, Temporal, and Bit Rate Scalability

H.264 provides scalable source coding for multiple resolutions. With slice coding, each frame is subdivided into one or more slices and the order can be modified when an error resilience method such as flexible macroblock ordering (FMO) is used. The slice is given increased importance because it is the basic spatial segment. Each slice is independently coded from its neighbors. In this way, errors or missing data from one slice cannot propagate to any other slice within a frame. This in turn, increases the flexibility to extend frame types (I, P, B) down to the level of slice types. Redundant slices are also permitted. Bits below and above the slice layer are related to VCL and NAL respectively. A recent extension provides layered (hierarchical) coding, which comprises a base layer and two enhancement layers. The media may be scaled down by dropping the enhancement layers when bandwidth resources become scarce or the channel becomes lossy.

Slice coding in VC-1 is restricted to only one slice group (SG) as shown in Figure 3.6. Several slices can be implemented but the length of each slice is restricted to a single row of MBs in the frame. The slice takes only a rectangular shape. Clearly, this is less flexible than H.264 where several SGs can be used by FMO and in addition, the length of the slices can be less than a row of MBs. Granularity can be a single MB if desired (Figure 3.7).

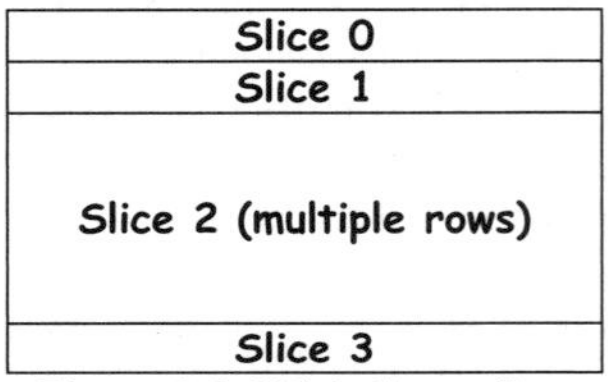

Figure 3.6: VC-1 slice coding.

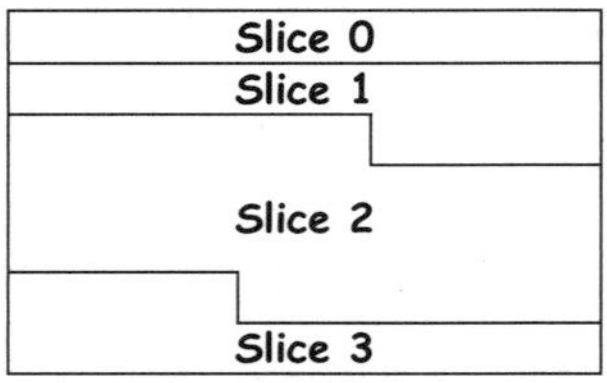

Figure 3.7: H.264 slice coding.

3.6.2 Error Resilience

H.264 offers a number of error resilience methods, which are key to TV broadcasting applications. These methods typically require additional overhead bits for implementation but make the H.264 standard more error resilient than prior MPEG standards. The NAL allows the same video syntax to be used in many network environments. The SPS and PPS provide more robustness and flexibility than prior designs. The SPS and PPS, together with the instantaneous decoding refresh (IDR) slice, enable a bitstream to be decoded from any random point. If errors are unavoidable, frame numbering allows H.264 to detect and localize frame losses. It also allows the creation of subsequences and enables temporal scalability by optional inclusion of extra frames between other frames.

The more powerful error resilience methods spread the errors at the receiver, allowing EC methods to perform effectively. This minimizes the visual impact of losses on the actual distorted image. Contrast the use of interleaving at the packet or bit level, which help spread burst errors such that forward error correction

(FEC) methods can perform more effectively. Two popular methods are data partitioning (DP) and FMO. DP allows the separation of coded video information into different levels of importance. Each level is encapsulated in a single packet and is therefore independently decodable. Higher-priority syntax elements (e.g., sequence headers) can be separated from lower-priority data (e.g., B frame transform coefficients). Thus, transmission of each level can be prioritized or accorded unequal protection, which may in turn minimize loss rates for important data and add further robustness. Each video frame can be divided into three partitions namely, A, B, and C. Partition A assumes the highest importance since it contains the control and sequence header information along with QP values and motion vectors (MVs). Thus, Partition A data should not be dropped or left unprotected during transmission, otherwise Partitions B and C become useless. Partition B contains intracoding information while Partition C (biggest partition) contains intercoding information. Since intracoding information can prevent error propagation, Partition B is more important than Partition C. This means more frequent Partition B updates are needed, leading to larger B partitions. RFC 3984 provides some NRI values for the different partitions, as shown in Table 3.10.

Table 3.10: NRI Values for Coded Slices and Data Partitions of Primary Coded Reference Frames [9]

NALU Type	Content of NALU	NRI (Binary)
1	Non-IDR coded slice	10
2	Coded slice data partition A	10
3	Coded slice data partition B	01
4	Coded slice data partition C	01

Higher-priority partitions (e.g., Partition A) in DP can be protected with stronger FEC to ensure that all users, including multicast users, receive the video at the minimum quality with no artifacts. Lower-priority partitions (e.g., Partitions B and C) can be accorded with standard FEC to reduce the network overheads. This is a much more scalable solution compared to designing a video headend that adapts its transmission to suit many individual receivers simultaneously.

As discussed previously, H.264 allows the possibility of dividing a video frame into regions or partitions called SGs. The partitioning is specified by the MB to SG map. Each SG can also be divided in one or several slices with each slice representing a sequence of consecutive MBs (processed in raster scan order–left to right and top to bottom) within the same (unique) SG. Thus, a slice can be decoded independently. However, the MBs in a slice are not necessarily in raster scan order within a frame. Similarly, the MBs in an SG are not necessarily in raster scan order. FMO may subdivide a video frame into 2 to 8 SGs. H.264 limits the number of SGs in a frame to eight to prevent complex allocation schemes.

FMO rearranges the coding order of MBs to reduce the probability that a packet loss will affect a large region of the frame. It enables more effective EC by ensuring that neighboring MBs will be available for prediction of a missing MB. The MBs are arranged in different SGs using an MB allocation map (MBA map). The MBA map consists of an identification number that specifies to which SG each MB belongs. This ordering exploits the spatial redundancy in frames. For example, the SGs may be designed such that no MB is surrounded by other MBs from the same SG (i.e., error accumulation in a limited region is avoided). In such cases, even if an entire SG is lost, it is possible to construct the MBs using the surrounding MBs corresponding to a different slice. If FMO is deactivated, each video frame will comprise a single slice with the MBs in raster scan order. The use of FMO is compatible with any type of interframe prediction.

Arbitrary slice ordering (ASO) is a complementary tool to FMO. ASO allows decoding of slices from an SG in any order. Like FMO, this can prevent entire regions of a frame from being affected when burst packet losses occur. In addition, ASO reduces decoding delay when NALUs are not delivered in sequence. ASO and FMO can be used independently. When FMO is not used, the slices of a unique SG of the frame are not sent in raster scan order. The redundant slices (RS) feature sends an extra representation of a picture region (typically at lower fidelity) that can be used if primary representation is corrupted or lost, thereby improving the robustness of the video transmission. The RS mode requires the decoder to be informed about its activation. Since an MB is contained only in one slice, RS allows secondary representations of an MB to be coded in the bitstream.

3.6.3 Error Concealment (EC)

EC at the H.264 or VC-1 decoder improves the VQ by correcting any observable or perceptible artifacts caused by errors associated with coding, decoding, or transmission, which can be detected with the help of the coded syntax/semantics. Although its effectiveness improves when combined with error resilience methods, error resilience is not mandatory for EC to function. In some cases, EC can be fairly effective when operated independently and therefore, additional overheads due to error resilience can be avoided. However, the use of error resilience methods mandates the use of EC—once the errors are spread using say FMO, it makes sense to have them concealed. Since EC targets the video bitstream rather than PHY layer bitstream, its effectiveness is not limited to bit errors and single packet losses. In many cases, it can deal with multiple (burst) packet losses that comprise a video frame and are therefore more effective and efficient than FEC. Like FEC, it does not rely on channel feedback. However, unlike FEC and error resilience methods, EC does not introduce any overhead bits although it may incur additional processing time. EC typically incurs more processing time than error resilience. EC can be made adaptive. For instance, EC is not needed when sufficient bandwidth or a good channel is available. EC changes the types of video artifacts and the frequency of occurrence of these artifacts becomes substantially lower. Figure 3.8 illustrates the effectiveness of EC with and without FMO. As can be seen, EC improves the VQ even without FMO and is especially effective with minor errors affecting one or two MBs. However, EC may fail to stop or effectively reduce propagation errors, due to the use of VLC and that frames corrected by EC may not serve as good references for future predicted frames.

EC can be applied to any coded video bitstream (i.e., its activation is codec agnostic) and are generally based on spatial/temporal interpolation from the adjacent areas of the same frame or the reference frame. This requires detection of missing MBs after decoding in order to locate the damaged areas of the image. While H.264 does not define any EC schemes, the Joint Model (JM) reference software [6] provides code to some of these schemes for handling partial and entire frame losses. In the "frame copy" method, each pixel value of the concealed frame is copied from the corresponding pixel of the previously decoded frame. This concealed frame is displayed and stored in the reference frame buffer for decoding subsequent frames [10]. In the "MV copy" algorithm, the MVs and reference indices of the colocated MBs in the previously coded reference frames are copied and used to conceal the missing MBs. Thus, the corrupted frame is assumed to have the same motion as its reference frame.

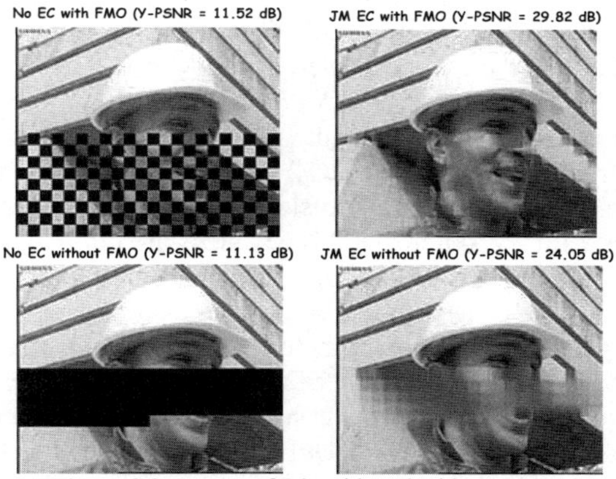

Figure 3.8: Impact of EC, with and without FMO.

The "MV copy" algorithm generally performs better than the "frame copy" algorithm. The "frame copy" method cannot protect the first frame or frames with scene changes and may be prone to error propagation. Its performance also depends on the scene complexity. Scenes with fast motion or rapid scene changes are difficult to conceal with this method. With the exception of movie trailers, most video content do not have abrupt scene changes and fast motion occurring simultaneously. Methods based on interpolation do not suffer these disadvantages since the pixel values of received or concealed neighboring MBs are interpolated from the same frame. Interpolation is employed when the frame to conceal does not resemble the previous frame (e.g., a scene change), but the method may not always perform well. For I frames, the JM EC algorithm uses a weighted sample average of the pixel values of the surrounding MBs. The process is applied for the luma and chroma components. In the case of P frames, motion compensated EC can be performed to repair the damaged part of the frame. Assuming that the motion is smooth and continuous in a frame, the MVs of the lost MBs are predicted from the surrounding decoded MBs.

Actions performed after EC (e.g., deblocking filtering) may change the value of the pixels of the MB edges depending on the smoothness of the MB edge and its neighbor. However, performing deblocking filtering with concealed blocks may corrupt the pixels of the correctly received MB. The authors in [11, 12] refine the idea of spatial EC by using the deblocking filter and data hiding, respectively. These spatial approaches conceal the missing blocks without using other frames and are justified when the I frames do not resemble the preceding frames (e.g., in a scene change). They give poor results for high error rates and are generally not efficient for concealing large areas of the frame.

3.7 H.264 versus MPEG-2, VC-1, and VP8

When compared to MPEG-2, there are higher computational demands in H.264 decoding due to the entropy coding, smaller block size, and in-loop deblocking (Figure 3.9). In general, H.264 decoder is about two times more complex than the MPEG-4 Advanced Simple Profile (ASP) decoder (used in DivX and Xvid). The H.264 coder is about 10 times as complex as the MPEG-4 ASP coder. Table 3.11 compares H.264, MPEG-2, VC-1, and VP8. In MPEG-2, an 8 × 8 discrete cosine transform (DCT) is used to transform the predicted error blocks from the spatial domain to the frequency domain. DCT is known to have favorable energy compaction properties on the intra and residual video data. In H.264, three types of integer transforms with 4 × 4 and 2 × 2 block sizes are defined, as shown in (3.2). An 8 × 8 transform is defined for the high profile (see Chapter 4). Unlike MPEG-2, H.264, and VP8, VC-1 defines rectangular transforms, in addition to square transforms. In general, smaller transforms perform better in areas with discontinuities and block boundaries because they produce fewer ringing artifacts. $T_{4\times4}$ is applied to all luma and chroma components. $H_{4\times4}$ is a secondary transform (a Hadamard transform) applied to all 16 DC coefficients if the 16 × 16 luma MB intraprediction method is chosen. $H_{2\times2}$ is used to transform the 4 DC coefficients of each 4:2:0 chroma component. For 4:4:4 chroma components, $H_{4\times4}$ is used. For 4:2:2 video, both $H_{4\times4}$ and $H_{2\times2}$ are used to perform a 2 × 4 chroma DC secondary transformation. The computation of the integer transform and its inverse transform requires only a few simple shift, add, and subtract operations. For orthonormal Hadamard transforms, add and subtract operations are sufficient operations. The integer operations avoid mismatches of the inverse transform. All transform coefficients are quantized by a scalar quantizer. Unlike MPEG-2, which requires a 32-bit (or floating point) transform, a 16-bit transform is sufficient for 8-bit video data using H.264 and is cheap to implement on 16-bit processors (which form the bulk of commonly used digital signal processors).

$$T_{4\times4} = \begin{bmatrix} 1 & 1 & 1 & 1 \\ 2 & 1 & -1 & -2 \\ 1 & -1 & -1 & 1 \\ 1 & -2 & 2 & -1 \end{bmatrix} \quad H_{4\times4} = \begin{bmatrix} 1 & 1 & 1 & 1 \\ 1 & 1 & -1 & -1 \\ 1 & -1 & -1 & 1 \\ 1 & -1 & 1 & -1 \end{bmatrix} \quad H_{2\times2} = \begin{bmatrix} 1 & 1 \\ 1 & -1 \end{bmatrix} \qquad (3.2)$$

3.7.1 Entropy Coding

Entropy coding is a lossless process that compresses the transform coefficients, prediction modes, MVs, and so forth, into the final output file. The H.264 entropy coding schemes incorporate context modeling to offer a high degree of statistical adaptivity when coding the source content (see Section 4.2). VC-1 employs multiple VLCs for entropy coding.

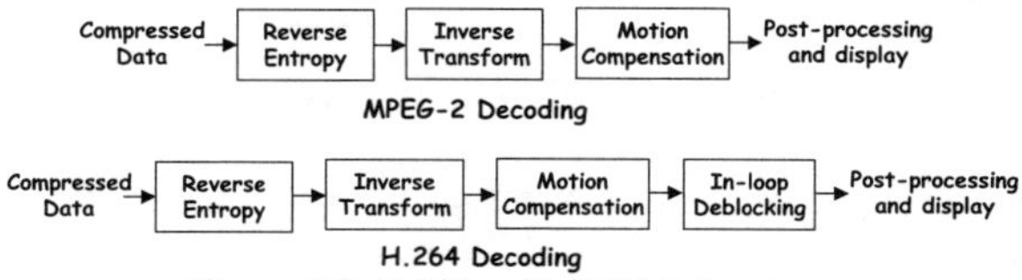

Figure 3.9: H.264 and MPEG-2 decoders.

Table 3.11: High-Level Comparison of MPEG-2, H.264, VC-1, and VP8

	H.262 (MPEG-2)	H.264 (MPEG-4 AVC)	VC-1 (WMV-9)	VP8
Intracoding	DC predictor	4 × 4, 16 × 16 spatial predictor	DC predictor (mandatory) AC predictor (optional)	4 × 4, 16 × 16 spatial predictor
Motion Vector Precision	Full Pel Half Pel	Full Pel Half Pel Quarter Pel	Full Pel Half Pel Quarter Pel	Full Pel Half Pel Quarter Pel
In-Loop Filters	None	Deblocking filter	Deblocking filter, overlap transform smoothing	Non-adaptive: fast and normal modes
Entropy Coding	Non-adaptive VLC (Huffman)	CAVLC (with UVLC) CABAC	Adaptive VLC	Non-adaptive arithmetic coding
Transform Coding	8 × 8 DCT	4 × 4, 8 × 8 integer transform	4 × 4, 8 × 4, 4 × 8, 8 × 8 integer DCT	4 × 4 integer DCT

3.7.2 Block Size

Many coding standards, including MPEG-2 and H.264, treat portions of the video image in blocks, which are often processed in isolation from one another. Regardless of the number of video pixels in the image, the number of blocks has an effect on the computational requirements. MPEG-2 employs a fixed block size of 16 × 16 pixels, H.264 permits simultaneous mixing of 7 different block sizes (down to 4 × 4 pixels) while VC-1 only allows two (16 × 16 and 8 × 8). Thus, H.264 can accurately represent fine details with smaller blocks while not having to "waste" small blocks on coarse detail. For example, patches of blue sky may use large blocks, while details of a forest may be coded with smaller blocks.

3.7.3 In-Loop Deblocking

Deblocking is a computationally intensive postprocessing step (performed after MB decoding) that adaptively smoothes the edges between adjacent blocks, thereby reducing blocking artifacts and improving the decoded VQ. It is particularly useful for lower quality video, which is more susceptible to blocking artifacts. In prior coding standards (e.g., MPEG-4 ASP), deblocking is an optional decoding step, only enabled when it was possible to perform the function in real time. In-loop deblocking is mandatory in H.264. The deblocked image is buffered and used to predict the motion of future frames in a coding loop. This improves VQ because past reference frames used for motion compensation will be filtered versions of the reconstructed frames. Clearly, the effectiveness of the adaptive filter in smoothing the video has an important effect on the overall sharpness and this is why H.264 deblocking defines 3 levels. In addition, the threshold for blockiness detection needs to be chosen properly and this is dependent on the quantizer. When the quantization step size is small, the effect of the filter is reduced. The filter is turned off when the quantization step size is very small.

3.7.4 Motion Compensation, Estimation, and Prediction

Advanced motion estimation is a key strength of many new video coding standards. When sufficient temporal correlation exists, MVs may be accurately predicted and only the prediction error (the residual) is transformed and quantized. Clearly, if the prediction is accurate, less data needs to be stored compared to coding each frame's motion. Because objects tend to move between neighboring frames, detecting and compensating motion are essential for accurate prediction. Such techniques also help partition and scale the bitstream with priority given to data that is more globally applicable. Thus, they not only improve the coding efficiency but also enhance the error resilience. A P frame predicts by displacing the reference frame and finding matching blocks similar to the current coded block. The difference in displacement in the horizontal and vertical directions with the best match is sent as a MV. B frames predict the MVs using spatial prediction (based on neighboring MBs from the current frame), temporal prediction (based on neighboring frames), or weighted prediction (based on reference frames).

Motion estimation employs algorithms to search for motion and compute the MVs. An accurate search algorithm produces better VQ but incurs more time for coding. Conversely, a fast search algorithm may compromise VQ, especially for sports videos. If the video is of higher resolution, more pixels should be used for motion estimation although a typical value of 16 is a good trade-off between speed and quality. Values of 24 or 32 can also be used for more complex motion search algorithms. Quarter-pel motion estimation is normally used for higher accuracy, often activated with rate distortion optimization (RDO). In this case, the reference image is generated by interpolation (via a 6-tap interpolation filter) in order to estimate and compensate fractional-pel displacements. By averaging the luma sample at integer and half-pel positions, the sample at the quarter-pel position is generated. Chroma motion estimation uses color information for motion detection. The chroma sample at a fraction-pel position is also obtained by averaging (horizontally, vertically, diagonally) via bilinear interpolation. In addition, VC-1 applies intensity compensation to a reference frame before motion estimation.

During coding, a frame is normally broken down into MBs or even smaller partitions called a block. The coder will then search for similar MBs in order to discard redundant data via motion estimation. The probability of finding a matching block increases when smaller block sizes and more reference frames are used. If an MB has motion characteristics that can be predicted from the motion of the neighboring MBs and contains no nonzero quantized transform coefficients, then it is flagged as skipped in the P frame. Nonzero MVs can be inferred in this way. A similar mode (direct mode) exists for B frames. In this mode, the MVs for an MB are not sent but derived by scaling the MV of a colocated MB in another reference frame or by spatially inferring motion from neighboring MBs.

In H.264, temporal MV search seeks matching MBs or blocks of variable sizes, which can range from the full MB size of 16 × 16, to an 8 × 8 block (also called a sub-MB), to a 4 × 4 block. Searches may also identify MVs associated with rectangular blocks of sizes 4 × 8, 8 × 4, 8 × 16, 16 × 8. A distinct MV can be sent for each sub-MB and sub-MB partition (i.e., 8 × 4, 4 × 8, 4 × 4). Variable block size motion compensation is a significant feature because it allows the MVs characterizing the displacement of blocks to model the actual motion with high accuracy. In MPEG-2, motion estimation is limited since it searches reference frames for a 16 × 16 set of samples that closely matches the current MB. In addition, matching a set of pixels must be within the reference frame. In general, analyzing the MB partitions can result in a more accurate prediction, and thus improve the visual quality. There are several options.

- 8 × 8, 8 × 16, and 16 × 8 P frame search enables 8 × 8 partitions on P frames.
- 8 × 8, 8 × 16, and 16 × 8 B frame search enables 8 × 8 partitions on B frames.
- 4 × 4, 4 × 8, and 8 × 4 P frame search enables 4 × 4 partitions on P frames, but VQ improvement is not significant.
- 8 × 8 I frame search enables 8 × 8 partitions on I frames and requires the 8 × 8 adaptive transform (see Section 4.1).
- 4 × 4 I frame search enables 4 × 4 partitions on I frames.

Using a finer video resolution results in higher granularity and more accurate prediction of the MVs. Consider a 720p video and 1080p video that have the same content. If the MVs are identical for both videos, then theoretically, the MV prediction will be faster for the 720p video. However, there are two main reasons why the MVs cannot be identical. First, a higher resolution implies more MBs and thus more MVs. Second, because the MVs are given in terms of samples, the same value for MVs derived from videos with different resolutions does not imply the same "magnitude." Nevertheless, the trajectory of the motion will be similar. For example, if a car is going from left to right, then many MVs will point in that direction independent of resolution in the corresponding area of the frame. Figure 3.10 shows the directions of the MVs (depicted by arrows) for a hand moving in the downward direction. More changes in the MVs are associated with the index finger and the edges of the hand whereas many background MVs remain unchanged.

Several motion search algorithms are listed below (based on x264 [13]):

- Diamond search: This is a four-sided shape analysis method. It is the fastest but may compromise VQ. Complexity is $O(n)$.
- Hexagonal search: This is a six-sided shape analysis method that provides reasonable quality at reasonable speed. Complexity is $O(n)$.
- Uneven multihexagon search: A more complex version of hexagonal search, produces high VQ at the expense of speed. Complexity is $O(n)$.
- Exhaustive search: A complete and extensive analysis method. This brute-force method is slow with only a small improvement in VQ compared to the uneven multihexagon search method. Complexity is $O(n^2)$.

- Hadamard exhaustive search: A more accurate version of exhaustive search method using Hadamard transform but is the slowest among all search methods. Computes the sum of absolute Hadamard transform differences on each MV candidate instead of the much faster SAD. Complexity is $O(n^2)$.

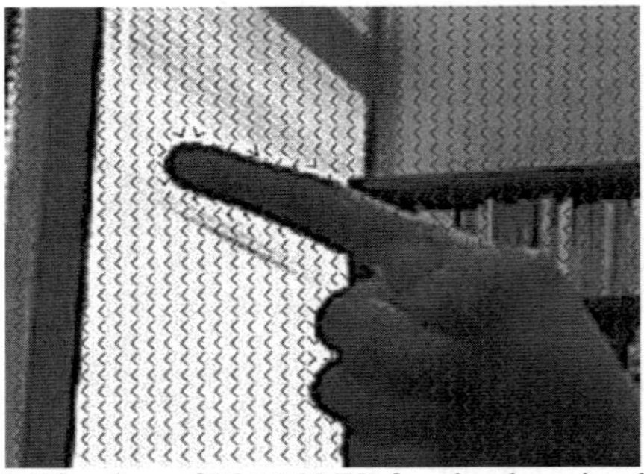

Figure 3.10: Directions of 16 × 16 MVs for a hand moving downwards.

Other methods include subblock split and search, weighted prediction, weighted biprediction, and implicit weighted biprediction. Weighted prediction in P frames employs a predictor with scaling and offsets and is especially effective for fade-in and fade-out scenes. After temporal prediction, the MVs and reference frame indexes are entropy coded. The median MV is first obtained from the left, top, and top-left or right MB or block. The difference between this median and the current MV is then entropy coded. For B frames, two MVs are allowed for each instance of temporal prediction. They can be from any reference frame in the past or future. The weighted average of the pixel values of the reference frames are used as a predictor. The predictor can further employ scaling and offsets to apply the weighted average for biprediction, thereby mitigating cross-fade scenes.

3.7.5 Multiple Reference Frames

Unlike MPEG-2, which restricts the reference frame to the immediate previous frame, H.264 provides additional flexibility for frames to reference up to 16 reference frames (i.e., frames that can be referenced by P and B frames for motion-compensated prediction). This may be any combination of past and future frames that provides opportunities for more precise interprediction and improved robustness to lost frame data. For instance, partial motion compensation for a P frame can be applied when one of its reference frames is missing or corrupted. Reference frames can also be switched via the switching-predictive (SP) and switching-intra (SI) mechanisms that allow identical reconstruction of frames even when different reference frames or a different number of reference frames are used in the prediction process. These mechanisms avoid transmission of an I frame and are useful for supporting trick modes, program channel switching, and bitstream decoding other than from the beginning. Since the SI frame is created without using any reference frame, it is more robust but less efficient than the SP frame. In comparison, VC-1 allows a maximum of 2 reference frames whereas VP8 supports a total of 3 reference frames: previous frame, alternate reference frame, and golden frame, but does not permit the use of B frames, just like H.264's baseline profile. The use of more reference frames typically leads to more efficient coding at the expense of an increase in coding time. However, using excessive reference frames may not improve coding efficiency significantly since the content of the reference frames that are farther away become very different from the frame to be predicted.

3.7.6 Multiview Coding (MVC)

Multiview 3D video bitstreams contain much redundancy yet they are sometimes coded separately (e.g., left-eye and right-eye views) to improve parallelism. This increases the bandwidth and storage requirements, especially for displays that support many simultaneous views (as many as 100) for wide-angle coverage. Although VC-1 and VP8 currently do not support MVC, H.264 includes an MVC extension that provides new techniques for reduced decoding complexity and scalable multiview operations, including marking of reference frames and efficient view switching (e.g., using different SPSs). An MVC bitstream comprises a base view and one or more nonbase views. To improve coding efficiency, a nonbase view can utilize other views for inter-view prediction using temporal motion tools such as disparity vectors. In addition, the parallel decoding SEI provides coded views with systematic constraints so that an MB in a specific view depends only on a subset of decoded MBs in other views. While MVC currently does not add new coding tools (e.g., new MB modes) on top of AVC, MVC does specify new high-level syntax for the NALU and slice headers. For example, a new NALU type called coded MVC slice is used for coding nonbase views. It consists of a new 4-byte header that includes the priority ID, temporal ID, anchor picture flag, and inter-view flag. Anchor frames can be decoded without previous frames and thus act as random access points. Random access at nonintra frames is also possible using gradual decoding refresh (GDR). Anchor and nonanchor frames can have different dependencies, which can be signaled in the SPS.

The efficient conversion of existing 2D videos to 3D has become a major component of 3D content production and delivery. One way to convert 2D to 3D is to generate a depth map or view dependency tree at the decoder. The received MV can be used to derive the displacement of objects over two consecutive frames. In doing so, the depth information need not be sent, thus conserving bandwidth. Alternatively, 2D-plus-depth or 2D-plus-difference methods can be applied. The left and right views can be temporally multiplexed into a single MVC bitstream by alternating frames or fields in a left-right-left-right sequential pattern, which preserves the full spatial resolution at the expense of compromising the temporal resolution (see Exercise 3.17). As for H.264 MVC video streaming, more information can be found in [14], which is an enhancement of RFC 3984.

3.8 Efficient Video Network Transport

The video coder generates data units containing the coded video bitstream, possibly stored in a buffer before transmission. The channel may delay, lose, or corrupt individual data units due to upstream contention among multiple STBs for access bandwidth, congestion in network routers, and retransmissions that lead to delays and additional losses. Overflows in the receiving STB buffer may also occur, resulting in data loss. There is a direct link between artifacts that occur during video playback and data loss. Instead of applying reactive measures at the application level, the use of preventive measures at the network level may be more desirable due to reduced latency. Packet fragmentation, congestion control, proper STB buffer dimensioning can all reduce these losses. Another important cause of video distortion is the lack of synchronization among different video frame types when errors occur in the bitstream. This is specific for coded digital video such as MPEG. In analog video, a separate synchronization subcarrier is used to eliminate the synchronization problem. This may not be possible in digital video.

3.8.1 Dealing with Packet Corruption

The fundamental method to avoid excessive data loss is to ensure that minimum packet loss rate of network standard is maintained. Although the loss rate differs for different network types, it is usually designed to be very low (much less than 1%). In a DOCSIS cable network, a packet loss rate above 1% will result in unacceptable voice over IP (VoIP) call quality. As another example, sometime in 1994, the author carried out an evaluation of a video conferencing system using the integrated services digital network (ISDN), an end-to-end digital network. Here there is no transport layer, only a link layer to perform retransmission if needed. VQ was good, even for a 64-Kbps link that connects Singapore and the United States. The packet loss rate may be low on single access link. As such, monitoring the delay and delay variation (jitter) may be more crucial when supporting real-time video. The packet loss rate may be increased by intentionally discarding video packets when they do not arrive on time. This is closely tied to STB buffer dimensioning, which provides a timing threshold for video packet discard (or retransmission). For example, a faster playback may require less buffer space for storing the video frames. The downside of doing this is the possibility of sacrificing VQ due to packet discard. Finally, channel errors and packet losses can be minimized without completely sacrificing VQ using EC, selective application layer retransmission, and low bit rate feedback channel for loss control and management messages (e.g., using RTCP).

3.8.2 Selective Information Dropping

Highly efficient codecs introduce a high degree of bit rate variability in the coded video stream, which may lead to packet losses due to buffer overflow, possibly impacting VQ. Thus, the use of fixed bit rate allocation may lead to severe underutilization of the channel bandwidth or degradation in the VQ during playback. One method is to reduce the bandwidth requirements via selective information discard, which may help reduce losses under congested situations. As we recall, MPEG, including H.264, encodes the video as a stream of I, P, and B frames. Each frame contains a varying amount of coded video information. I frames are still images that contain the most information. P frames contain predictive information that is used to reconstruct the B frames. B frames are typically the smallest of all 3 frame types and contain the least amount of information. In a GOP, there are typically more B frames than P and I frames. When present, the B frames tend to cause higher bit rate variability.

B frames may be dropped to conserve bandwidth resources and reduce bit rate variability, which is equivalent to smoothing the coded video bitstream. It can also mitigate the impact of an abrupt scene change. Dropping B frames is less harmful because subsequent frames following a B frame are typically not dependent on that B frame. In addition, predictive information from P frames (in both MPEG and H.264) can be used to reconstruct the dropped B frames. This is effective for a small number of dropped B frames. For a high number of B frame loss, EC can compensate for loss in MVs. B frames contain temporal information and so their loss only causes motion artifacts. Such artifacts may be difficult to notice unless the loss rate is very high. In some cases (e.g., low motion video), all or a large number of B frames in the entire video can be removed without introducing visible artifacts. This is in contrast to random frame loss, which can cause artifacts randomly in both temporal and spatial domains, and these artifacts may be more observable even at low loss rates. Note that the dropping of the B frames is performed after the video is coded with B frames. This is different from coding a video without B frames (i.e., using only I and P frames).

To evaluate the performance of DP under selective information discard, partitions B and C of all I, P, and B frames are dropped. The quality of the coded videos is observed and subsequently compared to the case when the B and C partitions from the B frames (rather than all the frames) are removed. Figure 3.11 shows snapshots of the playback of the H.264 videos using the mechanisms described above. Table 3.12 shows the file sizes. Note that by dropping the B and C partitions, one can improve the VQ (by lowering the QP value) for the same coded file size. Dropping the partitions may also reduce the bit rate variability of the coded bitstream since fewer video frames are now transmitted. As observed in Figure 3.11(a), dropping the B and C partitions for all frames causes much distortion in the video, which increases as the playback of the video progresses. Although this produces the smallest file size, the level of distortion observed is unacceptable. However, as shown in Figure 3.11(b), the second method does not show such distortions. This suggests that with DP, the partitions B and C may be dropped for the B frames without compromising VQ. Such selective removal of video information may be useful during peak periods or when the link quality is poor and retransmission becomes ineffective.

To evaluate the performance of FMO under selective information discard, we tested the same H.264-coded *Foreman* video with two or three SGs. In each case, the FMO type is varied from 0 to 5. For types 0 to 3, some form of blocking is observed when the B frames were dropped. However, for types 4 or 5, no artifacts or fast playback were observed in the decoded video even when entire SGs from all B frames (including header information) were removed. From Table 3.12, the bandwidth savings are higher compared to B frame dropping using DP (523/524 Kbyte versus 678 Kbyte using DP). The total bandwidth savings is roughly 30%.

(a) (b)

Figure 3.11: (a) Partitions B and C drop for I, P, B frames (b) Partition B and C drop for B frames.

Table 3.12: Bandwidth Conservation using DP and FMO

Method	File Size (Kbyte)
Partition B and C dropped for all frames	298
Partition B and C dropped for B frames	678
FMO Type 4, 2 SGs, B frames removed	523
FMO Type 5, 2 SGs, B frames removed	524

Foreman H.264 CIF video, QP = 26, 753 Kbyte original size.

3.8.3 Impact on Perceived Video Quality

The level of VQ can be affected by the loss rate, loss pattern (e.g., bursty versus random), delay, codec, video content (e.g., fast-motion versus slow-motion, abrupt versus gradual scene changes), and so forth. In contrast, voice quality is typically not affected by the voice content. Most VQ assessment models rely either on comparing the received and original videos, or on performing complex feature extraction on the received video. We conduct several subjective tests in order to determine the relative impact of different NALU losses on the VQ. This will in turn affect the retransmission decision and the prioritization of the video information. We employ x264 to encode the raw sequence of the *Foreman* CIF video. The impact of information loss on the decoded H.264 VQ is shown in Table 3.13. The subjective quality varies from undecodable to good. In all cases, error resilience and/or EC are not activated. When the SPS or PPS information is removed, the decoder cannot decode any video frame until it receives the next SPS or PPS. Thus, these NALUs should be set to the highest priority and heavily protected. If an entire intracoded IDR slice is removed, the decoded VQ is very poor. IDR slices allow switching from one video stream to another so that decoding can start from any point in time without spatial error propagation caused by lost reference slices in the bitstream. An FU-A NALU loss (belonging to a coded IDR slice) will also degrade the VQ. An NALU loss in the coded non-IDR slice (e.g., P slice) may cause discernible perceptual defects. However, an NALU loss in coded non-IDR slice (e.g., B slice) causes virtually no degradation in VQ. Note that an I frame normally contains multiple I slices, whereas a P or B frame is normally contained in one P slice or one B slice, respectively.

Table 3.13: Impact of Information Loss on Decoded H.264 Video Quality

Loss Description	Subjective Quality
Original H.264-coded bitstream	Good
Remove first SPS at 0x557	Undecodable
Remove first PPS at 0x587	Undecodable
Remove one coded slice IDR (I slice) at 0x595	Very bad
Random cuts within coded slice IDR (I slice)	Very bad
Remove one coded slice non-IDR (P slice) at 0x26892	Bad
Remove one coded slice non-IDR (B slice) at 0x630	Good
Replace bytes from 0x30080 to 0x30160 with zeros	Very bad
Remove bytes from 0x30080 to 0x30160	OK

3.9 H.264 Coding Parameters

This section focuses on the basics of video coding, using H.264 as an example. The parameters are based on several popular H.264 software codecs, including x264. YouTube and Facebook are known to use the x264 coder. A major drawback of the coder is that it does not support error resilience methods such as FMO. The following are typical parameters that can be varied. With multithreading, the coder can fully utilize the processing power of modern multicore processors. This is achieved by coding several frames in parallel.

- Number of frames in between I frames, including the IDR interval. An IDR frame, just like an I frame in MPEG-1, MPEG-2, and MPEG-4 ASP, allows all subsequent frames to make reference to the IDR frame. In other words, no frame after the IDR frame will refer to a frame before the IDR frame. Non-IDR I frames are rare, but since they cannot be ruled out, enforcing a minimal IDR interval can help improve coding of some high-motion scenes. In H.264, I frames within a GOP are possible but they are not seekable since a P frame after the I frame may reference a P frame before the I frame.
- Number of consecutive B frames between the previous and following I frame (or P frame). Allows B frames to compress more efficiently than P frames.
- Number of reference frames. H.264 B frames can act as a reference frame. Limited by the maximum decoded frame buffer size and the video resolution.
- Minimum GOP size controls the minimum number of frames between two IDR frames. Short GOPs may result in inefficient encoding. Generally, this value should equal the video frame rate.
- Maximum GOP size controls the maximum number of frames between two IDR frames. Long GOPs may degrade error resilience, which may be a problem for streaming media and Blu-Ray authoring.
- Number of frames, frame rate.
- Video resolution and display size (horizontal size and vertical size in pixels).
- CBR MBwise, CBR framewise, bit rate.
- Profile and level (number of reference frames and frame size must be consistent with the level).
- High profile chroma array type (monochrome, 4:2:0, 4:2:2, 4:4:4), bit depth for luma [8, 12], bit depth for chroma [8, 12].

- High profile auxiliary format idc, bit depth aux [8, 12], alpha increment flag: [0, 1], alpha opaque value [0, $2^{(\text{bit depth aux} + 9)} - 1$], alpha transparent value [0, $2^{(\text{bit depth aux} + 9)} - 1$].
- Start QP values for I, P, B frames.
- Deblocking filter alpha and beta values are dependent on the QP value. Alpha deblocking controls how much the deblocking filter will smooth the video. It can produce a sharper video or remove more details. Beta deblocking controls the threshold for block detection.
- Disable deblocking filter idc specifies whether the operation of the deblocking filter will be disabled across some block edges of the frame and specifies for which edges the filtering is disabled.
- Entropy coding mode (CAVLC, CABAC).
- The threshold for scene change detection allows the coder to insert an I frame at every scene change (instead of a P or B frame), which leads to better looking scene transitions. A lower threshold yields a more aggressive detection whereas a higher threshold will detect fewer scene changes.
- The hypothetical reference decoder (HRD) specifies constraints on the variability of the coded NALU or bytestream in order to enable cost-effective decoder implementations. It verifies the conditions of the video buffer to ensure compatibility among hardware devices. It also specifies the maximum buffer size and the initial buffer occupancy at the start of playback.
- The hypothetical stream scheduler (HSS) is a delivery mechanism for the timing and data flow of the input of a bitstream into the HRD. It is used to check for conformance of a bitstream or a decoder.

Just like metadata may be included in video containers to allow efficient search for stored H.264 videos, SEI can also be added on a frame-by-frame basis to allow the container to store "in band" data. SEI can be used for a variety of purposes. It is typically coded in a proprietary format since there is no standardized format for unregistered user data in H.264. By providing the current frame number, it can be used for stream positioning. It may indicate the stream bit rate or depth information for 3D video decoding. Stereo video SEI allows the coder to identify the use of the video (e.g., left and right views) on stereoscopic displays. Another use of SEI is to provide decryption information such as a decryption key or a seed for deriving the decryption key for the encrypted video stream. SEI can be employed to validate the frame using a checksum or Hash-based Message Authentication Code (HMAC). The newly created SEI NALU may itself be encrypted and/or signed (validated) so that information contained in it is not easily accessible to an unauthorized user and thus, the video content can be used for security purposes without being compromised.

3.10 Quantization

As described previously, quantization is the lossy part of video coding. Quantization controls the coded video bit rate. The transform coefficients are divided by the quantization matrix and then rounded off. Rounding errors can be minimized using integer transforms with exact-match inverse transforms. The flat matrix is the default quantization matrix of H.264 where all entries are simply filled with the value 16. The Joint Video Team (JVT) matrix is the alternative H.264 quantization matrix but may perform poorly compared to the flat matrix. Custom matrices can also be used to target a specific range of bit rates (e.g., ultrahigh or ultralow bit rates). The deblocking filter impacts the performance of the quantization matrix. In general, high-frequency components are rounded off because the human is less sensitive to these components. Trellis quantization can be used to improve VQ in DCT-based coding methods. It maximizes the peak signal to noise ratio (PSNR) relative to the bit rate for each block in the frame.

The QP value is a measure for the amount of data loss and can be changed for every MB. A higher value leads to more lossy coding and poorer visual quality but a smaller file. Conversely, a low QP value implies more data will be retained, hence better visual quality but leads to a larger file and longer coding and decoding times. Unlike MPEG-2, which uses 31 quantization levels, H.264 uses a quantizer scale that ranges between 0 and 51 (52 levels) when the video supports 8 bits per decoded sample. Although the step size doubles with each QP increment of 6, these parameters are not linearly related. A QP increment of 1 results in an increase in the data rate of about 12.5%. In general, a QP value between 16 and 32 gives satisfactory results for complex scenes. For less complex scenes (e.g., talking heads), a lower QP value (e.g., 10) can be used. For perceptibly lossless coding, a QP value of 0 is chosen to disable the quantizer. In Table 3.4, the H.264 videos are CABAC-coded with a QP value of 0.

In fixed quantization, the QP value is held constant, resulting in variable bit rate (VBR). As an alternative, an average QP value can be used instead. In this case, the QP values in fast-moving scenes increase since the loss may not be visible whereas in slower scenes, the QP values decrease. Thus, this mode may result in a similar subjective quality as the fixed QP mode, but usually achieves higher coding efficiency. Alternatively, one can lower the QP value slightly to give approximately the same file size as the QP mode but with better VQ.

In the average bit rate mode, the video is coded at the desired average bit rate. Therefore, the final file size can be predicted. A higher bit rate will result in better VQ at the expense of a bigger file and vice versa. A tolerance setting can be used to control how precise the coder will hit the target bit rate or target file size. In addition, the use of a maximum quantizer step can limit how much the quantizer can change between two consecutive frames. Typically, a value of 4 is chosen. The coder's ability to adjust the bit rate with respect to the video content is limited in this mode and this may in turn, compromise VQ, especially at medium and lower bit rates. The two-pass average bit rate mode can be used to enhance the VQ but requires twice the time of the single-pass mode described previously. In the first pass, the coder will perform a detailed analysis of the video and high motion scenes will be accorded a significantly higher bit rate than static scenes. In the second pass, the actual encoding takes place and the final file is created.

In general, the P frames should be quantized with a higher QP value than the I frames, as I frames serve as the initial reference of a scene and thus have a huge impact on the quality of the following frames. The B frames should receive stronger quantization compared to the P frames, as B frames are typically not referenced by other frames, while P frames serve as reference for the following frames. The chroma to luma quantizer offset controls how much stronger the color information will be coded compared to the brightness information. It is recommended that the color information is compressed more than the brightness information since data loss in the color information is less visible to the human eye than data loss in the brightness information.

The quantization curve compression (also called the qcomp) controls the level of bit rate fluctuation over the entire video. A 0% value produces a constant bit rate (CBR) while a 100% value results in constant quantization (VBR). A value of 60% gives good results for most videos.

Adaptive Quantization (AQ) allows each MB within the frame to choose different QP values dynamically depending on the complexity of the video content, instead of assigning the same value to all MBs within the frame. MB-level quantization can result in better VQ. AQ partially replaces the effect of qcomp and cannot be used with the constant QP mode. AQ allocates more bits for flat MBs by adaptively lowering the QP value. Without AQ, flat areas of the video image (e.g., sky, grass, walls) and dark backgrounds tend to show visible blocking or banding. Note that VC-1 allows the user to choose the QP value at the frame or MB levels. MB-level quantization is only available for the P and B frames.

3.11 Video Delivery Platforms

In addition to regular payTV service, several alternative video delivery modes exist. For instance, one could buy and own the video by mail or via a movie kiosk, and use it over and over. One could also rent the video or subscribe to premium video content online (e.g., movies, sports) with a monthly fee. These modes are commercial-free but may require the purchase of a console or proprietary STB. Finally, one could watch online videos for free. These videos normally include full episodes, TV shows, and selected movies that come with commercials.

Digital rights management (DRM) involves ownership, control, and distribution of stored media. DRM is easily managed with STBs. Thus, premium content is generally available if accessed via a hardware box. While a proprietary STB is required to receive payTV content, Internet TV can be accessed via a variety of devices, including broadband HDTV, game console, Blu-ray player, Roku set-top, TiVo box, Sling box, laptop, and PC. These hardware devices may support one-to-one personalized TV (where a user can watch TV shows on demand or videos of specific sports teams), social TV (where a user can invite friends to watch the same TV program), interactive TV, video recording (for playback at a later time), and ad management.

PCs or laptops may cause jerkiness and stalling in online video playback and sometimes this is due to the multitasking nature of the operating systems rather than a network or codec problem. In contrast, Internet-enabled TVs may provide a better viewing experience. Many HDTVs now come with good video cards, which may aid HD video playback. They can also be connected to a widget channel that extends Web-based services and applications directly to TV. Widgets are designed to pull selective content from the Internet to complement TV watching. For instance, users can buy products advertised on TV from online stores. Game consoles are TV-friendly and easily connected to the Internet, and are currently used by more than 90 million households worldwide.

Transcoding is equivalent to network streaming except that the output is sent to a file instead. Transcoding allows selection of the appropriate codec and bit rate for the delivery network, which is key to anywhere, anytime, any device delivery. This is because it is difficult to maintain 15 different versions of the same movie, which is common in some deployments. Transcoding software has matured and now produces a wide range of output formats, from Web and mobile standards like MPEG-4, Windows Media, QuickTime, and Flash, to VC-1 and H.264 for next-generation Blu-ray players.

3.12 Online versus PayTV Viewing

In online video viewing, multiple videos can be preloaded on the browser while the user decides on the selection and this reduces the channel change latency when the desired video is eventually selected. Normally, a commercial is streamed from a separate server before the start of the selected video, which allows for initial buffering of the selected video. In payTV, users are presented with a single channel (usually the last channel that was accessed), hence channel change may occur more frequently than online TV. Fast channel change is therefore more critical for payTV video viewing.

3.13 Video Quality Assessment

VQ metrics are used to benchmark or measure the quality of the video as perceived by the user. However, this is not easy since the reconstructed image is not meant to be identical to the original, especially with low bit rate coding, where perceptual quality assessment assumes a more important role. In lossy coding, perceptually irrelevant information will be discarded. What counts as "irrelevant" depends on the viewer's subjective response and three measures are normally used: resolution, noise, and overall impression. These metrics typically detect video artifacts caused by codec and packetization errors, and losses due to network transport, but have limited utility in detecting artifacts due to digital capture. Some useful tools for measuring VQ can be found in [15, 16]. The basic metrics can be classified under subjective and objective. ITU recommends longer sequences (10 sec where possible) for subjective viewing. For subjective comparisons, the sequence under test is presented side-by-side with a sequence generated by the JM reference software. VQ assessment may not be important with low-quality or highly compressed videos (e.g., H.264 videos coded with low QP values). In addition to the poorer quality, these videos tend to be smaller in size and rather than streaming video, users may prefer to download them instead, especially when a broadband connection is available. Note that assessing VQ may incur a fair amount latency because the video content for each frame must be decoded.

3.13.1 Subjective versus Objective Metrics

Subjective metrics assess actual distortions (e.g., blockiness, blurriness, ringing artifacts, added high frequency content, picture outages) perceived by the viewer using experiments that are performed in a controlled environment. These metrics cannot be measured rigorously using quantitative measures since they take into account the sensitivity and perceptivity of the human visual system (HVS), which is complex and may not be consistent across all video displays, resolutions, and human subjects. For instance, children, young adults, and seniors may have varying visual perceptivity. In addition, the manner in which many subjective tests are conducted to collect opinion scores may not lead to the detection of subtle artifacts or loss of detail in the video playback. Averaging subjective ratings of a panel of viewers via a single mean opinion score (MOS) is clearly restrictive and requires proper calibration. Even if an accurate human visual and perceptual system can be modeled, human intelligence is still required to prevent false alarms. For example, humans can intuitively decipher the age of the movie and an old movie is expected to have poorer quality. Humans can also distinguish between deliberate slow motion (e.g., sports replay), problems in playback, as well as deliberate blurring in background versus blurring or censorship on the subject. Watching a video up close and from afar can also lead to significant differences in subjective VQ. This led to some analysts predicting that great VQ with HDTV may not be as important as good choice in video content.

Objective metrics attempt to quantify the observed video distortions. These quantitative values are computed using the actual video and serve as an indicator for possible distortions or artifacts that may be observed in the video. These objective metrics can be further classified into content-independent and content-dependent metrics. Content-independent metrics do not depend on the type or content of the video and thus, have the same impact on all videos. For example, the number of frames affected by the loss of a P frame remains the same irrespective of the video under consideration. They play an important role in optimizing the network transport of videos. Content-dependent metrics, on the other hand, provide measures whose values depend on the actual video under consideration. Many content-dependent metrics require a reference for their computation. Using the original (perfect) reference video image and the received (possibly distorted) image, a relative comparison can be evaluated.

Most existing objective VQ approaches are known as full-reference, which compare the similarity of two images using an initial distortion-free image as reference. In reduced-reference quality assessment, the reference is only partially available, in the form of a set of extracted features made available as side information to help evaluate the quality of the distorted image. A no-reference or "blind" quality assessment approach does not require a perfect reference but may suffer in reliability and accuracy when evaluating VQ compared to the reduced-reference and full-reference metrics. No-reference metrics are also prone to error propagation. Full-reference metrics are normally computed at the video coder since it may be impractical to send an undistorted reference frame to the decoder for comparison. On the other hand, no-reference metrics are typically employed by a video decoder at various points in the network at the service provider side (e.g., the core or metro network connected to the video headend, which are usually point-to-point optical fiber links). In a point-to-multipoint access network, the usefulness of VQ metrics becomes very limited due to the enormous network overheads needed to track the quality of video playback at thousands of customer premise devices spread across multiple locations. In addition, collecting VQ measurements from these devices via an access network, and using these measurements to monitor and act on possible distortions at the video headend may be ineffective due to the latency in decoding the video to compute the VQ values. In a one-way broadcast access network (e.g., satellite broadcast), VQ measurements become impractical while in mobile or portable access (e.g., using DVB, WiMax), such metrics may not be updated in a timely manner. Unfortunately, most packet errors and losses occur during the transmission across the access network, which ultimately impacts end-user visual experience. The problem is exacerbated with multicast services when the same video stream is transported under varying link qualities to distributed end-users.

Tables 3.14 and 3.15 show some content-independent and content-dependent metrics, respectively. These metrics only identify possible degradation in VQ and are primarily designed for detecting artifacts in still images. Identifying the cause of these problems and providing remedies to correct these problems can be more challenging. For example, the VQ metrics included in ITU-T Recommendation J.247 have outperformed PSNR in validation tests, but many typical types of transmission artifacts related to packet losses are not covered by the tests.

Table 3.14: Content-Independent Metrics

TMDR	Time duration for which video is affected during a frame loss
FRAME SIZE	Size of the frame
FRAME TYPE	Type of frame (I, P, or B) that is lost/dropped

Table 3.15: Content-Dependent Metrics

MSE	Mean squared error in a frame
PSNR*	Peak signal to noise ratio in a frame
RELATIVE PSNR	PSNR as compared to highest quality video
SSIM*	Structural similarity index

* Full-reference metric that measures image quality using a distortion-free image as reference.

3.13.2 Peak Signal to Noise Ratio (PSNR)

A high PSNR indicates a less noisy signal. In general, an image or video frame with significant details or high scene complexity lowers the PSNR since it is more difficult for a lossy encoder to replicate the original frame. PSNR is governed by the mean squared error (MSE) and number of bits/sample (B). For two $m \times n$ monochrome images I and K, the PSNR is given by (3.3). For RGB color images, the same equation is valid but MSE is sum over all squared value differences divided by the image size and by three. Typical PSNR values range from 30 to 50 dB, where a higher value indicates better VQ. A PSNR above 40 dB normally results in transparent video (i.e., the coded video becomes indistinguishable from the original video). The PSNR is infinite when two images are identical since the

MSE will be zero. A PSNR of zero corresponds to the case when I is completely white and K is completely black (or vice versa). The luma or Y-PSNR is widely used since visual perception is most sensitive to brightness.

$$PSNR = 20\log_{10}\left(\frac{P_I}{\sqrt{MSE}}\right)$$
$$\text{where } MSE = \frac{1}{mn}\sum_{i=0}^{m-1}\sum_{j=0}^{n-1}\left\|I(i,j)-K(i,j)\right\|^2 \text{ and } P_I = 2^B - 1 \tag{3.3}$$

There are several limitations of PSNR although it must be pointed out that no VQ metric is perfect. For instance, PSNR applies only to videos with the same resolution. Identical video content with different resolutions may lead to a similar PSNR, as shown in Figure 3.12. This is true if the videos are coded with the same QP value. For low QP values (i.e., 1), the PSNR is higher, even for videos of lower resolutions, which is expected. Another example is seen in Figure 3.15, where a few video images with identical content are produced with similar PSNR but the VQ ranges from poor to good. Thus, a fairly high PSNR in the range of 30 dB does not necessarily imply good VQ. Note that it is not possible to recreate the same situation when the PSNR is high (e.g., over 40 dB). Yet another example, if the image is displaced or rotated slightly by say 2 pixels or 2 degrees, it can destroy the PSNR but the subjective VQ may still be good. If the received image is shifted in any way, most likely, the source is also shifted. From the last two examples, one can generally conclude that a high PSNR provides a reliable measure of good VQ whereas a low PSNR may not always result in poor VQ.

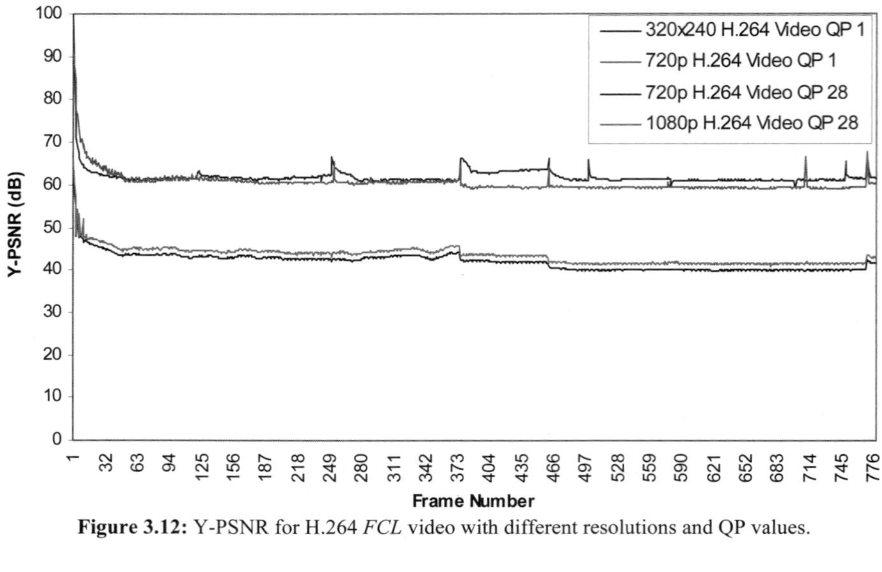

Figure 3.12: Y-PSNR for H.264 *FCL* video with different resolutions and QP values.

3.13.3 Structural Similarity (SSIM) Index

The SSIM index accounts for higher-level structural information in the video content. Unlike error-based metrics such as PSNR, which can be inconsistent with human eye perception, SSIM takes into account the human visual system. Some extensions include structural texture similarity index and color structural texture similarity index. SSIM is also commonly used as a method for testing the quality of lossy video encoders. For example, the popular open source H.264 encoder x264 is set by default to display an SSIM value at the end of an encoding operation. The SSIM index is a decimal value between 0 and 1. A value of 0 implies zero correlation with the original image, whereas a 1 means the exact same image. 0.95 SSIM, for example, would imply half as much variation from the original image as 0.90 SSIM. Through this index, image, and video coding methods can be effectively compared using three components (in contrast, PSNR uses one). Luminance comparison is given by (3.4) and is a function of the mean intensity. The contrast comparison, as a function of variance, is given in (3.5). The structure comparison, as function of covariance, is given in (3.6). SSIM is defined by (3.7). As can be seen from Figures 3.13 and 3.14, there is a strong correlation between luminance SSIM (Y-SSIM) and luminance PSNR (Y-PSNR).

$$l(x,y)=\frac{2\mu_x\mu_y+C_1}{\mu_x^2+\mu_y^2+C_1} \tag{3.4}$$

$$c(x,y)=\frac{2\sigma_x\sigma_y+C_2}{\sigma_x^2+\sigma_y^2+C_2} \tag{3.5}$$

$$s(x,y)=\frac{\sigma_{xy}+C_3}{\sigma_x\sigma_y+C_3} \tag{3.6}$$

where μ_x and μ_y, and σ_x and σ_y, represent the mean and standard deviation of the decoded and original images, respectively.

$$\text{SSIM}(x,y)=l(x,y)^{\alpha}c(x,y)^{\beta}s(x,y)^{\gamma} \tag{3.7}$$

3.13.4 Czenakowski Distance (CZD)

Like PSNR, CZD is a per-pixel quality metric that estimates the quality by measuring differences between pixels of the reference frame and pixels of the received frame. It measures the similarity among different samples and has a better correlation with subjective quality assessment than PSNR [17].

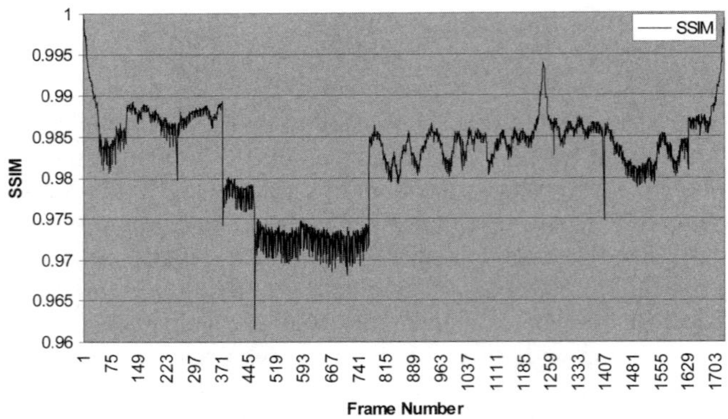

Figure 3.13: Y-SSIM for the *FCL* 1080p H.264 video.

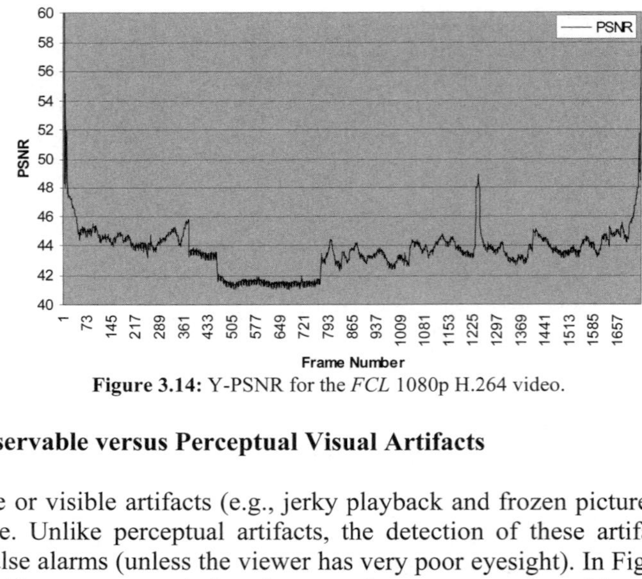

Figure 3.14: Y-PSNR for the *FCL* 1080p H.264 video.

3.13.5 Observable versus Perceptual Visual Artifacts

Observable or visible artifacts (e.g., jerky playback and frozen pictures) are more quantitative. Unlike perceptual artifacts, the detection of these artifacts are not prone to false alarms (unless the viewer has very poor eyesight). In Figure 3.15(b), the two artifacts are very obvious because they occur on the subject, and on the most important part of the subject—the eyes. The background artifact in Figure 3.15(c) is less visible. However, if temporal aspects are accounted for, the superposition of video frames in Figure 3.15(a) and (c) reveals an obvious artifact in the background. Note that there is no movement associated with the background, hence no change in MVs. Figure 3.15(d) shows a uniform degradation in VQ throughout the video frame using a higher QP value. However, the PSNR is similar to the frames in Figure 3.15(b) and 3.15(c), and more important, there are no observable artifacts (they may be perceptible depending on the sensitivity of the viewer). This shows the limitations of using the PSNR metric to assess VQ when the QP value is low. It also demonstrates the difficulty of assessing VQ. For instance, quantifying the degradation in VQ for the different number of artifacts occurring in Figure 3.15(b) and (c) is very challenging (they may well be parts of the frame that are intentionally censored). Some online streaming systems adjust the QP value and resolution dynamically to prevent observable artifacts (VQ and resolution degrade with losses). However, the loss of picture detail or fidelity may be visible for HD videos even though a lower resolution may not result in an artifact. To summarize, the practical measurement of VQ may not be accurate or precise. It is also difficult to quantify—the overall picture quality can be poor even if there are no observable artifacts. Conversely, the artifacts may be observable but the overall picture quality is good.

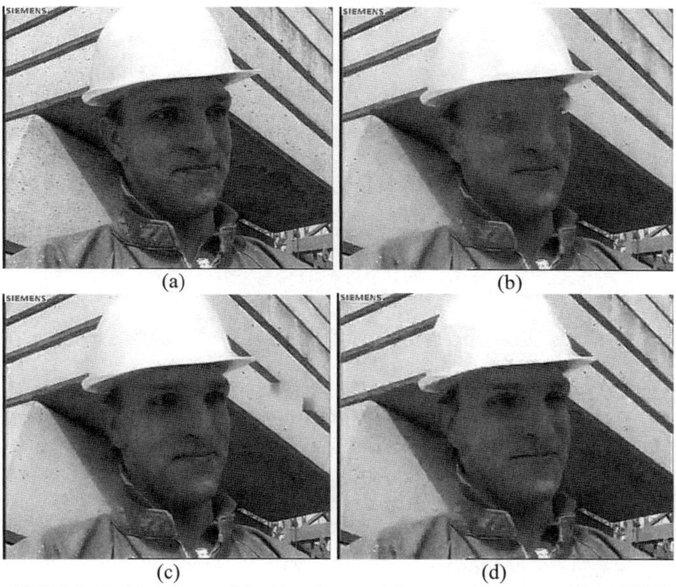

Figure 3.15: (a) Original video frame, (b) video frame with two artifacts on subject, PSNR = 32.10, QP = 30, (c) video frame with one artifact on background, PSNR = 32.09, QP = 30, and (d) video frame with no artifact, PSNR = 32.14, QP = 37.

3.14 CBR versus VBR Coding

The VBR (open loop) coding scheme reduces bandwidth and storage space requirements more efficiently than CBR (close loop) coding. VBR coding can be applied to many popular audio (MP3, WMA, AAC) and video (MPEG, VC-1, H.264) formats. VBR coding is not only favored for its smaller file sizes (thereby making it easier to store and stream), it also helps maintain VQ at a constant level because each video frame is coded using a fixed QP. This in turn simplifies the codec design and reduces coding and decoding time. In the CBR approach, frames with a large size (typically due to scene changes and/or high motion) are coded using a higher QP value in order to maintain a constant coded bit rate at the output. However, this degrades the VQ of the frame and rate RDO via a rate controller is normally applied to prevent the degradation of future frames, which increases encoder complexity and reduces coding efficiency. In addition, it is not possible to measure the overall VQ since this is variable and may change dramatically for each frame due to changes in the QP value and content complexity. The variation is more noticeable with HD as shown in Figure 3.16. A comparison of the VQ between VC-1 and H.264 under the CBR mode is shown in Table 3.16. As can be seen, the VQ of H.264 appears to be superior to VC-1 for a variety of videos but this is highly dependent on the decoder implementation.

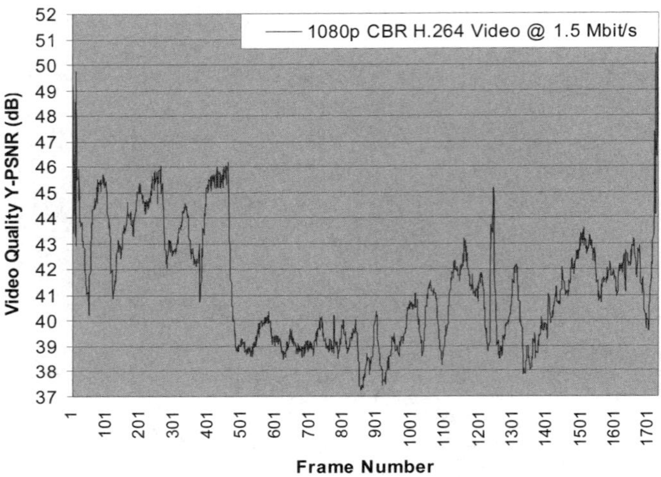

Figure 3.16: Variable video quality with CBR HD coding.

Table 3.16(a): Comparison of Y-PSNR (in dB) for H.264 and VC-1 under the CBR Mode

Bit Rate (Mbps)	*Dell* (480p)			*Parkrun* (720p)			*Shields* (720p)		
	VC-1	H.264 (CAVLC)	H.264 (CABAC)	VC-1	H.264 (CAVLC)	H.264 (CABAC)	VC-1	H.264 (CAVLC)	H.264 (CABAC)
2	32.39	34.34	35.2						
5	37.73	39.88	40.71				33.15	37.07	37.27
10	43.54	45.12	46.01	26.63	30.12	30.61	36.34	38.24	38.45
20				30.38	33.30	33.85	38.67	39.79	40.13
30				33.09	35.42	36.05	40.16	41.14	41.63
40				34.81	37.20	37.95	41.52	42.47	43.01

Table 3.16(b): Comparison of Y-PSNR (in dB) for H.264 and VC-1 under the CBR Mode

Bit Rate (Mbps)	*FCL* 1080p		*Parkrun* (1080i)		*Blue Sky* (1080p)	
	VC-1	H264 (CAVLC)	VC-1	H264 (CAVLC)	VC-1	H264 (CAVLC)
2	38.46	42.78				
5	43.97	46.38				
10	47.2	48.68	25.3	26.74	35.57	39.17
20	50.5	51.27	27.81	29.38	39.16	41.39
30			29.44	31.24	40.76	42.72
40			30.63	32.67	41.87	43.48

The coded H.264 frame sizes of some 1080p and 720p HD VBR videos are shown in Figures 3.17 to 3.19. Some frames exhibit larger sizes compared to the average. The peaks are normally caused by scene changes, which are intracoded. High motion may lead to large frame sizes but they do not correspond to the highest peaks due to motion prediction. For the highest peak value, the peak to average ratio of the frame sizes can approach 30 (Figure 3.20) or some high value (Figure 3.21). An interesting phenomenon arises when videos with the same content but different resolutions are coded with the same QP value. As shown in Figure 3.22, the peak to average ratios of the frame sizes for both videos are almost identical.

Streaming VBR video traffic is a special challenge due to the high dynamic range of the frame sizes that results in high bit rate variability. Smoothing the coded video stream reduces the bit rate variability. Figure 3.23 shows the bit rate variability of smoothed and unsmoothed VBR HD videos. The H.264-coded videos have the same file size. The smoothed video is sent at a rate that is equivalent to the average rate of the video (see Chapter 9 for more details). As can be seen in Figure 3.23, the smoothed video makes efficient use of available bandwidth (even when transporting video over CBR-SDV systems) and unlike the unsmoothed video, avoids peak rates that may cause packet losses over the network or buffer overflow at the receiver.

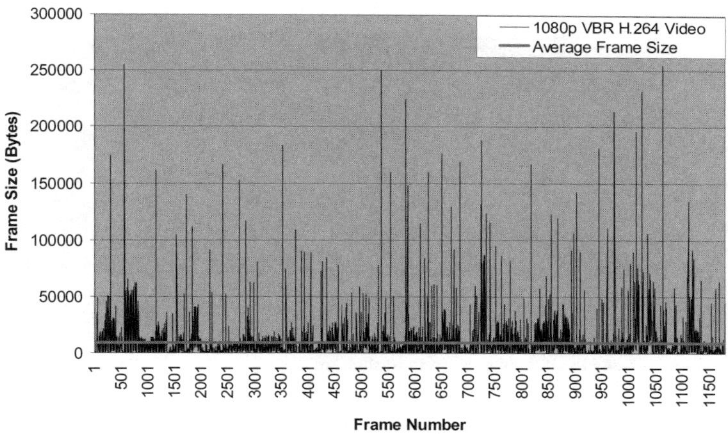

Figure 3.17: Frame sizes for *BBB* 1080p video.

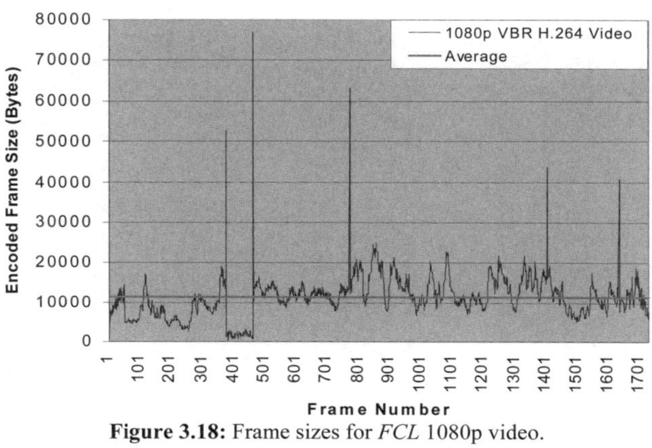

Figure 3.18: Frame sizes for *FCL* 1080p video.

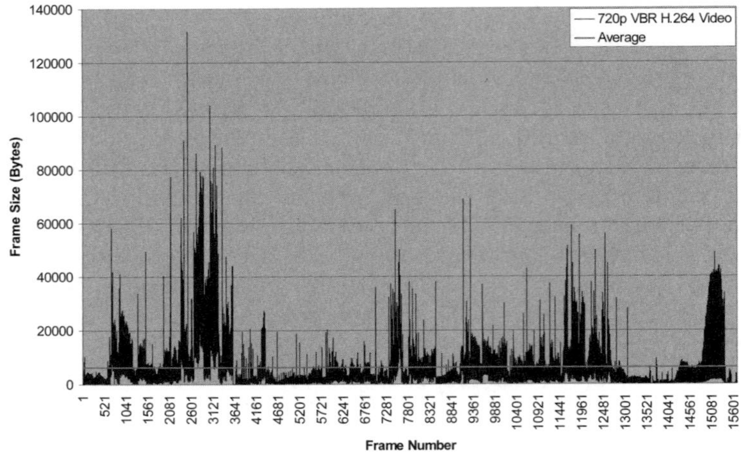

Figure 3.19: Frame sizes for *Elephant* 720p HD video.

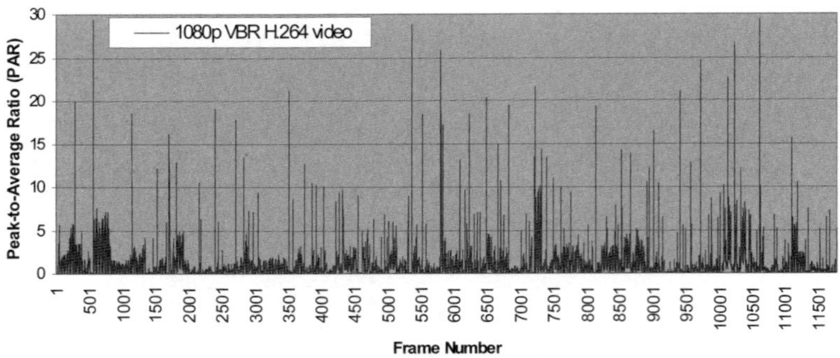

Figure 3.20: Peak to average ratio of the frame sizes for *BBB* 1080p HD video.

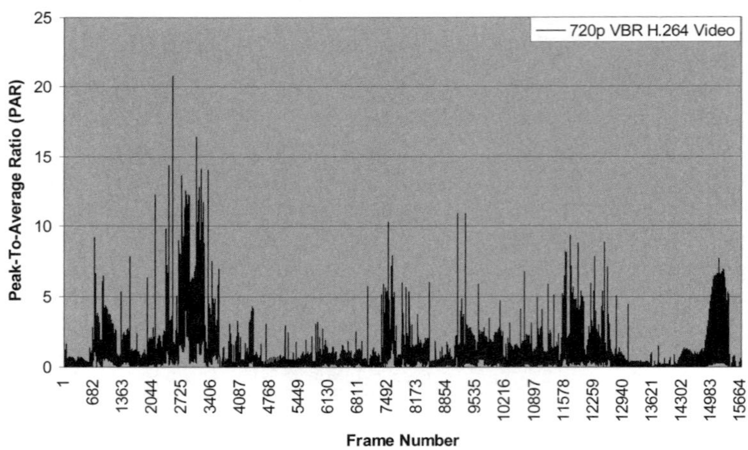

Figure 3.21: Peak to average ratio of the frame sizes for *Elephant* 720p video.

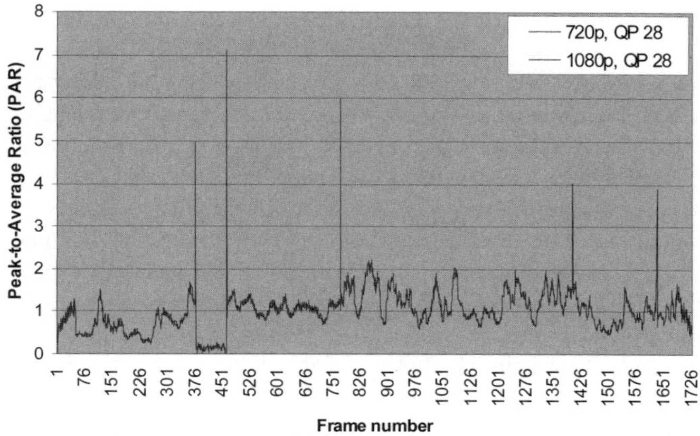

Figure 3.22: Peak to average ratio of the frame sizes for *FCL* 720p and 1080p videos.

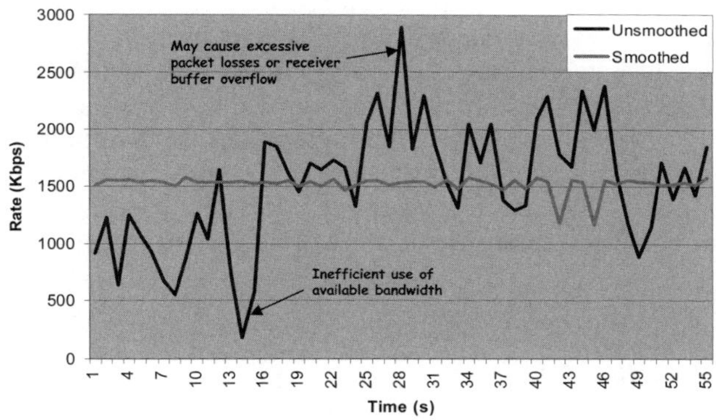

Figure 3.23: Comparison of smoothed and unsmoothed VBR video streaming.

3.15 Scalable Video Coding

SVC, as specified in Annex G of the H.264 standard [1], splits a video bitstream into subbitstreams. This allows lower resolution bitstreams to be reconstructed from partial bitstreams. SVC typically encodes the video once at the highest resolution but enables decoding of partial bitstreams, depending on the rate and resolution required by the applications operating over disparate networks, including 3D videos. For temporal bitstream scalability, the temporal sampling rate for the subbitstream can be lowered. High-level syntax and interpredicted reference frames are constructed accordingly. For spatial and quality bitstream scalability, the spatial resolution and quality can be lowered. NALUs are removed to derive the subbitstream and interlayer prediction is typically used for efficient coding. Although SVC provides more flexibility in transporting the coded video, it can also suffer reduced coding efficiency. For example, employing multiple layers of coded video with different levels of error protection may sometimes lead to less efficiency compared to a single-layer structure.

3.16 Conclusions

Compression reduces the cost of storing and transmitting video by converting the information to a lower bit rate. We have covered the main video coding features of the H.264, VC-1, and VP8 standards. We have also described methods to transport coded video efficiently and assess VQ. These methods are key to reducing the artifacts in online streaming and payTV systems, which vary greatly. For example, freeze frames are common in online streaming (which employ H.264) while video frame breakup is more commonly encountered in cable/satellite payTV systems that employ MPEG-2. H.264 offers several advantages such as a significant reduction in bandwidth requirements and a built-in ability to detect and track video artifacts, and collect statistics. H.264's variable length decoder can detect missing MBs in a video frame with no false alarms. EC can be employed to complement H.264's capabilities. EC requires no channel feedback and is applicable to broadcast, multicast, unicast networks, as well as end-to-end Internet streaming systems. EC may help maintain the video coding rate and VQ in the presence of burst or random losses and conceal errors caused by packet losses after video decoding. It can also help increase the frame rate for high-action videos or remove the need for recurring PHY layer error correction overheads. With powerful standards such as H.264 and VC-1, and with proper bandwidth management, the access network and the digital media industries are poised to be revolutionized by high quality, bandwidth-effective, and scalable delivery of digital content to heterogeneous home and enterprise networks, and user terminals.

References

[1] *Advanced Video Coding for Generic Audiovisual Services*, ITU-T Rec. H.264, March 2010, http://www.itu.int/rec/T-REC-H.264/e.

[2] ITU H.264 brochure, http://www.itu.int/itudoc/gs/promo/tsb/87066.pdf.

[3] *VC-1 Compressed Video Bitstream Format and Decoding Process*, SMPTE 421M, 2006, http://www.smpte.org.

[4] On2 VP8, http://www.on2.com/index.php?599.

[5] WebM project, http://www.webmproject.org.

[6] H.264 reference software, http://iphome.hhi.de/suehring/tml or http://www.itu.int/rec/T-REC-H.264.2.

[7] H.264 test sequences and video bitstreams, http://www.itu.int/rec/T-REC-H.264.1 or http://www.mpegif.org/resources.php.

[8] VC-1 reference decoder, http://wiki.multimedia.cx/index.php?title=Understanding_VC-1.

[9] S. Wenger, M. M. Hannuksela, T. Stockhammer, and D. Singer, "RTP Payload Format for H.264 Video," IETF RFC 3984, February 2005.

[10] S. Bandyopadhyay, Z. Wu, P. Pandit, and J. Boyce, "An Error Concealment Scheme for Entire Frame Losses for H.264/AVC," *IEEE Sarnoff Symposium*, 2006.

[11] A. Yilmaz and A. Alatan, "Error Concealment of Video Sequences by Data Hiding," *IEEE International Conference on Image Processing*, Vol. 2, pp. 679–682, December 2003.
[12] P. Nasiopoulos, L. Coria-Mendoza, H. Mansour, and A. Golikeri, "An Improved Error Concealment Algorithm for Intra-frames in H.264/AVC," *IEEE International Symposium on Circuits and Systems*, Vol. 1, pp. 320–323, May 2005.
[13] x264 encoder, http://www.videolan.org/developers/x264.html.
[14] Y. Wang and T. Schierl, "RTP Payload Format for MVC Video," IETF Draft, April 2010, http://www.ietf.org/id/draft-ietf-avt-rtp-mvc-00.txt.
[15] MSU video quality measurement tools, http://compression.ru/video/quality_measure/index_en.html.
[16] MPEG verification tests, http://mpeg.chiariglione.org/quality_tests.htm.
[17] D. Andreutos, K. N. Plataniotis, and A. N. Venetsanopoulos, "Distance Measures for Color Image Retrieval," *IEEE International Conference on Image Processing*, Chicago, 1998.

Exercises

3.1. Confirm that a raw 90-minute 480p video requires roughly 112 Gbyte of storage. Confirm that a 720p H.264 video comprises 3,600 MBs. Show that a 1080p (4:4:4) video at 60 Hz requires a raw bit rate of roughly 3 Gbps, which corresponds to the bit rate of 3G-SDI.

3.2. How is it possible that the lossless coding efficiency using H.264 is superior to data compression? By removing all P and B frames in an MPEG video, will this result in no compression? Note that in Dirac Pro (SMPTE VC-2) video coding (http://www.bbc.co.uk/rd/projects/dirac/diracpro.shtml), this mode is used for high quality and low latency applications (e.g., professional production).

3.3. Since the encoder is not defined in the H.264 standard, can a vendor build an MPEG-2 encoder that is more efficient than a poorly designed H.264 encoder?

3.4. When compared to SD, HD video requires roughly four times more bandwidth and storage space with MPEG-2. If H.264 is used instead, will this ratio change?

3.5. A H.264 encoder compresses a 7680×4320 video and a 4K video to 90 Mbps and 25 Mbps respectively. Compute the coding efficiency for both cases, assuming a progressive frame rate of 60 Hz and a color format of 4:4:4.

3.6. The DVB asynchronous serial interface (DVB-ASI) is similar to SDI and frequently used in satellite transmission. A DVB-ASI program stream utilizes 7 TS packets per IP packet. One stereo audio pair at 256 Kbps is included in the stream. Together with a 480i H.264 video (60 Hz) that is coded using the 4:2:2 color format with 0.25 bit/sample, compute the bit rate of the transport stream. How many program streams can be multiplexed to a 35 Mbps MPTS?

3.7. The mismatch in progressive frame rate between film (typically 24 Hz) and television (typically 25/30 Hz) creates a need for "pulldown" methods that stretch the frame rate by exploiting the nature of 60 Hz interlaced (also called 60i) videos. Besides frame repetition and insertion of null frames, devise other pulldown methods to achieve this objective. It is common that 30 Hz and 24 Hz videos are played back at 29.97 Hz (i.e., 30/1.001 Hz) and 24.98 Hz (i.e., 24/1.001 Hz) respectively. How can synchronization problems be avoided for these cases?

3.8. Compare the raw bit rates for a 720p video versus a 1080i video. Assume that 60 Hz videos are used (60p or 60i). How will the rates compare to a 1080i 60 Hz video that is deinterlaced before being transmitted? Compare the rates for a 1080i video versus a 1080p video at 60 Hz. What can you conclude about the bandwidth requirements? Why do some ATSC and payTV channels prefer 720p over 1080i? Will 720p provide better VQ than 1080i when playing high-motion sports videos?

3.9. In many HDTVs, it is common to have several display modes, for example, wide zoom, zoom, normal, full, and so forth. Suppose you set the HDTV to 1080p with a display aspect ratio of 16:9. Explain whether this ratio will change when different display modes are selected. Assuming square pixels, compute the display aspect ratio for 1080i and compare with 720p and 1080p. Compute the display aspect ratio for a 720p 3D movie (Note: there are 2 answers to this question). Explain why most computer display resolutions are 4 by 3 (e.g., 640 × 480, 1024 × 768) and 8 by 5 (e.g., 1280 × 800, 1440 × 900). How will a 1080p video appear on a 720p HDTV and a 1280 × 800 computer display? Will the pixel to aspect ratio and the display to aspect ratio change?

3.10. H.264 uses 52 quantization levels. How many bits are required to represent each level? Compare with the number of levels employed by MPEG-2.

3.11. For a video with successive black frames, will information be coded? List the factors that may affect the VQ for videos coded at the same rate. In general, an encoder must include a decoder to perform prediction and compensation. A H.264 encoder is shown in the figure below (common modules in bold). The decoder conceptually works in reverse, comprising the entropy decoder and the common modules. Will this imply that encoders are always more complex than decoders? Explain why in-loop deblocking filtering is performed at the encoder.

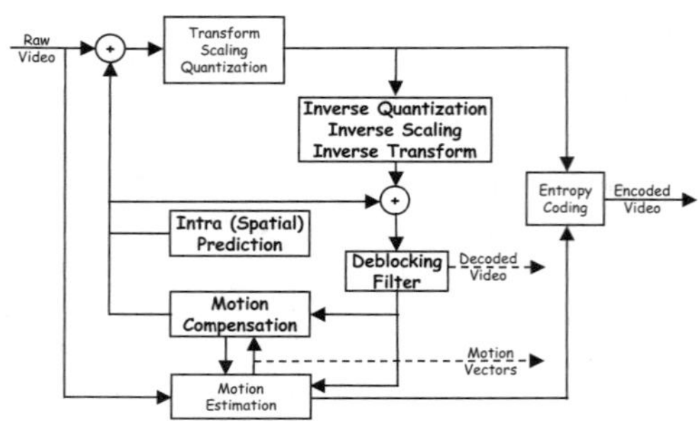

3.12. There are two general ways for checking the characteristics of a coded video. First, the frame sizes can be extracted and then summed up to give the overall video file size. Second, the natural video frame rate can be used to derive the frame duration. When multiplied by the total number of frames, this should match the duration of the video. Can the same methods be applied to 3D videos?

3.13. Since EC exploits inherent redundancy in the coded video and does not require extra overhead bits to conceal visual artifacts, can EC be used to improve the efficiency of video coding? Give a counter example to show that FEC and bit interleaving can perform better than error resilience and EC. Discuss whether packet interleaving can perform better than bit interleaving and error resilience. Explain whether DP can be combined with error resilience. Will DP with unequal error protection be useful if packet fragmentation is employed? FEC is normally embedded in each transmitted packet and this constitutes a recurring overhead. Consider a different method of FEC where a dedicated FEC packet is transmitted after a group of n packets. Will this method incur less overhead? What are the pros and cons? Can such a method be used to reduce bandwidth requirements? How can this method service real-time video traffic?

3.14. High refresh rate HDTVs such as 240-Hz HDTVs may employ interpolation between frames to create and insert additional frames into a lower frame rate video. To convert a 60p video to 240 Hz, how many additional frames must be inserted in one second? Explain whether this method is more complex than performing EC. Note that error resilience is not needed here. Can the pulldown method in Exercise 3.7 be used instead? Many movie projectors operate at 24 frames per second. However, each frame is illuminated two or three times before the next frame is projected using a shutter in front of its lamp. As a result, the projector has a 48 or 72 Hz refresh rate. Can this method be adopted to display a 24p video (commonly used in Blu-ray players) on a 240 Hz HDTV? Now consider partial and entire frame losses to the I, P, and B frames. Explain whether frame interpolation will still work for these losses. Online streaming systems normally use a progressive frame rate of 30 Hz and below. Explain whether higher frame rate videos can be enabled. In this case, packet losses associated with the network transmission must be taken into account, in addition to frame interpolation.

3.15. Dirac (http://diracvideo.org) is an open-source royalty-free video coding technology that employs a two-dimensional 4 × 4 discrete wavelet transform to remove spatial redundancies. The transform allows Dirac to operate on a wide range of video resolutions because entire frames, as opposed to smaller MBs or blocks, are used. Like block-based transforms such as DCT, it decorrelates data in a frequency-sensitive way but preserves fine details better and thus, achieves high coding efficiency. Dirac provides 1/8 pel MV precision and uses arithmetic coding but not in-loop filters. Explain why CBR encoding is not supported by Dirac.

3.16. Assuming an error-free link, discuss whether the picture quality in the free ATSC broadcast service can be better than satellite or cable payTV. Compare this quality with a DVD. Watch a sports video such as a tennis match and compare the VQ of 60 Hz and 240 Hz HDTVs when viewed at 2 ft and 15 ft. What can you conclude about raising the frame rate with respect to VQ? In general, what can you say about evaluating VQ at different distances from the display and using subjects with different levels of visual clarity? Visual artifacts tend to occur in high-action sports videos but they may not be perceptible. Would a slow-motion playback reveal these artifacts more succinctly? Although the HVS works with a continuous flow of photons, analyze the maximum number of frames per second the HVS can process. Suppose a tennis player serves at 140 mph on a 1080p video. Compute the minimum frame rate in order to avoid any visible artifact. Repeat the computation for a 3D 1080p video with two views. Note that even with a badminton shuttlecock served at 200 mph, the player can still retrieve. More information about the HVS is available at http://webvision.med.utah.edu.

3.17. Broadcasters and payTV providers employ side-by-side and top-bottom spatial multiplexing technologies as an alternative to the frame-sequential or simulcast (one channel per eye) method (used by active-shutter TVs and Blu-ray players) to deliver 3D videos to the home. These frame-compatible technologies consume the same bandwidth as standard 2D transmissions but only half of frame-sequential technology, which requires the decoder to operate *n* times faster to support *n* views. It splits each frame into two images, one for each eye. This is achieved by spatially reducing the horizontal or vertical resolution of each image by half to fit into a single video frame as a side-by-side image or as a top-bottom image (see figure below). The STB doubles the length of each image and then displays those images sequentially for the shutter glasses. While not as dense or rich as frame-sequential images, it requires only a simple firmware update and no new STB hardware. The technology can be adapted to broadcast interlaced or progressive videos and was adopted by ESPN to deliver the 2010 World Cup and by DirecTV for its 3D programming. How does it compare to methods involving 2D to 3D conversion (Section 3.7.6), EC, column-by-column/row-by-row line interleaving, and interlacing (or vertical decimation)? Which of these methods has the consequence of reducing the overall spatial resolution in order to include stereoscopic depth (e.g., horizontal disparity)? How can the VQ of 3D videos be assessed effectively? Are PSNR and SSIM adequate metrics in addressing MV corruption, including disparity vector corruption? With 3D on the rise, shouldn't providers consider migrating to H.264/MVC STBs for more efficient delivery of 3D-HD services, instead of H.264/AVC STBs?

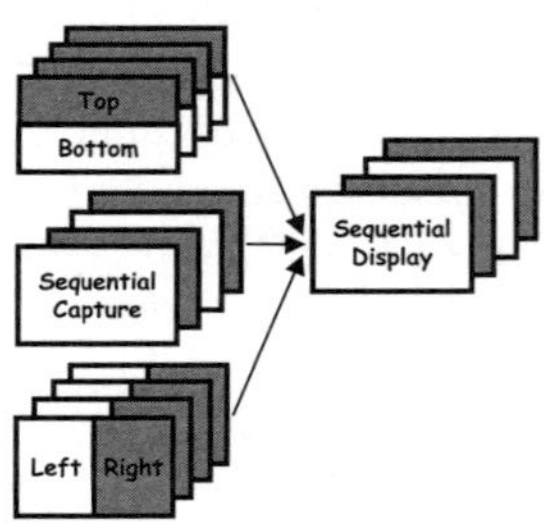

3.18. A small rotation (say 1 degree) in the video frame may not result in poor visual quality although the PSNR metric may indicate otherwise. Suppose there is a larger rotation of say 10 degrees. Design a VQ metric that can assess the quality of the rotated video reliably. How is the quality threshold determined (e.g., is a 3-degree rotation acceptable compared to a 5-degree rotation)? How can a rotated video image occur during encoding and network transmission?

3.19. Some VQ metrics attempt to assess the temporal and spatial aspects of a video. For example, the ITU-T P.910 recommends the use of spatial and temporal information based on the deviation of luminance samples of consecutive frames while other metrics evaluate the MVs in the encoded sequence. Discuss the effectiveness of these methods in the context of Figure 3.15(a) and (c). Can PSNR detect motion artifacts in successive frames?

3.20. Many service providers try to optimize the VQ and bandwidth consumption by adjusting the coding parameters before sending the coded video across the access network. In this case, no visual artifacts are introduced by losses or errors attributed to the network. Explain whether full-reference VQ metrics are sufficient for this purpose. Given that the user will watch video on the best available screen, how can the VQ of the same content be streamlined over different screen sizes?

3.21. PSNR has been criticized as a poor VQ metric but it has been used extensively, including assessing the VQ of MPEG standards 1, 2, and 4 that have evolved over the years. Should PSNR be replaced by "better" VQ metrics, given that there is no perfect VQ metric? Is the accurate VQ measurement (or even visual inspection) of videos with low PSNR a necessity, given that these videos are mostly poorly coded or highly corrupted? Or should methods that identify the cause of the low PSNR assume a more important role (e.g., evaluating the link quality of the network, assessing the types of missing MBs)? Note that these methods do not require the measurement of VQ and can function independently.

3.22. Given that high PSNR works well and that EC improves VQ, would PSNR suffice in measuring the VQ of a video with frames that have been operated upon by EC? Justify the following statement: In general, no-reference VQ metrics cannot perform as well as full-reference VQ metrics.

3.23. In perceptual coding, quantization noise is shaped so that it is masked by the signal energy. The technique has been applied successfully in voice/audio compression. Explain whether a similar technique can be used to improve VQ.

3.24. Which of the following videos (with the same content) will produce a higher PSNR? (a) a raw 720p video or (b) a 1080p video coded to a bit rate that is equivalent to the raw 720p video.

3.25. MPEG-2 is known to be susceptible to as many as 70 different video artifacts while H.264 has a lot less. Given that MPEG-2 achieves a lower coding efficiency than H.264, explain whether EC in MPEG-2 is more effective than H.264 since there is more redundancy in the MPEG-2 video that can be exploited. With the next-generation video coding standard expected to provide a two fold improvement in coding efficiency over H.264, will EC be effective?

3.26. Achieving good VQ requires an understanding of the artifacts generated by coding, packetization, and network transport. This can be a challenge for legacy MPEG codecs because the network may not be able to decipher which frame within a GOP is lost and thus take appropriate measures. Fortunately, many coding artifacts in prior MPEG codecs have been addressed by H.264 (e.g., blockiness, frame synchronization, error propagation, color bands) and different frame types can be accorded different priorities by the network. Can digital (source) capture artifacts can be detected and corrected by H.264? How effective is an H.264 VQ meter that simply flags an artifact whenever a missing MB is detected?

3.27. Identical video content coded with the same QP value but with different resolutions may lead to a similar PSNR or peak to average ratio of the frame sizes. Will this imply a connection between PSNR and the peak to average ratio of the frame sizes? In other words, will a larger frame size lead to a higher peak to average ratio, and hence a higher PSNR? Take into account the fact that PSNR may not be consistent and that peak to average ratio is not an absolute metric.

3.28. In general, it is more difficult for a lossy encoder to code a video with fine details, including videos recorded at high resolutions. This may in turn lead to a lower PSNR because the decoded video may not be a good reproduction of the original. Refine this observation in the context of Figure 3.12. Which factor has a more dominant influence on PSNR: QP or video resolution? Explain whether it is possible to devise methods to improve the PSNR of the source capture. For a video coded with a fixed QP value, is constant PSNR achievable?

3.29. When evaluating the observable and perceptible artifacts in Figure 3.15, what type of VQ metrics is used: subjective or objective?

3.30. Explain whether the following pictures reveal a visual or physical artifact. Hint: Detection of visual artifacts should be instantaneous.

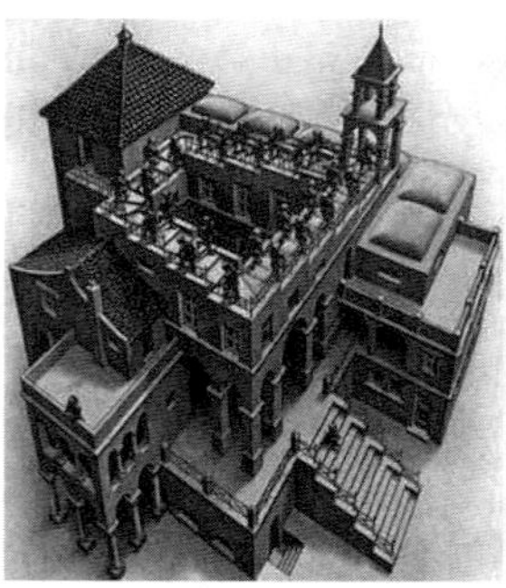

Chapter 4

The H.264 Standard

H.264 is excellent for transmitting rich interactive media over narrow and broadband networks and can be easily integrated into existing and future networks. Coding efficiency is improved by at least a factor of two over prior MPEG standards. H.264 provides bit rate adaptivity and scalability, which addresses the needs of different applications when transmitted over heterogeneous networks. This is also necessary when the receiving device is not capable of displaying full resolution or full quality images. H.264 is future-proof since new extensions can be added as technology improves. This chapter provides experimental insights into the key features of the H.264 standard and illustrates how the strengths of different capabilities can be harnessed in a complementary fashion. Among the H.264 attributes that were tested include profile type, entropy coding methods, rate distortion optimization, and the effectiveness of flexible macroblock ordering under varying levels of packet loss. In all the experiments, we employ the JM software version 12.4. Since the JM software only provides a test platform for normative features of the H.264 standard, it may not be optimized in terms of coding and decoding times. However, the results presented are based on a relative comparison of these times.

4.1 Profiles and Levels

The H.264 profiles are defined as follows:

- Profiles and levels specify restrictions on the bitstreams and hence limit the capabilities needed to decode the bitstreams.
- Each profile specifies a subset of algorithmic features and limits that shall be supported by all decoders conforming to that profile.

Three basic profiles (baseline, extended, main) are defined (Table 4.1). The high profiles are called fidelity range extensions (FRExt) and may achieve further coding efficiency than MPEG-2, as much as 3:1. They include all main profile coding tools. In MPEG-2, 4 profiles (simple, main, high, 4:2:2) and 4 levels are defined. A comparison of the profile, application, decoder complexity, and efficiency of H.264 and MPEG-2 is shown in Table 4.2. A profile defines which set of H.264 coding tools (or subset of the bitstream syntax) are enabled or disabled. Profiles are needed to ensure that the coded video plays properly on the target decoder without unnecessary functions. For example, a video coded using the main profile will decode properly on any main profile decoder and high profile functions are not needed. If the same coding parameters are chosen, the profile has no impact on the video quality (VQ), coding time, or coded video file size. Unlike software decoders, hardware decoders are more restrictive on the profile they support. This is not surprising since the number of implemented features impacts decoder complexity and cost. H.264 also defines 16 levels that are specified together with the profiles (Table 4.3). While profiles define coding capabilities, levels restrict other properties of the coded video such as limits on the resolution, bit rate, frame rate (for a given resolution), and number of reference frames. These constraints imposed restrictions on values of the syntax elements in the bitstream. To decode the H.264 video, both the video profile and level must be supported. For example, Apple's iPod Touch supports the baseline profile with level 3.0 (up to 1.5 Mbps) whereas the iPad and iPhone supports the main profile with level 3.1. The Internet Streaming and Media Alliance (http://www.isma.tv) defines H.264 profiles and levels for streaming IP video.

Table 4.1: H.264 Profiles and Capabilities

Parameter	**Baseline**	**Main**	**Extended**	**High**	**High 10**	**High 4:2:2**	**High 4:4:4**
Use of I and P frames	X	X	X	X	X	X	X
Use of B frames		X	X	X	X	X	X
Use of SI and SP frames			X				
Interlaced coding (PicAFF, MBAFF)		X	X	X	X	X	X
Data partitioning			X				
Redundant slices	X		X				
Flexible MB ordering	X		X				
Arbitrary slice ordering	X		X				
Multiple slice groups	X		X				
8-bit sample depth	X	X	X	X	X	X	X
9- to 10-bit sample depth					X	X	X
11- to 12-bit sample depth							X
Transform bypass							X
Quantization scaling matrices				X	X	X	X
Weighted prediction for P and SP frames		X	X	X	X	X	X
CABAC		X		X	X	X	X
CAVLC	X	X	X	X	X	X	X
8 × 8 transform decoding				X	X	X	X
Predictive lossless coding							X
Separate picture scaling				X	X	X	X
Separate C_B and C_R QP control				X	X	X	X
Residual color transform							X
Monochrome format				X	X	X	X
4:2:0 Color format	X	X	X	X	X	X	X
4:2:2 Color format						X	X
4:4:4 Color format							X

Table 4.2: H.264 versus MPEG-2

Profile	Typical Application	Additional Decoder Complexity over MPEG-2	Typical Efficiency over MPEG-2
Baseline	Low delay applications	2.5 times	1.5 times
Extended	Mobile streaming	3.5 times	1.75 times
Main	Broadcast video	4 times	2 times

Table 4.3: H.264 Levels

Level No.	Max MBs per second	Max frame size (MBs)	Max video bit rate (VCL) for Baseline, Extended, and Main Profiles	Max video bit rate (VCL) for High Profile	Max video bit rate (VCL) for High 10 Profile	Max video bit rate (VCL) for High 4:2:2 and 4:4:4 Predictive Profiles	Examples of max resolution @ frame rate (max stored frames) in Level
1	1485	99	64 Kbps	80 Kbps	192 Kbps	256 Kbps	128 x 96 @ 30.9 (8) 176 x 144 @ 15.0 (4)
1b	1485	99	128 Kbps	160 Kbps	384 Kbps	512 Kbps	128 x 96 @ 30.9 (8) 176 x 144 @ 15.0 (4)
1.1	3000	396	192 Kbps	240 Kbps	576 Kbps	768 Kbps	176 x 144 @ 30.3 (9) 320 x 240 @ 10.0 (3) 352 x 288 @ 7.5 (2)
1.2	6000	396	384 Kbps	480 Kbps	1152 Kbps	1536 Kbps	320 x 240 @ 20.0 (7) 352 x 288 @ 15.2 (6)
1.3	11880	396	768 Kbps	960 Kbps	2304 Kbps	3072 Kbps	320 x 240 @ 36.0 (7) 352 x 288 @ 30.0 (6)
2	11880	396	2 Mbps	2.5 Mbps	6 Mbps	8 Mbps	320 x 240 @ 36.0 (7) 352 x 288 @ 30.0 (6)
2.1	19800	792	4 Mbps	5 Mbps	12 Mbps	16 Mbps	352 x 480 @ 30.0 (7) 352 x 576 @ 25.0 (6)
2.2	20250	1620	4 Mbps	5 Mbps	12 Mbps	16 Mbps	352 x 480 @ 30.7(10) 352 x 576 @ 25.6 (7) 720 x 480 @ 15.0 (6) 720 x 576 @ 12.5 (5)
3	40500	1620	10 Mbps	12.5 Mbps	30 Mbps	40 Mbps	352 x 480 @ 61.4 (12) 352 x 576 @ 51.1 (10) 720 x 480 @ 30.0 (6) 720 x 576 @ 25.0 (5)
3.1	108000	3600	14 Mbps	14 Mbps	42 Mbps	56 Mbps	720 x 480 @ 80.0 (13) 720 x 576 @ 66.7 (11) 1280 x 720 @ 30.0 (5)
3.2	216000	5120	20 Mbps	25 Mbps	60 Mbps	80 Mbps	1280 x 720@ 60.0 (5) 1280 x 1024@ 42.2 (4)
4	245760	8192	20 Mbps	25 Mbps	60 Mbps	80 Mbps	1280 x 720 @ 68.3 (9) 1920 x 1080 @ 30.1 (4) 2048 x 1024 @ 30.0 (4)
4.1	245760	8192	50 Mbps	62.5 Mbps	150 Mbps	200 Mbps	1280 x 720 @ 68.3 (9) 1920 x 1080 @ 30.1 (4) 2048 x 1024 @ 30.0 (4)
4.2	522240	8704	50 Mbps	62.5 Mbps	150 Mbps	200 Mbps	1920 x 1080 @ 64.0 (4) 2048 x 1080 @ 60.0 (4)
5	589824	22080	135 Mbps	168.75 Mbps	405 Mbps	540 Mbps	1920 x 1080@72.3 (13) 2048 x 1024@72.0 (13) 2048 x 1080@67.8 (12) 2560 x 1920@30.7 (5) 3680 x 1536@26.7 (5)
5.1	983040	36864	240 Mbps	300 Mbps	720 Mbps	960 Mbps	1920 x 1080@120.5 (16) 4096 x 2048@30.0 (5) 4096 x 2304@26.7 (5)

The baseline profile is a subset of the extended profile. Both profiles address video transmission in error-prone environments. The baseline profile minimizes complexity but offers the lowest coding efficiency. The extended profile supports more error resilience techniques and is the only profile that supports SI and SP frames. It reduces temporal correlation using B frames but is computationally more complex. However, the extended profile is not a popular choice in deployment. The baseline profile is used when short coding and decoding times are desired (e.g., video conferencing applications). For this reason, B frames are not allowed in the baseline profile. The main profile provides the best VQ among the basic profiles at the expense of even higher decoder complexity, primarily due to the use of B frames and CABAC.

The high profiles employ more than 8 bits per sample and allow higher resolution color formats (i.e., 4:2:2 and 4:4:4). For each additional bit of source video accuracy, the quantization step expands by 6. A decoder conforming to the high 4:4:4 profile will be capable of decoding a bitstream coded with high 4:2:2, high 10, high, or main profiles. Similarly, the high 4:2:2 profile decoder is capable of decoding the high 10, high, and main profiles. These profiles do not offer tools for error robustness or resilience and are mainly designed for storage or for broadcasting. They achieve higher coding efficiency using weighted prediction for P frames (Section 3.7.4), 8×8 luma prediction with low-pass filtering to improve predictor performance, and adaptive transform coding for the intracoded frames. In adaptive transform coding, 4×4 blocks may be used for more detailed parts of the frame while 8×8 blocks are applied to areas with less details. The 8×8 transform in (4.1) is more complex than the 4×4 transform in (3.2) but simpler than the 8×8 DCT transform used in MPEG-2.

$$T_{8\times 8} = \begin{bmatrix} 8 & 8 & 8 & 8 & 8 & 8 & 8 & 8 \\ 12 & 10 & 6 & 3 & -3 & -6 & -10 & -12 \\ 8 & 4 & -4 & -8 & -8 & -4 & 4 & 8 \\ 10 & -3 & -12 & -6 & 6 & 12 & 3 & -10 \\ 8 & -8 & -8 & 8 & 8 & -8 & -8 & 8 \\ 6 & -12 & 3 & 10 & -10 & -3 & 12 & -6 \\ 4 & -8 & 8 & -4 & -4 & 8 & -8 & 4 \\ 3 & -6 & 10 & -12 & 12 & -10 & 6 & -3 \end{bmatrix} \tag{4.1}$$

Adaptive transform coding is supported by perceptual-based quantization scaling matrices that take into account the visibility of a specific frequency associated with each transform coefficient to tune the quantization fidelity. The quantized coefficients are then scanned in the zig-zag or field fashion [8] before being entropy encoded. The scan ordering is designed to order the highest-variance coefficients first and to maximize the number of consecutive zero-valued coefficients appearing in the scan. The higher profiles target more complex videos using more chroma samples per luma sample (e.g., 4:4:4 versus 4:2:0) or finer QP values (up to 12 bits per sample for high 4:4:4). Thus, the use of higher profiles is justified if video is already in good quality. As such, the high profiles are normally deployed in broadcast SD and HDTV, and optical storage media.

Table 4.4 shows the profiles and levels of VC-1. VC-1 defines 3 profiles. The simple and main profiles define the sequence, frame, MB, and block layers. The slice and entry point layers are defined in the advanced profile. The entry point layer is similar to the GOP, providing random access to the bitstream. The simple profile is designed for low bit rate video applications, the main profile for good-quality online video streaming, and the advanced profile for entertainment-quality DTV and HD-DVD applications.

Table 4.4: VC-1 Profiles and Levels

Profile	Level	Maximum Bit Rate	Resolutions by Frame Rate
Simple	Low	96 Kbps	176 x 144 @ 15 Hz (QCIF)
	Medium	384 Kbps	320 x 240 @ 24 Hz (QVGA) 352 x 288 @ 15 Hz (CIF)
Main	Low	2 Mbps	320 x 240 @ 24 Hz (QVGA) 352 x 288 @ 30 Hz (CIF)
	Medium	10 Mbps	720 x 480 @ 30 Hz (480p) 720 x 576 @ 25 Hz (576p)
	High	20 Mbps	1920 x 1080 @ 25 Hz (1080p) 1920 x 1080 @ 30 Hz (1080p)
Advanced	L0	2 Mbps	352 x 288 @ 25 Hz (CIF) 352 x 288 @ 30 Hz (CIF) 352 x 240 @ 30 Hz (SIF)
	L1	10 Mbps	704 x 480 @ 30 Hz (NTSC-SD) 720 x 576 @ 25 Hz (PAL-SD)
	L2	20 Mbps	704 x 480 @ 60 Hz (480p) 1280 x 720 @ 25 Hz (720p) 1280 x 720 @ 30 Hz (720p)
	L3	45 Mbps	1280 x 720 @ 50 Hz (720p) 1280 x 720 @ 60 Hz (720p) 1920 x 1080 @ 25 Hz (1080i) 1920 x 1080 @ 30 Hz (1080i) 1920 x 1080 @ 25 Hz (1080p) 1920 x 1080 @ 30 Hz (1080p) 2048 x 1024 @ 30 Hz
	L4	135 Mbps	1920 x 1080 @ 50 Hz (1080p) 1920 x 1080 @ 60 Hz (1080p) 2048 x 1536 @ 24 Hz (Digital Cinema) 2048 x 2048 @ 30 Hz

4.2 CABAC versus CAVLC

There are two lossless entropy coding methods in H.264: context-adaptive variable-length coding (CAVLC) and context-adaptive binary arithmetic coding (CABAC) [1]. Both methods employ advanced entropy coding techniques but CABAC provides better efficiency than CAVLC (typically 5 to 10%) and achieves higher bit-rate savings under the VBR mode. However, CABAC demands more computational power than MPEG-2's entropy decoding, typically about 4 times. At higher quantization levels, the efficiency of CABAC improves. The arithmetic coder in CABAC permits a noninteger number of bits per symbol. CABAC employs frequency statistics of dominant symbols to perform efficient coding and this also allows the coder to adjust to changing symbol statistics (Figure 4.1). If CABAC is disabled, then the less efficient but faster CAVLC can be used. Unlike CABAC, which is used to code the entire bitstream, only the transform coefficients are coded adaptively in CAVLC. In this case, 32 different VLCs are used. Note that the transform coefficients take up the most bandwidth. CAVLC is combined with another entropy coding technique, universal variable length coding (UVLC), which employs exponential Golomb codes to code the headers (i.e., syntax elements). These are special Huffman VLCs with a regular construction scheme that favors small numbers by assigning them shorter codes. These codes have the generic form of {K zeroes, 1, K-bit data} where data is a binary representation of an unsigned integer. Each code can be decoded as Code Number = 2^K + integer(data) – 1, as shown in Table 4.5. We refer CAVLC to the joint use of CAVLC and UVLC. CABAC requires a main or high profile and more decoder processing power compared to the other algorithms [2]. For the baseline or extended profiles, CAVLC must be used. CAVLC is sometimes used to improve the performance of slower devices.

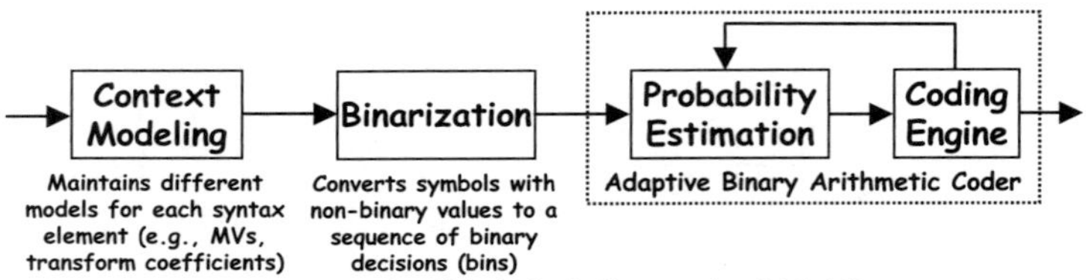

Figure 4.1: Simplified block diagram for CABAC.

Table 4.5: Examples of Exponential Golomb Codes

Code Number	Code
0	1
1	010
2	011
3	00100
4	00101
5	00110
6	00111
7	0001000
8	0001001
…	…

4.2.1 CABAC and CAVLC under VBR Mode

We employ videos with an IBPBPBPBPBPB GOP structure and a QP value of 30. The VBR mode implies no constraint on the coded bit-rate and VQ is maintained across all frames. Figure 4.2 shows identical Y-PSNR values obtained from the decoder for one of the videos (*Blue Sky* HD), which implies that the choice of the entropy coding method has no impact on the VQ. This is expected since the entropy coding method is lossless and only changes the coding of the quantized transform coefficients, not the QP value. Table 4.6 shows that CABAC achieves higher coding efficiency (i.e., a smaller coded file size) than CAVLC for all video resolutions (i.e., QCIF, CIF, HD). Although CABAC is more efficient for higher quality videos, the decoding times are almost identical. Thus, the increased processing time required by CABAC is compensated by smaller bitstreams. For videos with the same size, the decoding time for CABAC will be longer.

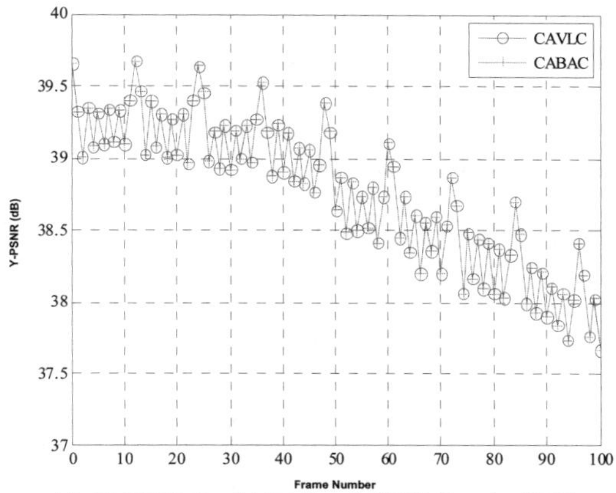

Figure 4.2: Y-PSNR for CABAC and CAVLC under VBR mode.

Table 4.6: CABAC and CAVLC Decoding under VBR Mode

Video	Entropy Method	Coded File Size (Bytes)	Decoding Time (s)
Foreman QCIF	CAVLC	235,649	27.6
(400 frames)	CABAC	223,133	27.4
Foreman CIF	CAVLC	554,249	47.3
(300 frames)	CABAC	511,711	48.2
Coastguard CIF	CAVLC	1,052,730	49.8
(300 frames)	CABAC	943,413	49.9
Mobile CIF	CAVLC	7,575,666	69.2
(300 frames)	CABAC	7,031,726	69.2
Blue Sky HD	CAVLC	7,372,239	583.6
(217 frames)	CABAC	6,551,358	584.3

4.2.2 CABAC and CAVLC under CBR Mode

The CBR mode adapts the QP value of each coded frame to fit a targeted bit rate and only allows for I and P frames (no B frames). To achieve CBR, the coder dynamically adapts the QP value of each frame according to previously coded frames and the GOP structure, and limits the variations of the QP value in subsequent frames so that perceptible disruptions in the VQ are minimized [3, 4]. However, in some situations such as a scene change, the CBR mode may result in temporary but noticeable degradation in the VQ. To show that the bit-rate savings provided by CABAC allow lower distortions in the CBR video, the *Foreman* video (CIF resolution, 300 frames) video is coded with fixed rates of 300 Kbps and 1 Mbps. The first frame of the coded video is an I frame while the rest are P frames. From Table 4.7, we observe that the coding time is significantly shorter for CABAC (21 sec less with a bit rate of 300 Kbps and 43 sec less with a bit-rate of 1 Mbps) whereas the decoding time is reduced slightly for CAVLC (1.2 sec less with a bit rate of 300 Kbps and 1.5 sec less with a bit rate of 1 Mbps). The CAVLC coding time increases with the specified bit rate but decreases for CABAC. Unlike the VBR case (where the higher amount of processing demanded by CABAC is compensated by lower rate bitstreams), the difference in decoding complexity is visible in the CBR mode. This difference is more pronounced with higher-quality and higher-resolution videos. Figure 4.3 shows that Y-PSNR for CABAC is consistently better than CAVLC on each coded frame. Due to the bit-rate savings provided by a more efficient coding technique, a lower QP value can be used for CABAC, which improves the VQ.

4.3 Rate Distortion Optimization (RDO)

The basic goal of lossy coding is to provide good rate distortion. As described in Chapter 3, H.264 macroblocks (MBs) can be intracoded (in the spatial domain) or intercoded (in the temporal domain). Depending on the intracoding mode, the luma component of each MB can be uniformly predicted or subdivided into 4 × 4 blocks, which is useful for coding highly detailed regions of the frames. In both cases, several prediction techniques are available. The intercoding modes also subdivide the MBs for more accurate motion-compensated prediction.

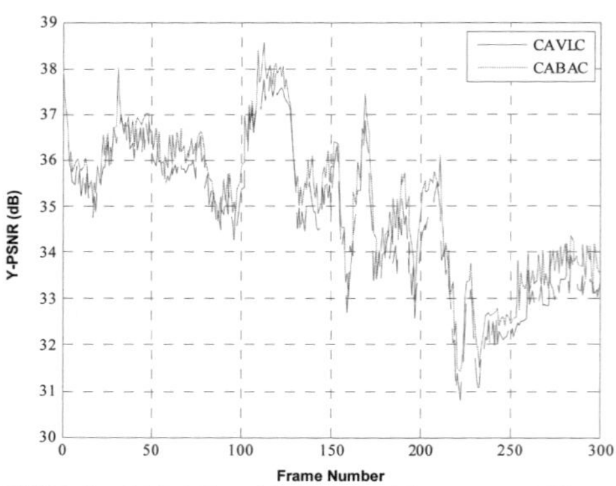

Figure 4.3: Y-PSNR for CABAC and CAVLC with a constant bit rate of 300 Kbps.

Table 4.7: CABAC and CAVLC under CBR Mode

CBR Rate	Entropy Method	Coding Time (s)	Decoding Time (s)	Average Y-PSNR (dB)
300 Kbps	CAVLC	1040	41.1	34.69
	CABAC	1019	42.3	35.13
1 Mbps	CAVLC	1045	50.9	39.69
	CABAC	1002	52.4	40.12

RDO allows a vendor to differentiate its encoder products. RDO algorithms essentially choose the coding modes to achieve the best trade-off between low distortion and low bit rate, based on specific metrics. Both distortion and rate can be minimized using Lagrangian optimization techniques. Psychovisually optimized RDO keeps the complexity of a coded block similar to the complexity of the original block, producing an image that looks much sharper and more detailed in many cases but may render classical full-reference VQ metrics, such as PSNR and SSIM, ineffective. RDO can be combined with Trellis quantization. A comprehensive coverage of RDO can be found in [3, 4].

There are two flavors of RDO, namely, fast high complexity and high complexity. The fast high complexity mode uses a simplified algorithm with fewer computations and thus, reduces the coding time. However, this is done at the expense of the VQ. High complexity is chosen in the following experiments, which employ the *Foreman* and *Coastguard* CIF videos to test the behavior of RDO under VBR and CBR coding.

4.3.1 RDO under VBR

The videos are coded with and without RDO using two GOP structures and different QP values. The results are presented in Table 4.8 and Figure 4.4. Several observations can be made. First, the file sizes are bigger when RDO is activated for both GOP structures. Although this is generally the case, other tests indicate that the impact of RDO on the size of the coded video is dependent on the video content. Second, as the QP value decreases, the coding time increases slowly without RDO whereas the increase is more significant with RDO. Thus, the difference in coding time is more pronounced for high-quality videos. Third, RDO improves the average Y-PSNR. In addition, a lower QP value (which corresponds to higher Y-PSNR) results in a higher gain with RDO. By comparing Figure 4.4(a) and (b), we observe that the gain is achieved on each coded frame and is independent of the specific temporal sequences in the video such as the GOP structure or the use of B frames. The improvement in visual quality is uniform across the video sequence and no perceptible variations can be observed from one frame to the next. To sum up, RDO in the VBR mode provides better VQ but the coded video is typically larger in size and the coding time is longer. Thus, RDO is particularly well suited for broadcasting of prerecorded high-quality videos but should be avoided for low delay applications.

4.3.2 RDO under CBR

This mode determines if the desired VQ can be maintained when the size of the coded videos are the same. CBR and RDO can be considered as complementary tools. If the bit rate is fixed, RDO simply employs the best prediction mode depending on the complexity of the scene. However, before providing a QP value, the rate control algorithm requires information that is only available after the RDO algorithm has completed its prior computations. In fact, to simplify the coding process, the results of the RDO algorithm are predicted based on the complexity of the previous scenes in the video sequence. More details on the trade-offs between RDO and CBR are available in [5].

We employ an IPPPPPPPPP GOP structure to test the RDO coding time and the average Y-PSNR for different CBR bit rates. From Figure 4.5(a), we observe that the coding time increases and then stabilizes as the bit rate increases without RDO whereas it increases steadily with the bit rate when RDO is used. Because higher Y-PSNR can be achieved with a higher bit rate (Figure 4.5(b)), the QP values computed by the rate control algorithm decrease when the available bit rate increases, resulting in a longer coding time due to the smaller QP values.

Table 4.8: RDO under VBR Mode

GOP	**IPPPPPPPPP**						**IBPBPBPBPB**	
QP value	**10**		**20**		**30**		**20**	
RDO mode	Off	On	Off	On	Off	On	Off	On
***Foreman* CIF**								
Coding time (s)	601	1,428	588	1,157	587	971	641	1,283
File size (KByte)	9,500	11,066	2,556	2,960	589	606	2,533	2,778
Average Y-PSNR (dB)	49.78	51.98	41.83	42.79	35.19	35.62	41.85	42.73
***Coastguard* CIF**								
Coding time (s)	664	1,503	644	1301	630	1,107	688	1,476
File size (KByte)	11,819	12,776,	4,440	5,027	1,164	1,292	4,312	4,768
Average Y-PSNR (dB)	49.60	51.98	40.92	42.41	33.16	33.83	40.93	42.42

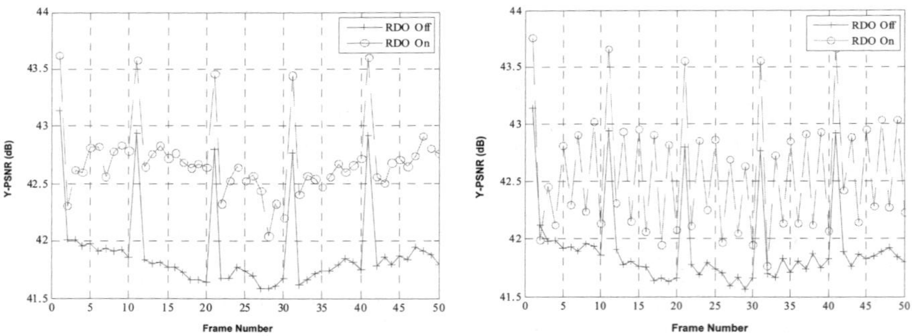

(a) IPPPPPPPPP GOP (b) IBPBPBPBPB GOP

Figure 4.4: Y-PSNR for QP=20 (first 50 frames of *Foreman* CIF video).

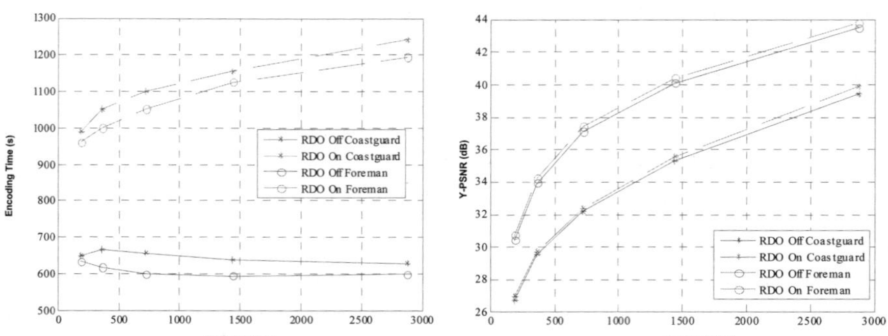

(a) Coding time (b) Y-PSNR

Figure 4.5: RDO under CBR mode (*Foreman* CIF and *Coastguard* CIF).

4.4 Flexible Macroblock Ordering (FMO)

H.264 defines seven FMO types: type 0 to type 6 (Figure 4.6). They contain a fixed pattern that can be exploited to reduce the overhead needed for reordering the MBs during decoding. In FMO type 0, each slice group (SG) contains a series of MBs in raster scan order before another SG starts. Only the length of each SG is transmitted. The number of SGs and SG identity are coded in the PPS. FMO type 1 can help improve the privacy of the coded video. In this case, the frame is divided into 2 SGs with the MBs rearranged in a checkerboard style. Each SG is sent in a different packet. The location of the MBs in the SG is given in (4.2). The MBA map is not needed since the formula is known to both encoder and decoder.

$$\text{MB number} \rightarrow \text{SG number } i \rightarrow (i \bmod w) + ((i/w)n)/2) \bmod n \qquad (4.2)$$

where n = number of SGs (coded in a PPS)
w = frame width in MBs (coded in an SPS)
"/" denotes integer division with truncation

FMO type 2 uses one or more rectangular SGs and a background SG. This separation allows the partition of different regions of interest within a frame. For example, a face can be coded with more bits whereas the background can be coded with lower quality to maintain the total bit rate. Only the top left MB number and the bottom right MB number of each foreground SG are needed for decoding. Types 3, 4, and 5 employ an SG configuration (coded in the PPS) that can change in every frame by making use of periodic patterns. The SG cycle change is included in the header of each slice to keep track of the current position within the cycle of changes. The cycle change is not included in the PPS to reduce overheads and to reflect possible changes in every frame. Type 6 is the most flexible. It allows any MBA map, which must be explicitly coded into the bitstream.

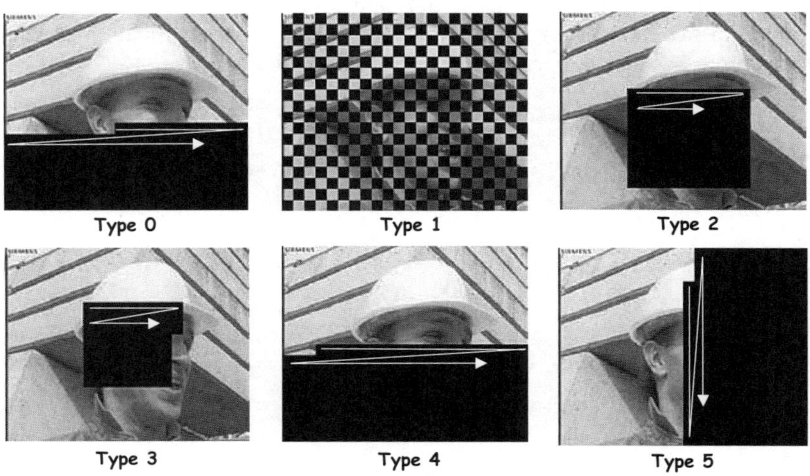

Figure 4.6: FMO types with two slice groups.

We evaluate the performance of each FMO type under different levels of packet losses using the Extended profile (B slices with CAVLC). We employ the *Foreman* CIF video with a GOP length of 12 and QP values of 20 and 30 for all frame types. For FMO types 3, 4, and 5, the number of MBs per SG is chosen to be equal to the size of the slice. Thus, an integer number of slices per frame is maintained. We consider slices of 33 MBs (12 slices per frame for all FMO types), 66 MBs (6 slices with 2 SGs for FMO types 3, 4, 5, and 1 with 2 SGs and 6 slices with 4 SGs for FMO type 1), and 99 MBs (4 slices for all FMO types). Only I, P, and B slices are removed and not the PPS or SPS. The slice-loss pattern is based on the uniform distribution. Loss percentages of 3%, 5%, and 10% are applied to each coded video. The resulting average Y-PSNR is shown in Figures 4.7 to 4.9.

4.4.1 Overheads

The relative overheads compared to a video coded without FMO (coded with only one slice per frame) are presented in Table 4.9. The relative overhead ranges from 1.1% for FMO type 5 (with a QP value of 20, an IPPPPPPPPP GOP, and a slice size of 99 MBs) to 51.3% for FMO type 1 (with 4 SGs with a QP value of 30, an IBBPBBPBBPBB GOP, and a slice size of 33 MBs). The overheads for FMO type 1 is considerably larger compared to other types (at least two times larger for 2 SGs and three times larger with 4 SGs in most cases). The first reason for this increased overhead is the reduction of the coding efficiency when more SGs or more slices, are used. This phenomenon can be observed when reading Table 4.9 column-wise: the overhead decreases when the slice size increases (i.e., the number of slices decreases). In H.264, each slice is designed to be decodable without other slices of the frame to minimize error propagation. This implies intraprediction and MV prediction are not allowed across slice boundaries, which degrades coding efficiency. The use of more than one slice creates an overhead that is difficult to predict in most situations. For example, although FMO types 3, 4, 5, and type 1 with 2 SGs have the same number of slices, the percentage difference in overheads ranges from 7.8% to 24.4%. This overhead is present for all FMO types because they require the creation of slices.

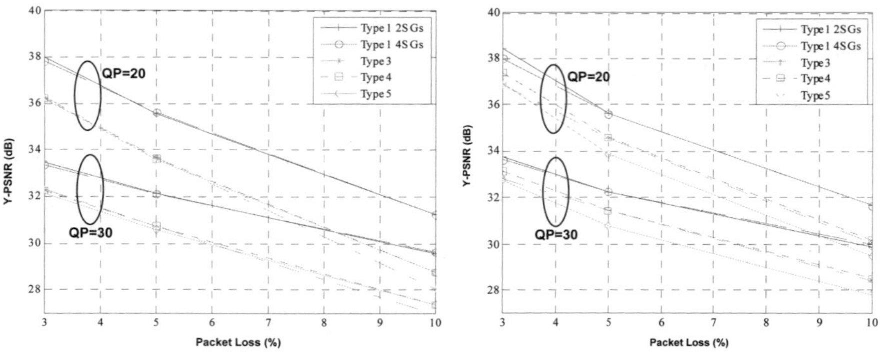

Figure 4.7: (a) 33MBs IBBPBBPBBPBB and (b) 33MBs IPPPPPPPPPPP.

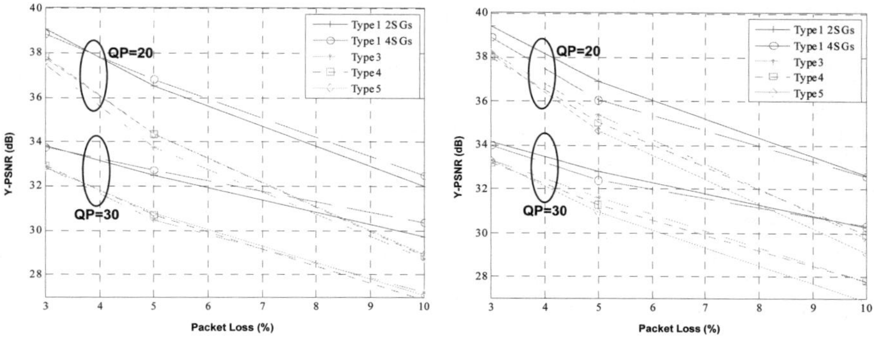

(a) (b)

Figure 4.8: (a) 66MBs IBBPBBPBBPBB and (b) 66MBs IPPPPPPPPPPP.

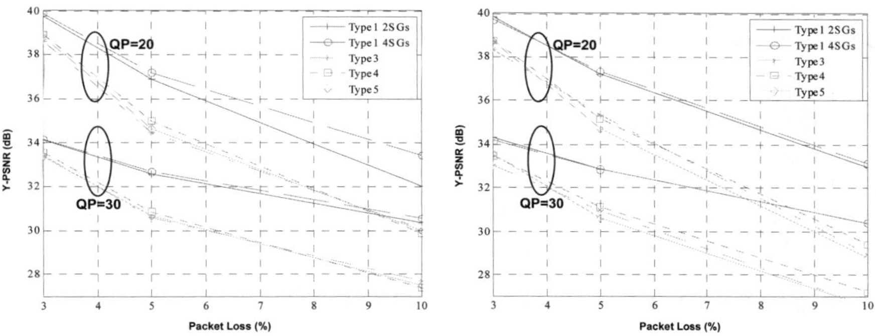

(a) (b)

Figure 4.9: (a) 99MBs IBBPBBPBBPBB and (b) 99MBs IPPPPPPPPPPP.

Table 4.9: Relative Percentage Overheads (Size of Reference File Shown in Parentheses)

GOP	QP	MBs per Slice	Type 1 2 SGs	Type 1 4 SGs	Type 3	Type 4	Type 5
IPPP	20 (2,484 KBytes)	33	12.2	15.9	4.4	4.6	4.3
		66	10.15	15.3	2.1	2.0	2.0
		99	9.4	14.7	1.3	1.3	1.1
	30 (587 KBytes)	33	34.7	41.7	15.4	16.1	15.2
		66	28.7	39.33	7.2	7.3	6.8
		99	26.7	36.9	4.5	4.7	4.0
IBBP	20 (2,520 KBytes)	33	16.25	21.8	5.7	5.8	5.7
		66	13.3	21.3	2.7	2.7	2.5
		99	12.2	20.7	1.7	1.7	1.7
	30 (586 KBytes)	33	39.7	51.3	15.7	16.1	16.2
		66	31.7	49.0	7.5	7.4	7.2
		99	29.0	46.7	4.8	5.1	4.6

FMO type 1 incurs significant overheads. When two slices are used, only the upper-right and upper-left neighbors can be part of the same slice. Thus, no neighboring MBs in the original frame are coded in the same slice. With four SGs, the difference is even more pronounced because MBs of the same slice are farther from each other than for 2 SGs. For FMO type 1 with two and four SGs, we also observe that the difference is the biggest when SGs of 66 MBs are used (31.7% and 49.0% with IBBPBBPBBPBB GOP and QP value of 30). In this case, the use of four SGs results in the creation of 3 extra slices and the coding efficiency degrades. Furthermore, the efficiency of the entropy coding methods (i.e., CABAC and CAVLC) depends on the elements that have already been coded. When they are correlated to the next elements, entropy coding becomes more efficient. In the case of several slices, the elements of an SG already coded are not used to choose the coding of the elements. Thus, the use of several slices breaks the entropy coding. Another reason for the increased overhead is due to extra slice headers and extra PPSs. However, four FMO types do not require extra PPSs:

- A FMO type 1 MB at a given location always belongs to the same SG.
- FMO types 3, 4, 5 contain a fixed cycle of MBs, thus a single PPS suffices.

Other FMO types may require extra PPSs. For instance:

- For type 0, the location of the first MB of each SG is coded in the PPS.
- For FMO type 2, the location of the top left MB and bottom right MB of each SG is coded in the PPS.
- For FMO type 6, the MBA map is coded in the PPS and is much bigger than the other FMO types. If the MBA map changes for different frames, new PPSs are coded in the bitstream.

The first 24 bits of a PPS are presented in Table 4.10. The bits related to the FMO syntax elements are:

- 3 bits for the number of SGs (2 SGs in this example);
- 5 bits for the FMO type (type 5 in this example);
- 1 bit to keep track of a new cycle of MBs in the SG that are added in raster scan order (without inserting a new PPS);
- 11 bits to indicate the number of MBs (33 in this example) that must be added to the SG with each new frame.

The remaining bits of the PPS do not carry FMO information and are the same as the bitstream not coded with FMO. Thus, the FMO overheads account for only 20 bits, which are minor compared to the size of a coded sequence.

Table 4.10: Initial Bit Pattern of a PPS

Bits	Parameter
1	pic_parameter_set_id = 0
1	seq_parameter_set_id = 0
0	entropy_coding_mode_flag = 0
0	pic_order_present_flag = 0
010	num_slice_groups_minus1 = 1
00110	slice_group_map_type = 5
0	slice_group_change_direction_flag = 0
00000100001	slice_group_change_rate_minus1 = 32

The use of B frames or B slices reduces temporal correlation, resulting in higher coding efficiency and hence a higher relative FMO overhead. For example, the overheads are smaller when an IPPPPPPPPPPP GOP structure is used compared to an IBBPBBPBBPBB structure. Note that the relative overhead of any FMO type is smaller as the QP value decreases. For example, the overhead for a QP value of 20 is 21.8% compared to 51.3% for a QP value of 30. This is because the size of the video coded without FMO is considerably bigger (2.5 Mbytes for a QP value of 20 versus 0.59 Mbytes for a QP value of 30) whereas the FMO overheads remain fixed. Thus, the relative cost of FMO is more acceptable for higher-quality videos. The main results are summarized in Table 4.11. In all cases, FMO type 1 (with two SGs or four SGs) provides the best VQ after EC.

Table 4.11: Main Results for FMO Overhead Evaluation

Parameter	Impact on Relative Overhead
FMO type	Largest for FMO type 1
Small slices	Increases relative overhead
Use of B slices	Increases relative overhead
Low QP values	Reduces relative overhead

4.4.2 FMO Operating under CBR

The use of CBR with FMO provides insights into the loss of visual quality when the file sizes are the same. As shown in Table 4.12, FMO type 5 exhibits the least sacrifice in Y-PSNR under the CBR mode. The price to pay for the improved robustness in FMO type 1 is a larger relative overhead. Thus, in error-free conditions, FMO type 1 will incur the most overheads. The results for FMO type 1 with four SGs are identical because the choice of 99 MBs per slice corresponds to the maximum size of the slices. As expected, the use of more SGs generates more overheads and thus, lowers the Y-PSNR. The degradation is more significant with lower bit rates because the relative overhead using FMO becomes higher.

Table 4.12: Average Y-PSNR in dB for CBR Coded Videos (*Foreman* CIF)

Data Rate (Kbps)	No FMO (1 SG)	99 MBs per Slice			1 Slice per SG		
		Type 1 (2 SGs)	Type 1 (4 SGs)	Type 5 (2 SGs)	Type 1 (2 SGs)	Type 1 (4 SGs)	Type 5 (2 SGs)
240	32.01	29.16	28.76	31.28	29.45	28.76	31.87
800	37.54	36.72	36.30	37.45	36.83	36.30	37.54
2400	42.59	42.11	41.84	42.52	42.16	41.84	42.58

4.5 Conclusions

We have evaluated several key features of the H.264 coding standard. The main results are summarized in Table 4.13. In the VBR mode, a higher Y-PSNR is provided by RDO for higher quality videos. Under the CBR mode, coding time using CABAC is superior over CAVLC, whereas the degradation in decoding time is insignificant. The use of more SGs increases the overhead for FMO. FMO type 1 is more suited for operation in high-loss transmission environments.

Table 4.13: Main Results

	Mode	**Results**
CABAC	**VBR**	• Smaller files. • Shorter coding time. • Slightly longer decoding time. • Better VQ compared to CAVLC.
	CBR	• Shorter coding time. • Slightly longer decoding time.
RDO (High Complexity)	**VBR**	• Larger files. • Relative overhead increases with low QP values (not desirable for high-quality videos). • Coding time increases with low QP values (not desirable for high-quality videos). • Better VQ (especially with low QP value) compared to the case without RDO.
	CBR	• VQ improvement is achieved for each coded frame. • Better VQ with higher relative gain for low bit rates. • Longer coding time compared to the case without RDO.
FMO Type 1	**VBR**	• Larger files. • Relative overhead increases with number of SGs. • Relative overhead decreases with low QP values (desirable for high-quality videos).
	CBR	• Better VQ in presence of packet losses than types 3, 4, and 5. • VQ decreases in lossless environments compared to the case without FMO.

References

[1] D. Marpe, H. Schwarz, and T. Wiegand, "Context-based Adaptive Binary Arithmetic Coding in the H.264/AVC Video Compression Standard," *IEEE Transactions on Circuits and Systems for Video Technology*, Vol. 13, No. 7, July 2003, pp. 620–636.

[2] S. W. Lee, and C. C. Kuo, "Complexity Modeling of H.264/AVC CAVLC/UVLC Entropy Decoders," *IEEE International Symposium on Circuits and Systems*, May 2008, pp. 1616–1619.

[3] G. Sullivan, T. Wiegand, and K. P. Lim, "Joint Model Reference Encoding Methods and Decoding Concealment Methods," JVT-I049, September 2003.

[4] T. Wiegand, H. Schwarz, A. Joch, F. Kossentini, and G. Sullivan, "Rate-Constrained Coder Control and Comparison of Video Coding Standards," *IEEE Transactions on Circuits and Systems for Video Technology*, Vol. 13, No. 7, July 2003, pp. 688–703.

[5] Pixel Tools, 2003, http://www.pixeltools.com/rate_control_paper.html.

[6] P. Lambert, W. De Neve, Y. Dhondt, and R. Van de Walle, "Flexible Macroblock Ordering in H.264/AVC," *Elsevier Journal of Visual Communications and Image Representation*, Vol. 17, 2006, pp. 358–375.

[7] J. Ostermann, J. Bormans, P. List, D. Marpe, M. Narroschke, F. Pereira, T. Stockhammer, and T. Wedi, "Video Coding with H.264/AVC: Tools, Performance, and Complexity," *IEEE Circuits and Systems Magazine*, Vol. 4, Issue 1, 1Q 2004, pp. 7–28.

[8] G. Sullivan, P. Topiwala, and A. Luthra, "The H.264/AVC Advanced Video Coding Standard: Overview and Introduction to the Fidelity Range Extensions," *SPIE Conference on Applications of Digital Image Processing*, August 2004.

Exercises

4.1. Justify whether the following statement is true. In general, the high profile provides better coding efficiency than the main profile, which in turn provides better coding efficiency than the extended and baseline profiles. Note that the baseline and extended profiles do not support CABAC.

4.2. Some video conferencing vendors claim to achieve a 1,000-fold coding efficiency using the baseline profile and a 2,000-fold efficiency using the main profile. Are these claims valid?

4.3. Compare the capabilities of the high and main profiles and explain why the high profile adds minimum complexity to the main profile.

4.4. Observe from Table 4.1 that many H.264 profiles support the 4:2:0 color format. Explain why other color formats (i.e., 4:2:2 and 4:4:4) are only supported by the high profiles. In comparison, VP8 and VC-1 only support the 4:2:0 color format. Note that in the 4:2:0 color format, each MB comprises 16×16 luma samples and two 8×8 chroma samples whereas in 4:2:2 and 4:4:4, the chroma samples are 8×16 and 16×16 respectively. Compute the raw bit rate for a 1080i (4:4:4) video at 60 Hz using 12 bits per sample. Compute the encoded bit rate if the high 4:4:4 profile is used, assuming an efficiency of 0.25 bit/sample.

4.5. Video is usually captured in RGB but the color components may be highly correlated. In addition, the HVS is better matched to YC_BC_R. To transform RGB to YC_BC_R, the following equations can be applied with $K_R = 0.2126$ and $K_B = 0.0722$.

$$Y = K_R R + (1 - K_R - K_B)G + K_B B\,;\; C_B = \frac{1}{2}\left(\frac{B-Y}{1-K_B}\right);\; C_R = \frac{1}{2}\left(\frac{R-Y}{1-K_R}\right).$$

Since the samples are represented using integers, rounding errors may be introduced in the forward and reverse transformations. In addition, the coefficients such as 0.2126 and 0.0722 increases complexity in digital compression and may compromise coding efficiency. Can the first problem be solved by adding two extra bits of precision to the samples? The high profiles in H.264 overcome the second problem by introducing a new color space called YC_GC_O, where C_G and C_O are the green and orange chroma components respectively.

$$Y = \frac{1}{2}\left(G + \frac{R+B}{2}\right);\; C_G = \frac{1}{2}\left(G - \frac{R+B}{2}\right);\; C_O = \frac{R-B}{2}.$$

The high profiles overcome the first problem by introducing two sets of equations for the forward and reverse transformations, where >> denotes an arithmetic right shift operation. Verify that these equations provide exact integer inverses and that an additional bit of sample accuracy is still required.

$$C_O = R - B;\ C_G = G - B - C_O >> 1;\ Y = B + C_O >> 1 + C_G >> 1.$$
$$G = Y - C_G >> 1 + C_G;\ B = Y - C_G >> 1 - C_O >> 1;\ R = B + C_O.$$

For 4:4:4 video, the high profiles employ residual color transformation to remove the conversion error without significantly increasing the complexity. This requires the use of the RGB domain.

4.6. When coding a video with predictable motion, say a plane flying in straight path, will motion JPEG2000 be more efficient than H.264? Motion JPEG2000 basically takes a series of JPEG frames and plays them in order, without motion compensation or interprediction. Hint: Compare JPEG2000's intracoding efficiency versus H.264's intracoding efficiency. JPEG2000 employs the discrete wavelet transform that works on the entire image (as opposed to MBs). This can significantly improve coding efficiency. Like CAVLC, JPEG2000 also uses context models in arithmetic coding. Note that a lossless MB prediction mode can be enabled in the High 4:4:4 profile. In this mode, values of the samples are interpredicted but sent without transformation (i.e., transform bypass).

4.7. Explain why an integer number of 16 × 16 MBs cannot be supported by a 1080p video. Devise methods to evaluate the performance of entropy coding under the CBR and VBR modes for these videos.

4.8. Explain why the VQ improves more with RDO when operated under the CBR mode than the VBR mode. Suppose an RDO algorithm minimizes only the distortion at the expense of a very high coded rate. In other words, only the distortion is optimized. Explain whether such an algorithm could still be useful.

4.9. The best prediction mode in RDO is chosen by minimizing the distortion and the number of bits for coding the block or MB. Both parameters are dependent on the chosen QP value and the prediction mode. The distortion is measured as the squared differences between the original block and the reconstructed block. Explain the relationship between the QP value and the distortion.

4.10. Will FMO be useful for a channel with bursty losses that corrupt entire frames? Is its use justified if the packet loss rate is low, say below 1%? Note that short packets carrying short slices are less prone to errors and easier to conceal. Explain why FMO overheads are higher if there are fewer slices in a frame (i.e., many MBs per slice). Can FMO be applied to videos encoded using the high profiles, even if these profiles do not support error resilience? Hint: See Chapter 7.

Chapter 5

Short-Term H.264 Bandwidth Prediction

H.264 compression gives rise to high traffic rate variability, which must be addressed in order to optimize bandwidth and buffering resources. In this chapter, we analyze the bandwidth requirements of H.264 video traffic by analyzing the traffic pattern of the group of pictures of the video, the autocorrelation of the trace of B frames, and the cross-correlation between the trace of B frames and the trace of I and P frames. We then describe a method for predicting H.264 bandwidth in the short term and evaluate the accuracy of two traffic prediction models. The first predicts the size of the B frame while the second predicts the size of the group of pictures (GOP). To quantify the prediction accuracy, a model for the prediction error is also described. It will be shown that the model provides highly accurate prediction for a wide variety of video types, in particular, high action and regular movies encoded in high-quality resolution. This is an important benefit since high-motion H.264 videos demand the highest bandwidth and generate the widest range of bit rate requirements. Although the prediction models are very simple to implement (a key requirement for short-term prediction), we also present two methods to simplify its operation further without sacrificing prediction accuracy. It will be shown experimentally that the method can provide accurate prediction for a variety of video traces using minimum buffering resources. We then enhance both models using a scene change detector. By allowing bandwidth to be predicted on a per frame or per GOP basis, a video headend can provision video bandwidth efficiently or reduce bit rate variability and bandwidth requirements via selective frame dropping, thereby minimizing buffering requirements and packet losses at the set-top box (STB).

5.1 Introduction

Typical characteristics of a broadband cable network include a dedicated STB buffer, fixed channel allocation, and an architecture that naturally lends itself to broadcasting, which is in contrast to the general Internet setting of point-to-point connections. In addition, the need to support packetized video breaks traditional Internet bandwidth utilization models since the video traffic is real-time constant bit rate (CBR) or variable bit rate (VBR) with peak viewing hours. Hence, current Internet traffic models cannot be applied directly to the video bandwidth allocation problem in cable networks. Moreover, in SDV implementations, the total bit rate is preset to a fixed value for each video stream. However, highly efficient video compression standards such as H.264 present high bit rate variability when operated in the VBR mode. Thus, if the fixed bit rate is set too low, it can lead to packet losses that impair the video quality dramatically. Conversely, if the fixed bit rate is set too high, then channel bandwidth is underutilized.

In order to optimally provision bandwidth, an algorithm has to depend on the underlying traffic model. A number of models have been proposed for MPEG-2 video traffic. In this chapter, we analyze and develop a practical traffic model for H.264 video traffic, based on predicting the frame size of the incoming video traffic. This predictive model draws certain insights and observations from previous work on traffic modeling [1]. The main insights relate to the design of a good traffic model that must be simple, be able to precisely model the traffic, and be able to capture the marginal error distribution effectively. These criteria will be used as benchmarks to evaluate our model.

The B frames may occupy the most number of frames (and typically, the most bandwidth) in a group of pictures (GOP). In addition, as will be shown in Section 2, the variability of the sizes of the B frames tends to be higher than the I and P frames in the video sequence. Hence, the trace of the B frames tends to be more bursty than the trace of the other frames, which implies the bandwidth allocation for the B frames is more challenging than the other frame types. On the other hand, B frames are the least important compared to the I and P frames, and can sometimes be dropped to conserve bandwidth resources and reduce bit rate variability. Reducing the bit rate variability is equivalent to smoothing the encoded video bit stream. In some instances, all or a large portion of the B frames in the entire video can be removed without compromising the quality of video playback. Thus, the ability to predict the size of the B frames can either allow a cable headend to provision bandwidth efficiently or reduce STB buffer requirements and packet losses under periods of peak network utilization.

For efficient bandwidth allocation, many dynamic schemes have been proposed in the literature [2]. More specifically, prior work on linear prediction of MPEG video traffic has been reported. In [3], a model is proposed to predict the changes of frame sizes between adjacent GOPs. In [4], the authors propose a model that partitions the vector of MPEG frames according to scene changes to improve the accuracy of the prediction. However, most of these schemes were designed based on the traffic characteristics of prerecorded videos encoded using the older MPEG-2 or MPEG-1 standards.

In this chapter, we present real-time algorithms to predict the size of B frames and the size of the GOP for H.264 video traffic. The B frame prediction model can be enhanced using a new scene change detector. We first perform a careful bandwidth analysis on the variability of the different frame types of the GOP as well as their importance in the GOP's structure. Then we analyze the traffic pattern of the GOPs, the autocorrelation of the B frames, and the cross-correlation between the B frames and P and I frames. This leads to the design of an accurate model for the B frame size prediction. In addition, the study of the autocorrelation of the GOP and the cross-correlation of the GOP with the I frames leads to the design of an accurate model for the GOP size prediction. A model for the error is also proposed.

5.2 Statistical Characteristics of H.264 Coded Videos

An H.264 video stream typically presents higher bit rate variability compared to MPEG-2, due to the higher efficiency in H.264 compression. This rate variability can sometimes be represented by the coefficient of variability (CoV) of the frame sizes, which is computed as follows. For a video sequence consisting of M frames encoded with a given quantization level, if X_m ($m = 1, 2, \ldots, M$) denotes the sizes of the encoded video frames, then the CoV of the encoded video is defined as (5.1). In this case, σ is the standard deviation and $\overline{X}$ is the mean of the frame sizes.

$$CoV = \frac{\sigma}{\overline{X}} = \frac{\sqrt{\frac{1}{(M-1)}\sum_{m=1}^{M}\left(X_m - \frac{1}{M}\sum_{m=1}^{M}X_m\right)^2}}{\frac{1}{M}\sum_{m=1}^{M}X_m} \tag{5.1}$$

The CoV of 21 H.264 video traces encoded in low quality (LQ) and in high quality (HQ) obtained from [5] are shown in Table 5.1. For videos encoded in LQ, the CoV varies from 1.12 to 2.87 whereas for the videos encoded in HQ, the CoV varies from 0.41 to 1.09. Clearly, LQ encoded videos exhibit a higher CoV than HQ encoded videos. This can be explained from (5.1). For LQ videos, the frame sizes are smaller, which results in a smaller mean and as a result, the CoV tends to be higher. The standard deviation tends to increase when the video quality increases (e.g., from LQ to medium quality (MQ) to HQ) but this increase can sometimes be offset by a higher average bit rate in the higher-quality videos.

From Table 5.1, it can be observed that for all H.264 videos, the CoV for the B frames is higher than the I and P frames while the CoV for the P frames is higher than the I frames. This is, in general, attributed to the fact that there are more B frames than P frames and I frames, and more P frames than I frames, for a GOP of 12 with 2 B frames (i.e., IBBPBBPBBPBB).

Table 5.1: CoV for I, P, and B Frames

	B frames			P frames			I frames		
	Std dev	**Mean**	**CoV**	**Std dev**	**Mean**	**CoV**	**Std dev**	**Mean**	**CoV**
Silence of the Lambs HQ	1929.8	2430	0.794	2407	3203	0.751	2678.4	5471	0.490
Silence of the Lambs MQ	774.5	635	1.220	1044	997	1.048	1574.8	2436	0.646
Silence of the Lambs LQ	338.5	247	1.371	710	642	1.107	1573.8	2436	0.646
Aladdin HQ	1342.4	1694	0.792	1674	2592	0.646	1894.9	4795	0.395
Aladdin LQ	238.7	247	0.966	480	616	0.779	946.9	2213	0.428
Ski HQ	1895	3596	0.527	2346	4716	0.498	2762.5	6974	0.396
Simpsons HQ	2116	5685	0.372	2526	7189	0.351	1957.5	11073	0.177

From Table 5.2, it is clear that if the B frames are removed from the entire encoded video, the CoV will be reduced. Intuitively, this is expected since the remaining I and P frames will lead to less variation in the frame size. The CoV reduction is consistent across all H.264 encoded videos and is equivalent to smoothing the encoded video bitstream. For the video without the B frames, the mean and standard deviation are computed using 4 frames (1 I and 3 P frames) and not 12 frames. If averaged over 12 frames (i.e., including the B frames of zero size to simulate the dropped B frames), the mean values become 1256.8, 452.2, 363.4, 1047.7, 338.5, 1760.1, 2719.9, which are roughly of the order of two times lower than the corresponding means of the original video that includes the B frames. Moreover, as can be seen from the standard deviation and the mean, HQ and high-motion videos (such as the *Ski* sports video) demand the highest bandwidth and generate the widest range of bit rate requirements.

Table 5.2: Impact on CoV when B Frames are Dropped

	Entire Video with B Frames			**Entire Video without B Frames**		
	Std dev	**Mean**	**CoV**	**Std dev**	**Mean**	**CoV**
Silence of the Lambs HQ	2291.1	2876.4	0.7965	2664.8	3770.4	0.7068
Silence of the Lambs MQ	1060.1	875.4	1.2110	1350.9	1365.5	0.9959
Silence of the Lambs LQ	876.3	528	1.6597	1265.2	1090.1	1.1606
Aladdin HQ	1723.7	2177	0.7918	1977.3	3143	0.6291
Aladdin LQ	678.9	503.3	1.3489	935.6	1015.6	0.9212
Ski HQ	2314.1	4157.4	0.5566	2644.2	5280.2	0.5008
Simpsons HQ	2683.7	6510.2	0.4122	2927.6	8159.8	0.3588

To further determine the amount of possible bandwidth savings associated with B frame dropping, we compare the sizes of the B frames against the size of the I frames. We observe, from Table 5.3, that the total size of all B frames in a video can exceed the total size of all I frames. This is especially true for HQ movies. The mean size of the B frames per GOP is also higher than the mean size of the I frames for HQ movies. As expected, the average size of the B frame is less than the average size of the I frame for all videos. These observations indicate that substantial bandwidth savings can be achieved in the removing of B frames, confirming our observations from Table 5.2.

Table 5.3: B Frame and I Frame Size Comparison

Movie	**Total Size of I Frames**	**Total Size of B Frames**	**Mean Size of I Frames**	**Mean Size of B Frames**	**Mean Size of B Frames per GOP**
Silence of the Lambs HQ	4.10×10^7	1.46×10^8	5471	2430	19436
Silence of the Lambs MQ	1.83×10^7	3.81×10^7	2436	635	5079
Silence of the Lambs LQ	1.83×10^7	1.48×10^7	2436	247	1976
Aladdin HQ	3.60×10^7	1.02×10^8	4795	1694	13552
Aladdin LQ	1.66×10^7	1.48×10^7	2213	247	1978
Ski HQ	5.23×10^7	2.16×10^8	6974	3596	28768
Simpsons HQ	2.80×10^7	1.15×10^8	11073	5685	45483

To summarize, for a high-action and/or HQ movie, the standard deviation is an important metric that determines the frame size variation and hence, the range of bit rate requirements. For such movies, in addition to the high average bit rate requirement, the range of bit rate variation is also high, as can be seen from the standard deviation in Table 5.2. Conversely, for LQ and/or animated movies (e.g., cartoons) that typically require low average bit rates, the CoV is a good metric that describes the bit rate variation. In this case, a high CoV may not necessarily result in a wider range of bit rate variation when compared to high-action or HQ movies, as can be seen from Table 5.2.

Since bandwidth allocation for videos with high bit rate variability is very challenging, one solution is to transmit HQ videos (with small CoV) whenever possible but reduce the bandwidth requirements and smooth the encoded video bit stream using selective B frame dropping. Here, the benefits are threefold: better-quality video, and reduced bandwidth requirements, and CoV. As an example using Tables 5.1 and 5.3, it is possible to transmit a low CoV but HQ *Aladdin* movie with B frame dropping and yet achieve comparable bandwidth requirements when compared to the LQ *Aladdin* movie with high CoV and no B frame dropping. It will be shown in Section 5.4 that in general, the B frame size prediction method is more accurate for HQ than LQ videos. This provides another incentive for sending HQ videos.

5.3 Problem Formulation

We consider an encoded sequence of the GOP comprising 12 frames (i.e., G12/B2 with 1 I frame, 3 P frames, and 8 B frames) as shown in Figure 5.1. The method that we propose can be extended to other GOP patterns and similar results should be obtained. Figure 5.2 illustrates the transmitted sequence. Note that for the final GOP, frames B-7 and B-8 are not transmitted.

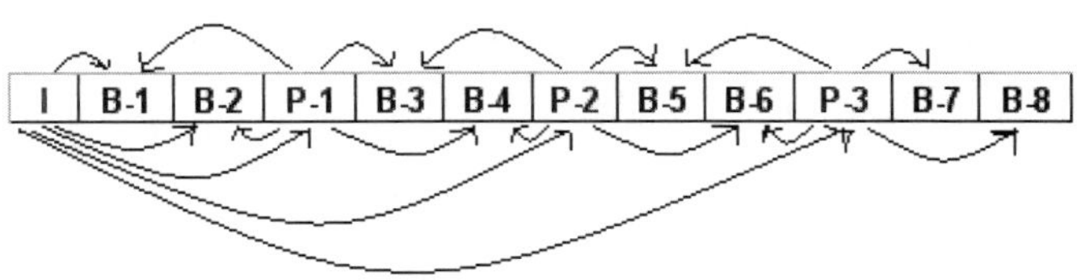

Figure 5.1: Encoded sequence of a GOP with G12/B2.

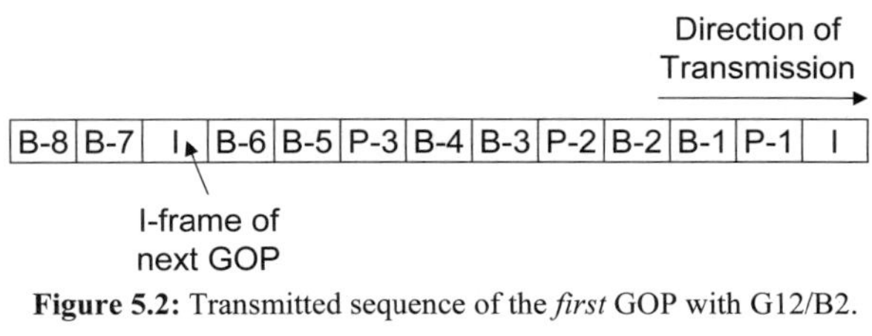

Figure 5.2: Transmitted sequence of the *first* GOP with G12/B2.

The B frames are constructed based on the reference frames, namely, the P and I frames, as shown in Figure 5.1. Further details on the construction are described in [1]. As a consequence, the size of the B frames may be strongly correlated with the size of the reference frames. This short-term correlation is obvious by observing the first 50 lags (in frames) of the cross-correlation between the B frames and some of their reference frames for three common movie types (regular, animated, and high-action) as illustrated in Figures 5.3 to 5.5.

In order to locate the most relevant correlation in the short term, we compute the coefficient of correlation ($\rho_{X,Y}$) between each B frame (denoted as the variable X), and each I or P frame (denoted as the variable Y) in a GOP using (5.2). Here, σ_X and σ_Y represent the standard deviation of X and Y respectively.

$$\rho_{X,Y} = \frac{E(XY) - \overline{X}\,\overline{Y}}{\sigma_X \sigma_Y} \tag{5.2}$$

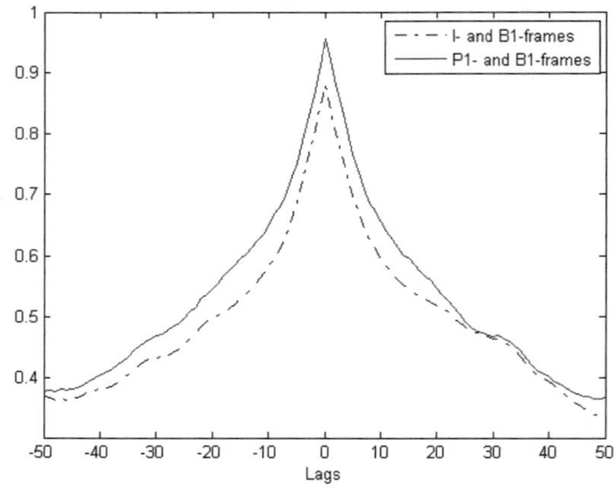

Figure 5.3: Cross-correlation of frame sizes of *Silence of the Lambs* (HQ) with G12/B2 GOP.

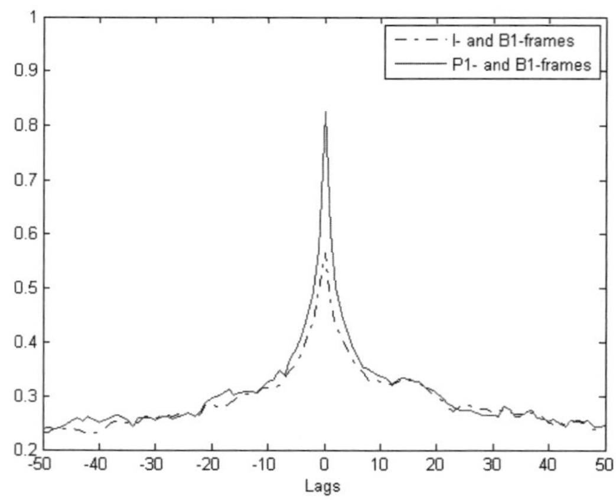

Figure 5.4: Cross-correlation of frame sizes of *Aladdin* (HQ) with G12/B2 GOP.

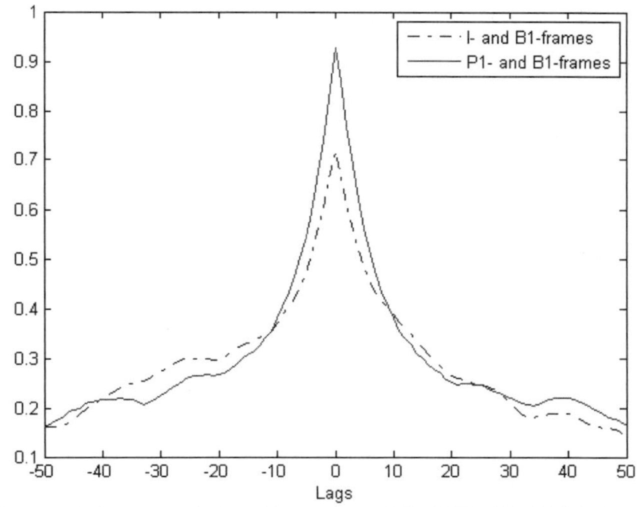

Figure 5.5: Cross-correlation of frame sizes of *Ski* (HQ) with G12/B2 GOP.

The results for several videos are presented in Table 5.4. The coefficients in bold for *Silence of the Lambs* HQ show a strong correlation between a B frame and the closest P frame. The correlation between the B and I frames is weaker. It is also confirmed by the plots in Figure 5.3. Even if some of the B frames (the first two B frames of each GOP) are encoded with an I frame as reference, because the manner in which they are encoded is very different, their coefficient of correlation is lower. This leads us to consider the size of the P frame when predicting the size of the B frames.

Table 5.4: Correlation Coefficients for Different Frames of Videos with Different Quality.

Video	Frame Type	B-1	B-2	B-3	B-4	B-5	B-6	B-7	B-8	I
Silence of the Lambs HQ	I	0.8796	0.8794	0.8715	0.8766	0.8652	0.8579	0.8438	0.8384	
	P-1	**0.9558**	**0.9612**	**0.9563**	0.9545	0.9450	0.9407	0.9312	0.9234	0.8888
	P-2	0.9366	0.9379	0.9524	**0.9559**	**0.9561**	0.9513	0.9382	0.9306	0.8734
	P-3	0.9220	0.9261	0.9328	0.9373	0.9512	**0.9627**	**0.9496**	**0.9404**	0.8619
Silence of the Lambs LQ	I	0.6449	0.6553	0.6635	0.6784	0.6672	0.6660	0.6151	0.6141	
	P-1	0.8919	0.9042	0.8982	0.8965	0.8837	0.8810	0.8572	0.8479	0.7490
	P-2	0.8595	0.8659	0.8991	0.9041	0.8943	0.8911	0.8567	0.8467	0.7537
	P-3	0.8384	0.8483	0.8678	0.8750	0.8970	0.9133	0.8678	0.8525	0.7531
Aladdin HQ	I	0.5670	0.5732	0.5675	0.5679	0.5331	0.5242	0.5130	0.4891	
	P-1	**0.8254**	**0.8523**	0.7783	0.7558	0.7268	0.7065	0.6493	0.6196	0.6699
	P-2	0.6752	0.7136	**0.8507**	**0.8634**	0.7742	0.7430	0.7413	0.7096	0.6417
	P-3	0.6563	0.6875	0.7261	0.7452	**0.8471**	**0.8615**	**0.7806**	**0.7389**	0.6143
Ski HQ	I	0.7165	0.7249	0.7066	0.7118	0.6943	0.6984	0.6716	0.6651	
	P-1	0.9272	0.9396	0.9359	0.9320	0.9144	0.9079	0.8860	0.8767	0.7767
	P-2	0.9026	0.9139	0.9345	0.9387	0.9352	0.9280	0.9043	0.8975	0.7686
	P-3	0.8712	0.8825	0.9002	0.9144	0.9265	0.9370	0.9238	0.9156	0.7502
Simpsons HQ	I	0.5496	0.5800	0.5307	0.5311	0.4838	0.4799	0.4870	0.4733	
	P-1	0.7762	0.8046	0.7226	0.6886	0.6658	0.6274	0.5611	0.5705	0.5472
	P-2	0.6366	0.6593	0.8005	0.7715	0.7356	0.6837	0.6321	0.6269	0.4636
	P-3	0.5790	0.5820	0.6475	0.6370	0.8047	0.7368	0.6824	0.6639	0.4494

Additionally, we have computed the autocorrelation of the trace of the B frames. The values of the curve for the movie *Silence of the Lambs* HQ in Figure 5.6 are very close to 1 for the first lags. This shows a strong correlation in the short term. Moreover, the curve decreases less exponentially. This shows a long-range dependence characteristic, which is the focus of Chapter 6. To confirm this observation, we plotted the autocorrelation of the B frames for the *Aladdin* HQ and the *Ski* HQ movies. The curves show a similar trend. These observations lead us to construct a model for predicting the size of the B frames using the trace of the closest P frames and the previous B frames.

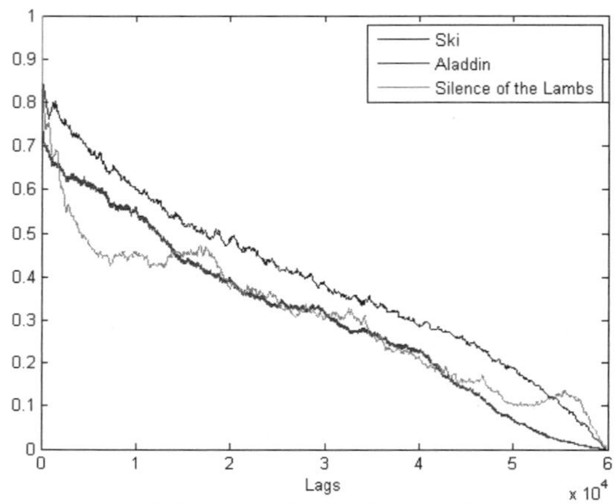

Figure 5.6: Autocorrelation of B frames for *Ski*, *Aladdin*, and *Silence of the Lambs* (all in HQ).

Additionally, we have computed the autocorrelation of the trace of the B frames using (5.3) with the same notations as before.

$$r(k) = \frac{E\left[\left(X_m - \overline{X}\right)\left(X_{m+k} - \overline{X}\right)\right]}{{\sigma_X}^2} \tag{5.3}$$

Coefficient of correlation for the I frames is weaker for the videos encoded in HQ than for the ones encoded in LQ or MQ (the I frames for LQ and MQ movies are the same except for the ski movie). That is unusual. For the B frames and the GOP the coefficients of correlation are stronger for the HQ videos than for the MQ and LQ quality.

5.4 Traffic Model for B Frame Size Prediction

According to the observations made in Section 5.3, the size of the next B frame can be predicted using the size of the previous B frame and the size of the closest P frame. Because all B frames of a GOP are not equivalent, we need to differentiate them. As shown in Figure 5.1, B-1 will be the first B frame of the GOP, B-2 the second, and so on. The same notation is used for the P frames.

As shown in Figure 5.1, the P-1 frame is encoded after the B-2 frame in the video sequence. However, for the need of the decoder and as shown in Figure 5.2, the P-1 frame is transmitted before the B-2 frame. This is due to the way the B frames are encoded. As a consequence, we can use the P-1 frame size to predict the B-2 frame size. The same method can be applied for the B-2, B-4, and B-6 frames.

We introduce the following notations. $B_{1,t}$ is the size of the *t*th B-1 frame. That is the size of the first B frame of the *t*th GOP of the encoded video. In the same way, we define the vectors B_2, B_3, B_4, B_5, B_6, B_7, B_8, P_1, P_2, and P_3 to correspond to the real trace. All these vectors have the same length (L), which is the number of GOPs in the encoded video. The vector B is the trace of all the B frames. $B_t = [B_{1,t}\ B_{2,t}\ B_{3,t}\ B_{4,t}\ B_{5,t}\ B_{6,t}\ B_{7,t}\ B_{8,t}]$. The length of B is $8L$.

We employ a linear model (denoted as 1M) for all t between 2 and L as described in (5.3). The model for the B-1 frames is slightly different because the B frame used is not part of the same GOP.

$$\begin{aligned}
\hat{B}_{1,t} &= \alpha_1 B_{8,t-1} + \gamma_1 P_{1,t} \\
\hat{B}_{2,t} &= \alpha_2 B_{1,t} + \gamma_2 P_{1,t} \\
\hat{B}_{3,t} &= \alpha_3 B_{2,t} + \gamma_3 P_{1,t} \\
\hat{B}_{4,t} &= \alpha_4 B_{3,t} + \gamma_4 P_{2,t} \\
\hat{B}_{5,t} &= \alpha_5 B_{4,t} + \gamma_5 P_{2,t} \\
\hat{B}_{6,t} &= \alpha_6 B_{5,t} + \gamma_6 P_{3,t} \\
\hat{B}_{7,t} &= \alpha_7 B_{6,t} + \gamma_7 P_{3,t} \\
\hat{B}_{8,t} &= \alpha_8 B_{7,t} + \gamma_8 P_{3,t}
\end{aligned} \tag{5.3}$$

The predicted size of the B frame can be used to provision the bandwidth for the next B frame. However, the value may not match the actual value exactly. Hence, it is useful to compute the marginal error of the prediction using the following equation:

$$\varepsilon_t = B_t - \hat{B}_t \tag{5.4}$$

The marginal error can be modeled by a G-and-H distribution, which can fit a long-tailed distribution.

$$K_{g,h}(x) = A + C\left(\frac{e^{gx} - 1}{g}\right)e^{h\frac{x^2}{2}} \tag{5.5}$$

where *A* and *C* are the location and scale parameters, respectively, *g* and *h* are the skewness and kurtosis parameters, and *x* is a random variable with standard normal distribution.

5.5 Results and Discussion

We employ a least-squares method, implemented on MATLAB, to find the coefficients α_j and γ_j with *j* ranging from 1 and 8. We perform our simulation on the movie *Silence of the Lambs* that was encoded using H.264 in HQ. The pattern of the GOP is as shown in Figure 5.1 (G12/B2). Table 5.5 presents the statistical characteristics of the encoded movie. The movie is 60 minutes in length and contains 7,499 GOPs. The values obtained by the least-squares method are presented in Table 5.6.

Table 5.5: Statistical Values of *Silence of the Lambs* (HQ) with G12/B2 GOP

	Entire Video	**I Frame**	**P Frame**	**B Frame**
CoV	0.7965	0.4895	0.7512	0.7943
Mean Frame Size	2900	5471	3203	2430
Minimum Frame Size	158	684	179	158
Maximum Frame Size	22239	18194	22239	17941
Standard Deviation	2280	2678	2406	1930

Table 5.6: Coefficients for Least-Squares for *Silence of the Lambs* (HQ) using 1M Model

	B-1	**B-2**	**B-3**	**B-4**	**B-5**	**B-6**	**B-7**	**B-8**
α	0.6053	0.7972	0.7433	0.8030	0.6935	0.7033	0.7954	0.9337
γ	0.3104	0.1586	0.1876	0.1521	0.2317	0.2312	0.1432	0.0503

We tested the model on six other videos that are encoded in H.264: *Star Trek: First Contact*, *Jurassic Park I*, *Simpsons*, *Susi und Strolch*, *Aladdin*, and *Ski*. The α_j and γ_j obtained with these movies are similar. Note that the values obtained for the B-1 and B-8 frames are quite different from the values obtained for the other B frames. This is due to the fact that the model used is quite different. In this case, the P frame used for the prediction of the B-1 and B-8 frames is two frames away from the B frames studied. As a consequence, the coefficients of correlation between these B and P frames are weaker – but are still the best.

After computing these coefficients, we use (5.3) to build the predicted vector of B frames. The plots for the original trace and the predicted trace are shown in Figures 5.7 and 5.8, respectively. They are very similar. This proves that our model fits well the real trace. In order to estimate the accuracy of our model, we study the error (ε) introduced by (5.4). The plot for the marginal error histogram is shown in Figure 5.9. We observe the long-tail characteristic of the curve. An algorithm can be used to fit a G-H distribution of (5.5) and the marginal distribution. A good (but not optimal) estimation of the parameters of the G-H distribution for the movie *Silence of the Lambs* HQ is: $A = 3.3674$, $C = 135$, $g = -0.03$, and $h = 0.4$.

The statistical values of the error introduced are provided in Table 5.7. The model introduces an average error of 7.47% for the movie *Silence of the Lambs*

HQ, which shows that the model is accurate and relevant. We repeated the experiments and tested the model on the six videos cited above. The model was highly accurate in most cases.

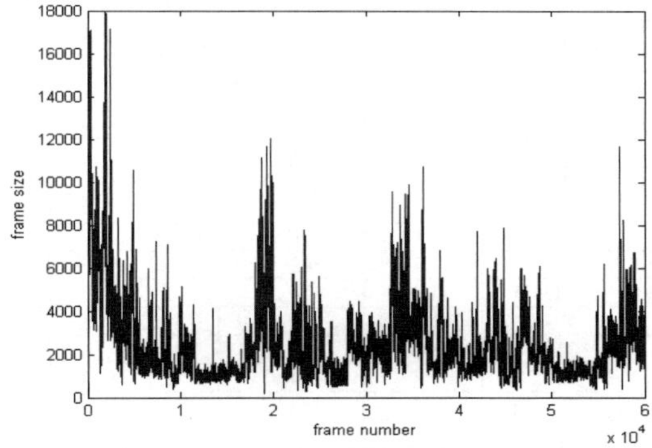

Figure 5.7: Original trace of B frames for *Silence of the Lambs* (HQ).

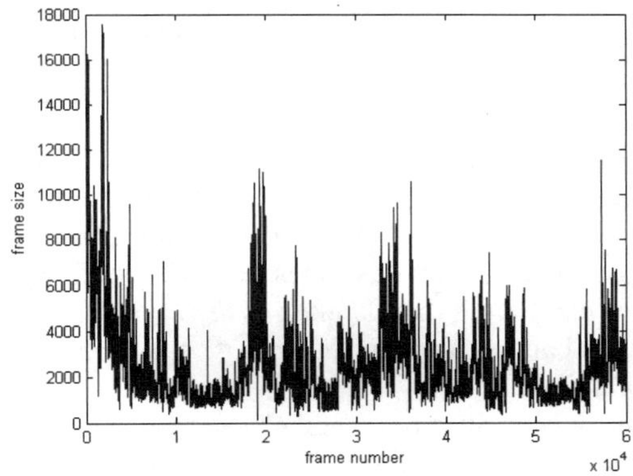

Figure 5.8: Predicted trace of B frames for *Silence of the Lambs* (HQ).

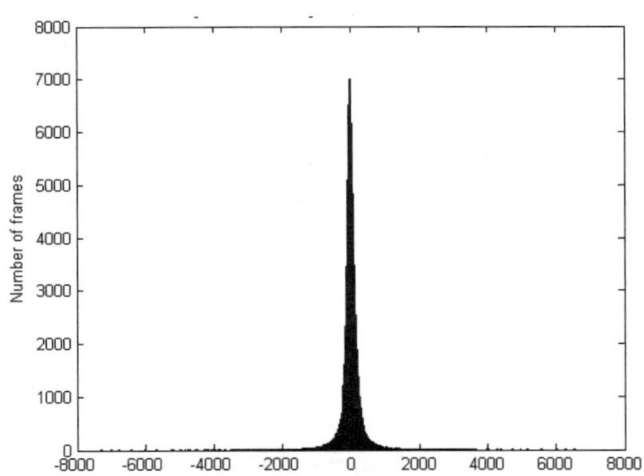

Figure 5.9: Histogram of marginal error for *Silence of the Lambs* (HQ).

Table 5.7: Marginal Error Statistics for Different Movies

Video	Method	Mean |ε|	Mean ε	Std Dev |ε|	Std Dev ε	Max ε	Min ε	RPE
Silence of the Lambs HQ	1M	181.72	-3.37	316.18	364.66	7326	-6536	7.47%
	2M	177.58	-4.96	310.07	357.28	7326	-8172	7.30%
	Arithmetic	185.98	-2.79	320.17	370.26	7303	-7149	7.65%
	Balanced	185.96	-4.76	320.23	370.28	7301	-7191	7.65%
	BB	192.72	15.62	130.86	121.94	8789	-8318	7.93%
Silence of the Lambs LQ	1M	42.23	-2.32	79.16	89.69	2411	-1951	17.10%
	2M	41.09	-4.22	77.35	87.49	2411	-2177	16.64%
	Arithmetic	43.16	-2.84	79.93	90.79	2377	-2056	17.48%
	Balanced	43.14	-2.93	79.94	90.79	2377	-2068	17.47%
	BB	43.96	4.87	2374.20	3346.60	2460	-2164	17.80%
Aladdin HQ	1M	379.85	6.14	544.72	664.06	9083	-8764	22.40%
	2M	350.92	-8.04	510.39	619.34	8390	-8764	20.72%
	Arithmetic	452.33	2.94	562.89	722.11	9692	-7713	26.70%
	Balanced	443.63	-1.68	574.37	725.75	9824	-8194	26.19%
	BB	481.74	121.19	169.69	146.90	10809	-9481	28.43%
Ski HQ	1M	254.88	0.98	350.87	433.67	7241	-6844	7.09%
	2M	250.20	-1.17	344.87	426.07	7241	-6844	6.95%
	Arithmetic	258.29	0.90	352.93	437.35	7248	-6940	7.18%
	Balanced	258.04	-0.34	353.15	437.37	7248	-6986	7.18%
	BB	270.06	19.98	147.59	141.15	7685	-8692	7.51%
Simpsons HQ	1M	612.94	25.42	1014.30	1184.90	30729	-25278	10.78%
	2M	541.86	33.76	909.72	1058.30	32739	-17803	9.53%
	Arithmetic	617.14	47.32	1008.30	1181.30	31039	-24204	10.85%
	Balanced	612.94	25.42	1014.30	1184.90	30729	-25278	10.78%
	BB	666.37	96.08	29153.00	26144.00	31588	-29872	11.72%

We now present the results obtained on two kinds of movies: an animated cartoon movie, *Aladdin* HQ, and a sports video, *Ski* HQ. These videos have been encoded in the same manner as the *Silence of the Lambs* HQ movie studied previously. These two videos allow us to test the 1M model on an animated movie (which requires less bandwidth than *Silence of the Lambs*) and a high-action video (which requires more bandwidth than *Silence of the Lambs*). The results obtained are presented in Table 5.7.

The model introduces an error of 22.4% for the *Aladdin* HQ movie, 17.1% for *Silence of the Lambs* LQ, and 7.09% for the *Ski* HQ movie. These results seemed to indicate that our proposed model performs best on HQ high-action and regular movies (such as *Ski* and *Silence of the Lambs*) than on animated or LQ movies. However, when we applied the model on another animated movie, *Simpsons* HQ, and the accuracy is comparable to the high action and regular HQ movies (10.3% error). Although the error introduced by the *Aladdin* HQ movie is higher, the model is still reasonably accurate. This higher percentage of error can be partly explained by the lower values for the correlation coefficients for the first lags for the *Aladdin* movie, as seen in Figure 5.4. Because our prediction is based on this correlation, it is logical that the prediction is less accurate if the coefficients of correlation are smaller.

5.6 Model Enhancements

In order to examine why the prediction model is less accurate with the *Aladdin* HQ movie, we compute the coefficients of correlation between each B frame and each

I and P frame of the GOP as we did for the movie *Silence of the Lambs* (HQ). The results are presented in Table 5.4.

As can be seen in the bold values, the strongest correlation appears to be between the B frame and the P frame placed just after in the GOP. The correlation between a B frame and its closest P frame is now smaller. This leads us to propose another model for the prediction of the B frames. The new model (denoted as 2M) described in (5.6) differs from the 1M model described by (5.3) only for the prediction of the B-3 and B-5 frames. As done previously, we use a least-squares method to find the best α_j and γ_j coefficients (Table 5.8) and compute the marginal error statistics (Table 5.7).

$$\begin{aligned} \hat{B}_{3,t} &= \alpha_3 B_{2,t} + \gamma_3 P_{2,t} \\ \hat{B}_{5,t} &= \alpha_5 B_{4,t} + \gamma_5 P_{3,t} \end{aligned} \tag{5.6}$$

Table 5.8: Coefficients for Least-Squares Method for *Aladdin* (HQ) using 2M Model

	B-1	**B-2**	**B-3**	**B-4**	**B-5**	**B-6**	**B-7**	**B-8**
α	0.0852	0.6964	0.1232	0.7431	0.1266	0.7330	0.0295	0.9241
γ	0.5929	0.2166	0.5905	0.1764	0.5842	0.1832	0.6185	0.0287

With this 2M model, the average error introduced represents 20.71% of the average size of the B frames. This value is smaller than the error obtained with the 1M model but is still higher than the other movies. Thus, although the 2M model enhances the 1M model slightly, it remains less accurate for the *Aladdin* HQ animated movie.

Table 5.7 also shows the results of the 2M model when applied to the movies *Silence of the Lambs* HQ and *Ski* HQ. We observe that the 2M model provides better results for these movies as well. However, the improvements achieved using this model are not significant. Based on the structure of the GOP, the 1M model seems more natural for these two movies. As a consequence, we retain the 1M model for the high action and regular movie types.

An improvement we can add to our model is to simplify its operation further. The least-squares method provides 16 coefficients to describe the different B frames. We would like to reduce this number to only one α and one γ to describe all B frames. The α will replace each α_j and the γ will replace each γ_j in (5.3). The values α and γ are determined using the values of α_j and γ_j determined previously. We propose two methods to compute them. In the arithmetic method, α and γ are simply the arithmetic average of the set of α_j and γ_j values. In the balanced method using (5.7), the values of α_j and γ_j are balanced by the inverse of the error they introduce into the model. We apply these two methods using the 1M model for different movies and the results are presented in Table 5.7.

The results obtained with these methods are nearly the same. The balanced method is only slightly more accurate than the arithmetic method but requires more processing. As such, the arithmetic method is preferred. Moreover, these results are very close to the values obtained without the arithmetic or balanced methods. As a consequence, the arithmetic method can be used to reduce the processing requirements of the predictive algorithm, which will in turn reduce the latency associated with its execution. Finally, we have devised yet another model to predict the size of the B frames using two previous B frames (denoted BB). The results are shown in Table 5.7. As can be seen, the BB method performs slightly worse than the 1M and 2M models.

$$\alpha = \sum_{i=1}^{8} \frac{\alpha_i}{\lambda_i}$$
$$\beta = \sum_{i=1}^{8} \frac{\beta_i}{\lambda_i} \qquad (5.7)$$
$$\frac{1}{\lambda_i} = \left|\hat{B}_i - B_i\right|$$

5.7 Results and Discussion

In [7], a linear model that predicts the size of the B frames according to the previous B frame and the closest B frame was presented. In this chapter, we wish to quantify the accuracy of this model so that we can observe the impact of the scene change detector to be introduced later. First, we define in (5.8) the marginal error as the difference between the predicted size of the frame and its actual size.

$$\varepsilon_m = \hat{X}_m - X_m \qquad (5.8)$$

From this marginal error, we can define the relative percentage error (RPE) as shown in (5.9).

$$\text{Relative Percentage Error} = \frac{\sum_{m=1}^{L} \varepsilon_m}{\sum_{m=1}^{L} X_m} \times 100\% \qquad (5.9)$$

The statistics on the marginal error and the RPE for the linear prediction model are shown in Table 5.8. In most cases, the RPE is low, implying that the model is accurate. However, a higher RPE for the *Aladdin* and *Silence of the Lambs* (encoded in LQ) movies is observed. The higher RPE for the *Aladdin* HQ movie can be explained by the lower values for the correlation coefficients for the first lags for the *Aladdin* movie (Figure 5.4). Because our prediction is based on this correlation, it is logical that the prediction is less accurate if the coefficients of correlation are smaller. The higher RPE for the *Silence of the Lambs* LQ movie is explained by our previous observation (in Section 5.3) that the autocorrelation of the B frames is typically weaker for a movie in LQ than for a movie in MQ or HQ. As a consequence, the linear prediction is less accurate. It is also worth pointing out that the marginal error presents a long-tail distribution.

Table 5.9: Marginal Error Statistics and RPE for B Frame Prediction

Video	Mean |ε|	Mean ε	Std Dev |ε|	Std Dev ε	Max ε	Min ε	RPE
Silence of the Lambs HQ	181.72	-3.37	316.18	364.66	7326	-6536	7.47%
Silence of the Lambs MQ	79.12	-6.10	149.74	169.24	4001	-3370	12.46%
Silence of the Lambs LQ	42.23	-2.32	79.16	89.69	2411	-1951	17.10%
Aladdin HQ	379.85	6.14	544.72	664.06	9083	-8764	22.40%
Aladdin MQ	141.10	5.93	217.17	258.91	4581	-3042	26.06%
Aladdin LQ	70.94	5.46	111.41	131.97	2684	-1761	28.70%
Ski HQ	254.88	0.98	350.87	433.67	7241	-6844	7.09%
Simpsons HQ	612.94	25.42	1014.30	1184.90	30729	-25278	10.78%

5.8 Traffic Model for GOP Size Prediction

We now propose a model to predict the size of a GOP. The size of a GOP is defined as the sum of the sizes of all frames in the GOP. Similar to the process for the B frame size prediction, we analyze the video trace to study the correlation statistics of the GOP size.

According to the statistics presented in Table 5.3, the biggest frame in a GOP is the I frame, which is also the first frame transmitted in a GOP. The cross-correlation between the I frame's size and the size of the GOP for some movies is shown in Figure 5.10. As we can see, the correlation is fairly strong.

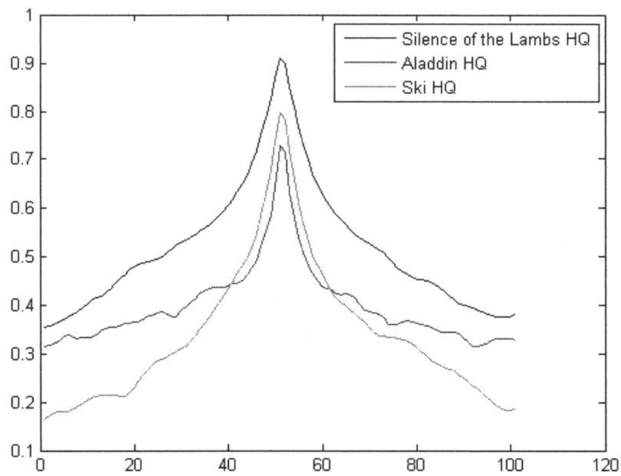

Figure 5.10: Cross correlation between I frame size and GOP size.

Figures 5.11 to 5.14 show the autocorrelation of the GOP size for different movies encoded in different quality. The autocorrelation is very strong for the first lags. As can be seen, the coefficient of autocorrelation for the first lags is more important than the coefficient of correlation between the size of the I frame size and the GOP. As a consequence, we can expect that the linear prediction of the GOP using two previous GOPs (5.10) will be better than the linear prediction of the GOP using the current I frame and the previous GOP (5.11). We use (5.10). Our results are presented in Table 5.10.

$$\hat{G}_t = \alpha G_{t-1} + \beta G_{t-2} \tag{5.10}$$

$$\hat{G}_t = \alpha G_{t-1} + \beta I_t \tag{5.11}$$

where G_t is the size of the tth GOP, I_t the size of the tth I frames and the coefficients α and β are computed using a least-squares method for all GOPs over the entire movie.

Table 5.10: Marginal Error Statistics and RPE for GOP Size Prediction Using (5.10)

Video	Mean |ε|	Mean ε	Std Dev |ε|	Std Dev ε	Max ε	Min ε	RPE
Silence of the Lambs HQ	3337.10	460.02	5474.10	6394.60	71190	-74841	9.67%
Silence of the Lambs MQ	1412.70	195.54	2320.00	2709.30	32810	-31248	13.45%
Silence of the Lambs LQ	860.44	111.45	1340.60	1589.10	19102	-17569	13.58%
Aladdin HQ	5764.10	1023.60	6839.80	8886.20	71932	-77313	22.06%
Aladdin MQ	2330.70	451.59	2812.60	3624.90	28650	-28764	25.07%
Aladdin LQ	1412.30	257.45	1657.00	2162.00	15679	-16047	23.38%
Ski HQ	5792.40	788.40	6878.00	8957.80	84554	-81882	11.61%
Simpsons HQ	8602.30	788.76	9941.30	13124.0	97495	-101580	11.01%

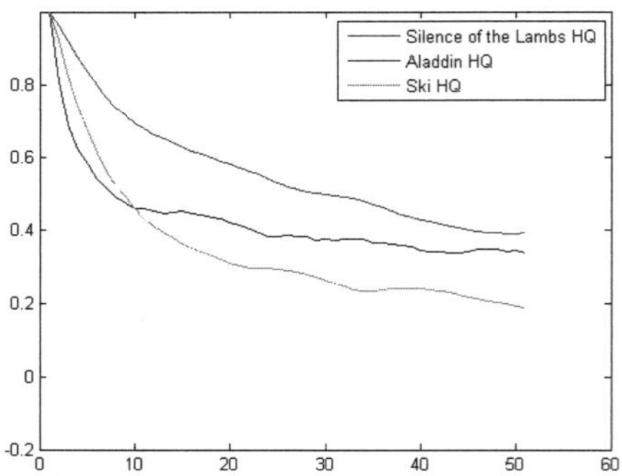

Figure 5.11: Autocorrelation of GOP size for *Silence of the Lambs*, *Aladdin*, and *Ski* (all in HQ).

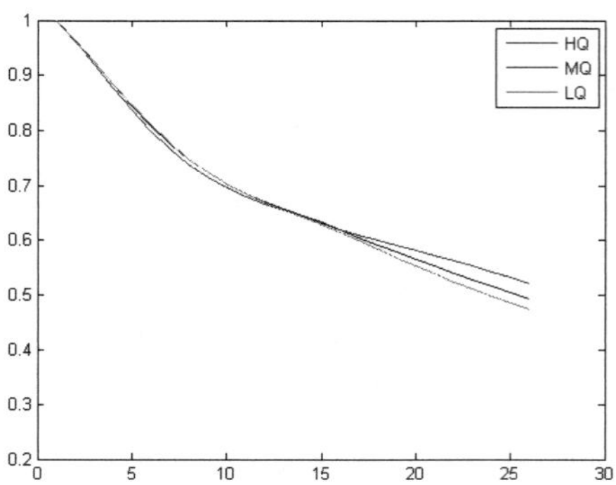

Figure 5.12: Autocorrelation of GOP size for *Silence of the Lambs* (LQ, MQ, and HQ).

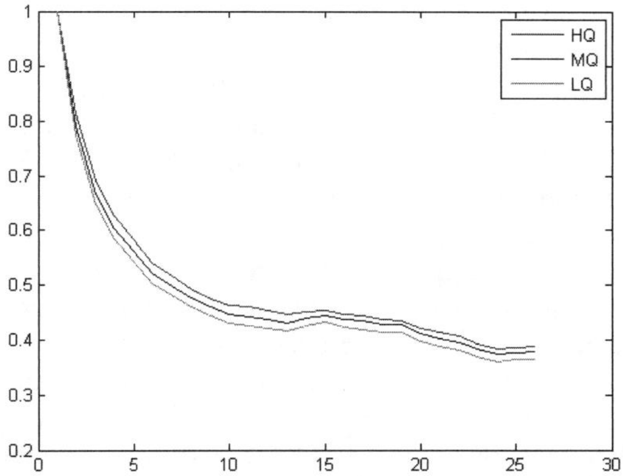

Figure 5.13: Autocorrelation of GOP size for *Aladdin* (LQ, MQ, and HQ).

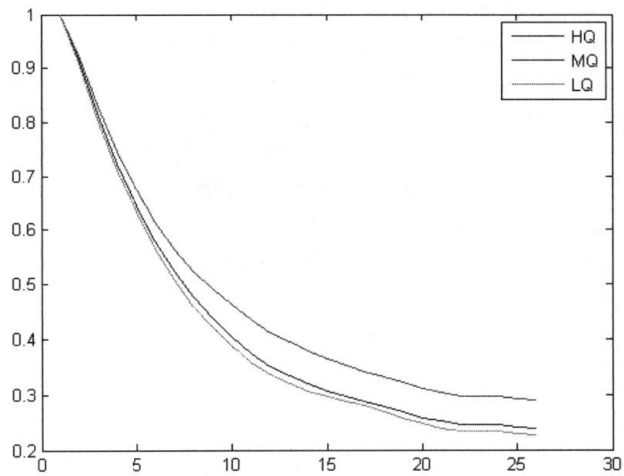

Figure 5.14: Autocorrelation of GOP size for *Ski* (LQ, MQ, and HQ).

The results in Table 5.10 prove that linear prediction is accurate for predicting the GOP's sizes. Moreover, we can make the same conclusions as we did for the high RPE for the *Aladdin* movie. In Figure 5.11, the autocorrelation for the first lags of the *Aladdin* movie is weaker than for the *Ski* video and the *Silence of the Lambs* movie. As a consequence, the RPE is bigger. This GOP prediction can be useful to allocate the required bandwidth in the longer term than allocating the bandwidth for the next B frames. However, as expected, the prediction can be less precise for the GOP prediction although it is still good.

5.9 Model Enhancement with Predicted Scene Change Detector

In order to find the best coefficient α and β, we employ the least-squares method for all the frames. However, during a movie, there may be scene changes, which typically lead to a significant change in the size of the frames (could be larger or smaller). This may impair the accuracy of the prediction method.

Scene changes can be detected easily via changes in the size of the I frame (from GOP to GOP) or the error in I frame size prediction. A scene change can be classified under sudden and gradual. Abrupt changes are easy to detect as two successive frames are completely uncorrelated. Gradual changes are used to enhance quality of video production and are more difficult to detect as the difference between frames corresponding to two successive scenes is small. Previous work on scene changes has been reported (e.g., [8]). However, the majority of these methods are proposed for prerecorded videos. In this section, we propose a new real-time algorithm to deal with scene changes. First, we use a simple metric to detect a scene change (e.g., I frame size). Then, whenever a scene change is detected, we change our algorithm to predict the sizes of the next frame. Each time a scene change occurs, our algorithm will "reset" and treats the frames like a new movie that was transmitted until the next scene change. The video trace is partitioned according to the scene changes. We are going to apply this algorithm for the B frame prediction and the GOP prediction. The models will be the ones presented in Sections 5.3 and 5.5, respectively.

In order for our algorithm to be executed in real time, it is necessary to develop a simple scene change algorithm. The algorithm we propose is given in Figure 5.15 (implemented in MATLAB). This new algorithm computes the difference in the size of two adjacent I frames. Each time the encoder receives an I frame (in every 12 frames of a GOP), it computes the difference in size with the previous I frame it has received. If the difference is bigger than a predefined threshold, the algorithm will start over. This metric is relevant because each scene change typically occurs at the beginning of a GOP. Moreover the I frames are computed without using any reference frame. As a consequence, the size reflects most of the scene changes.

```
nb_change=0;
place_change=[1];
for j=2:nb_GOP
    diff(j)=abs (I(j)- I(j-1));
    if diff(j)>Threshold
        if (j-place_change (end) > 2)
            place_change=[place_change i];
            nb_change=nb_change+1;
        end
    end
end
```

Figure 5.15: Scene change detection algorithm.

A key consideration is the choice of a good threshold. If the threshold is too high, the impact of this method may not be significant because the encoder will only detect a few scene changes, although these changes tend to be abrupt scene changes. On another hand, if the threshold is too low, the algorithm will have to start over very often, and this can prevent the transmission of the video in a real-time fashion but enables the detection of both gradual and abrupt scene changes. Therefore, the choice of the threshold is a trade-off between the efficiency of the algorithm and the real-time transmission. Note that even though a lower threshold enables the detection of gradual scene changes (in addition to abrupt changes), and hence better performance, this may not be necessary since our linear prediction algorithm can cope better with gradual scene changes than abrupt changes. For the linear prediction method to operate, every segment of the movie should be at least 2 GOPs in length for the B frame prediction and 3 GOPs in length for the GOP prediction. We applied this algorithm to several movies using several threshold values. The results are presented in Table 5.11. As expected, superior prediction is obtained when the threshold is low. However, this may incur more latency due to the need to restart the algorithm regularly. The prediction accuracy also improves for videos encoded with lower QP values (i.e., higher-quality videos).

Table 5.11: RPE for B Frame and GOP Size Prediction When Scene Change Model with Different Thresholds Is Applied on Different Movies

Video	Threshold	Number of Scene Changes	Mean Number of GOPs per Segment	B Frame RPE	GOP RPE
Terminator 2 QP = 10	4000	367	5	2.06%	0.12%
	10000	359	5	2.11%	0.21%
	20000	342	5	2.18%	0.52%
	50000	287	6	2.48%	1.34%
Terminator 2 QP = 28	4000	337	5	6.24%	1.21%
	10000	288	6	6.82%	2.65%
	20000	227	7	7.52%	4.34%
	50000	125	13	9.84%	8.12%
Terminator 2 QP = 48	4000	206	8	14.13%	6.17%
	10000	115	14	18.78%	10.22%
	20000	44	33	22.30%	13.79%
	50000	5	206	27.87%	16.26%
Sony QP = 10	4000	346	5	3.79%	0.22%
	10000	332	5	3.98%	0.30%
	20000	294	5	4.34%	0.75%
	50000	210	7	5.26%	1.54%
Sony QP = 48	4000	83	18	14.24%	4.30%
	10000	41	35	16.93%	5.35%
	20000	18	77	19.07%	6.56%
	50000	4	302	20.48%	6.85%
From Mars to China QP = 28	4000	990	5	4.88%	0.86%
	10000	894	5	5.25%	2.05%
	20000	774	6	5.82%	3.18%
	50000	568	8	7.06%	4.94%
Horizon Talk show QP = 28	4000	822	5	7.66%	2.64%
	10000	622	7	9.20%	4.81%
	20000	391	11	11.00%	7.07%
	50000	129	32	13.53%	9.41%

5.10 SAD Method for Scene Change Detection and Adaptation

Figures 5.16 and 5.17 show some correlation in the relative size of the I frame (corresponding to a scene change) compared to the previous or next P frame. This correlation suggests we can use the SAD method described in Section 3.5 to adapt the scene change threshold to the features of the scene. The threshold can be selected using a linear formula as follows:

$$T(n) = aX_{n-1} + bm_n + c\sigma_n \quad (5.12)$$

where $m_n = \frac{1}{N} \sum_{i=n-N-1}^{n-1} X_i$ and $\sigma_n = \sqrt{\frac{1}{N-1} \sum_{i=n-N-1}^{n-1} (X_i - m_n)^2}$

Figure 5.18 shows that with $a = 0$, $b = 1$, and $c = 2.5$, the first 200 frames of the video give 14 detected scene changes, 3 missed, 2 false positives, or a hit rate of 74%. Since this H.264 video contains very fast scene changes, the performance of the scheme should be better for most videos. The detection method can be extended to incorporate PSNR since the MSE used in PSNR is similar to SAD.

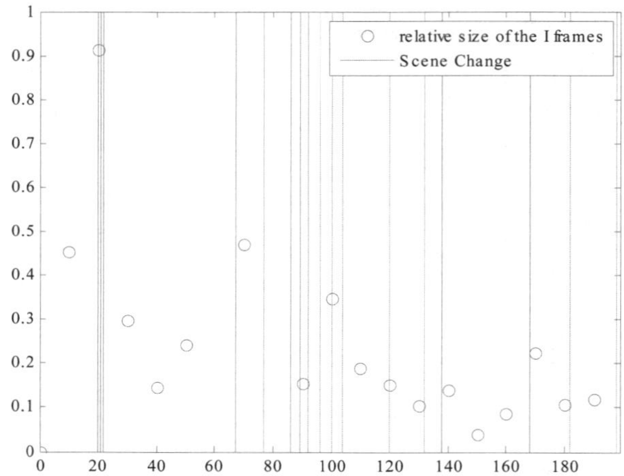

Figure 5.16: Relative size of the I frame compared to the previous P frame (480p *Dell* video).

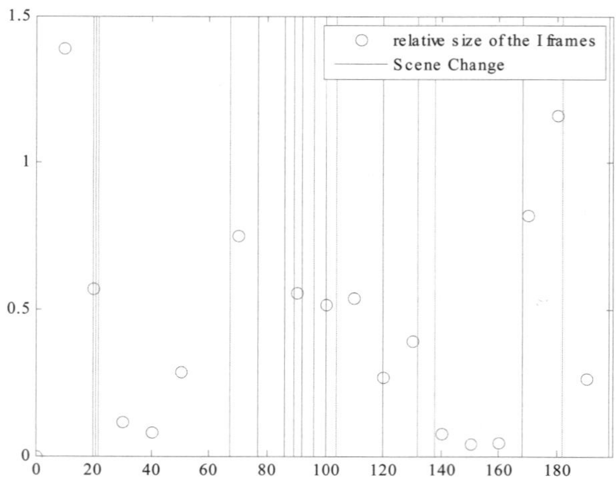

Figure 5.17: Relative size of the I frame compared to the next P frame (480p *Dell* video).

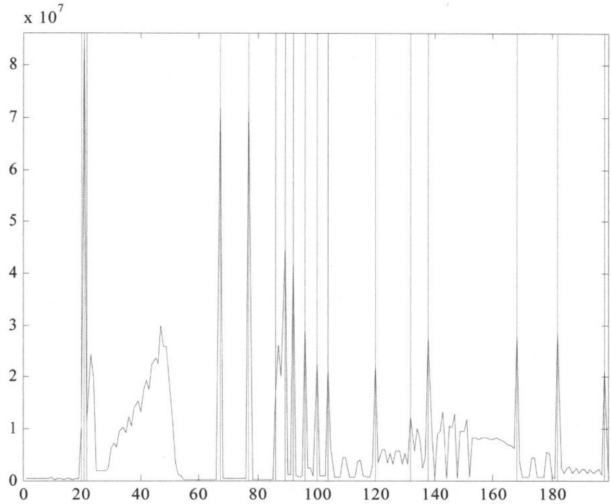

Figure 5.18: SAD computation.

5.11 Conclusions

We have analyzed the traffic requirements and presented a model that predicts, in the short term, the size of the B frames and the GOP of H.264 videos. It has been shown the model provides highly accurate prediction for a wide variety of videos, in particular, high action and regular movies encoded in high-quality resolution. Although the prediction model is very simple to implement, it can be simplified further using the balanced or arithmetic methods without sacrificing prediction accuracy. The same idea has been applied for the prediction of the size of P frames. We attempted to predict the size of the P frame using the previous P frame and the I frame of the same GOP. However the results are less convincing. We also improved the accuracy of the linear prediction model using a scene change detector. We have shown that the RPE can be lowered for any type of movie using this enhancement. We have also shown that the autocorrelation of the size of the B frames exhibits long-range dependence characteristics. This provides a foundation for developing a long-term prediction model. As we will see in the next chapter, the computation of the Hurst parameter using the R/S plot analysis confirms this observation. Clearly, a long-term prediction algorithm may suffer degraded prediction accuracy and the higher complexity may result in higher latency. However, this is offset by the additional time available for long-term prediction and the need to forecast bandwidth usage well ahead of time in order to minimize packet losses during periods of peak bandwidth demands.

References

[1] A. Lombardo, G. Morabito, and G. Schembra, "An Accurate and Treatable Markov Model of MPEG-Video Traffic," *IEEE INFOCOM*, 1998, pp. 217–224.

[2] Y. Dong, Z. Zhang, and D. Du, "Maximizing the Profit of VOD Service on Broadband Cable Networks," *IEEE Globecom*, pp. 2875–2879, December 2003.

[3] A. Dimou, O. Nemethova, and M. Rupp, "Scene Change Detection For H.264 Using Dynamic Threshold Techniques," *5th EURASIP Conference on Speech and Image Processing, Multimedia Communications and Service*, June 29–July 2, 2005.

[4] C. Lee, Y. Jung, S. Lee, Y. Oh, and J. Kim, "Real-Time Frame-Layer H.264 Rate Control for Scene-Transition Video at Low Bit Rate," *IEEE Transactions on Consumer Electronics*, Vol. 53, No. 3, Aug. 2007, pp. 1084–1092.

[5] Video Trace Resource, http://www-tkn.ee.tu-berlin.de/research/trace/trace.html.

[6] G. Van der Auwera, P. T. David, and M. Reisslein, "Traffic Characteristics of H.264/AVC Variable Bit Rate Video," submitted to IEEE Communications Magazine, 2007, http://www.fulton.asu.edu/~mre/COMMAG_06_00562_revised.pdf.

[7] W. Xu and A. G. Qureshi, "Adaptive Linear Prediction of MPEG Video Traffic," *IEEE ISSPA 1999*, pp. 67–70.

[8] S. Feng and R. Sankar, "Limitation and Improvement to Linear Prediction and Smoothing-based Bandwidth Allocation for VBR Traffic," *IEEE Globecom*, 1999, pp. 209–213.

Exercises

5.1. Derive a relationship between the CoV and the peak to average ratio of the frame sizes. Compute the CoV using the peak to average ratio for 10 frames of sizes 20, 10, 5, 5, 5, 5, 5, 20, 5, 20 Kbytes. Confirm your answer by computing CoV directly using (5.1). Now consider CBR frame sizes 10, 10, 10, 10, 10, 10, 10, 10, 10, 10 Kbytes, which give the same average as the previous case. Compute the CoV and the peak to average ratio.

5.2. Suppose a scene change occurs on a B frame. Will the frame be intracoded? Justify your answer in terms of abrupt and gradual scene changes. Note that in Table 5.5, the maximum frame size occurs on a P frame.

Chapter 6

Long-Term H.264 Bandwidth Prediction

In this chapter, we analyze the long-range dependency (LRD) or self-similarity of MPEG video traffic and evaluate the accuracy of a new long-term traffic prediction model. With this model, a video headend in a broadband network can forecast bandwidth usage well ahead of time and this can help minimize packet losses during periods of peak bandwidth demands or prefetching of video data. Although the model can be applied to a wide variety of compressed video traffic, we apply our model to several high-definition H.264 videos. First, we present three methods to estimate the Hurst parameter, which is used to test the LRD of the videos. This leads us to design a new algorithm to predict the size of the I, P, and B frames, and the group of pictures (GOP) in the long term. Although long-term prediction may potentially suffer degraded prediction accuracy compared to short-term prediction, we will show that this is not always the case, especially when the prediction is applied to videos encoded with high-quality resolution. In addition, we will demonstrate that the bit rate variability of multiplexed videos tends to smooth due a weaker LRD.

6.1 Introduction

The previous chapter focused on bandwidth analysis for short-term prediction of H.264 frame sizes. The model works well only when the frame sizes are predicted within a GOP. However, LRD is an inherent characteristic of VBR video traffic [1]. In contrast to short-range dependent traffic models (e.g., Markovian models), where the autocorrelation function decays very quickly in an exponential manner, long-range dependent traffic exhibits a much slower decay in correlations. In this chapter, we analyze the LRD characteristic of H.264 videos using the Hurst parameter and propose a long-term prediction model for at least a thousand frames or many GOPs in advance. First, we evaluate three methods to estimate the Hurst parameter. Then we present a new algorithm for the long-term prediction of different types of frames. We analyze the influence of the parameters of the model and the impact of the coding standard and video quality on the accuracy of the prediction. Finally, we apply the algorithm to multiplexed video streams.

6.2 Long-Range Dependency and Hurst Parameter

The LRD of H.264 videos can be verified by the long-term correlation and the estimated values of the Hurst parameter. For instance, as shown in Figure 6.1, the autocorrelation decays slowly, thus implying long-range dependence or self-similarity. Many videos exhibit this characteristic, which can be exploited for long-term prediction.

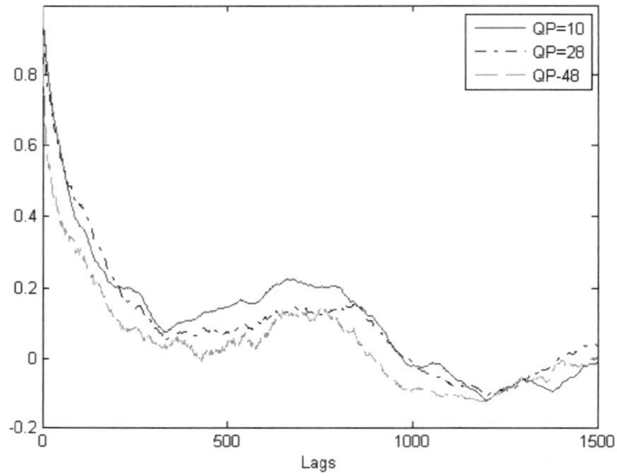

Figure 6.1: Autocorrelation of B frames for *Terminator 2* with G12/B2.

The Hurst parameter is a useful metric that verifies whether a time sequence of data presents LRD or not. Typically, the Hurst parameter values are between 0 and 1. If the Hurst parameter is above 0.5, then LRD between the frames is present. Further details on the Hurst parameter, its interpretation, and its calculation can be found in [2]. The Hurst parameter can be used to determine whether the long-term prediction (to be presented later) will be accurate. A higher value typically leads to more accurate long-term prediction.

There are several methods to estimate the Hurst parameter. The most common methods are rescaled range (R/S) analysis, variance analysis, and periodogram analysis [3]. These estimators can be used for different levels of aggregation. We describe the computational details of each estimator.

For the R/S estimator, we partition the video trace into different groups (Figure 6.2) and a R/S analysis is computed for each group using (6.1).

$$\begin{aligned} &\mathrm{W}(X_n)_k = \sum_{i=1}^{k} X_n(i) - k\overline{X}_n \\ &\frac{R_X}{S_X}(n) = \frac{range(X_n)}{\sigma_{X_n}} = \frac{max(\mathrm{W}(X_n),0) - min(\mathrm{W}(X_n),0)}{\sigma_{X_n}} \\ &\frac{R_X}{S_X}(n) \rightarrow c_X n^H \end{aligned} \tag{6.1}$$

where X_n is the vector of the first n frames of the group. σ_{Xn} and $\overline{X}_n$ are the standard deviation and the mean of X_n, respectively. H is the estimated value of the Hurst parameter and c_X is a constant. "→" means that $R_X/S_X(n)$ is asymptotically proportional to $c_X n^H$.

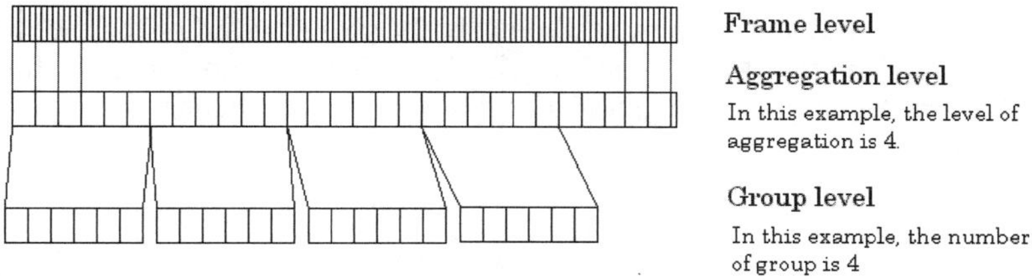

Figure 6.2: Steps of the R/S analysis.

Based on these values, we can generate the R/S plot with log(*R/S*) on the *y*-axis and log(*n*) on the *x*-axis. Using linear regression, we find the best fit to the data points (Figure 6.3). Finally, the value of the Hurst parameter is the average of the gradient found for each group. A tradeoff of this method is the choice of the values of *n*. For small *n*, short-term correlation dominates and the readings are not valid. For large *n*, there are fewer samples and the value of *R/S*(*n*) may not be accurate. A common choice is $N = [10\ 20\ 40\ 80\ldots]$ where N is a vector containing the values of n and $\max(N) < L$, where L is the length of the video trace.

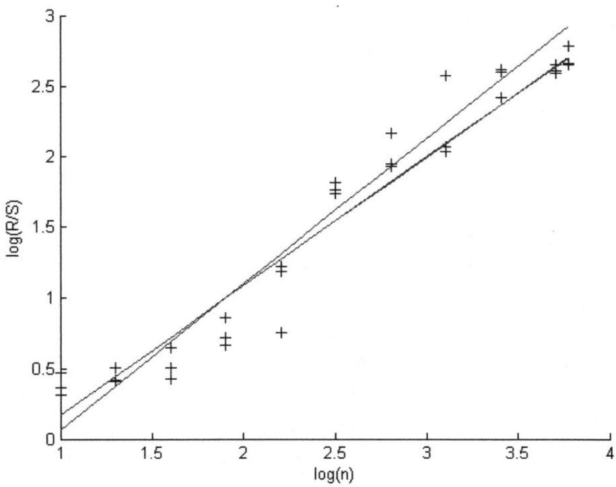

Figure 6.3: R/S plots for 3 different groups of a movie (aggregation level = 1).

The variance estimator splits the video frame into blocks of length *m*. Then the variance is computed using (6.2).

$$var\left(X^{(m)}\right) \rightarrow c_X m^{2H-2} \tag{6.2}$$

where $X^{(m)}$ is a time series derived from the original video trace X by aggregating it over blocks of size m. The sample variance $var(X^{(m)})$ is asymptotically proportional to $2H - 2$ for large values of L/m and m. We then plot $\log(var(X^{(m)}))$ versus $log(m)$. Using linear regression, we find the best fit to the data points and determine the gradient $(2H - 2)$ of the plot. We deduce H from this value. A trade-off for this method is the choice of the values of m.

In the periodogram method, we compute the periodogram of the video trace using (6.3) and plot $\log(I(\lambda))$ versus $\log(\lambda)$. Using linear regression, we find the best fit to the data points. The gradient of the straight line obtained is $1 - 2H$. We then deduce H.

$$I(\lambda) = \frac{1}{2\pi L}\left|\sum_{j=1}^{L} X_j e^{ij\lambda}\right|^2 \rightarrow c_X \lambda^{1-2H} \tag{6.3}$$

where L is the length of the video trace and λ is the frequency.

The R/S analysis and the variance analysis are time-domain estimators. The periodogram analysis is a frequency domain estimator. These estimators may not be accurate due to the nonstationarity of the scaling exponent. Moreover, the frequency-domain estimator is easily affected by any strong short-range dependency of the video traces. This can lead to estimations of the values of the Hurst parameter to be greater than 1 [4].

In Table 6.1, we compare the results obtained with the three estimators. An aggregation level of a means that the sizes of a consecutive video frames are averaged. Note that the range of the values as well as the variation trends are similar for the three estimators. The periodogram estimator often exhibits values above 1 due to the strong short-range dependency characteristic of the video traces. Therefore, this estimator is not reliable in the case of video traces presenting strong short-term correlation. The R/S estimator exhibits two values slightly above 1 for the *Terminator 2* movie encoded with quantization parameter (QP) of 10. This movie presents very strong short-term correlation. Note that an estimate of the LRD cannot be completely independent of the short-term dependency. This leads us to conclude that the R/S and variance estimators as more reliable estimators of the Hurst parameter.

Table 6.2 presents the estimation of the Hurst parameter using the R/S estimator for different movies encoded with G12/B2. These values are all above 0.5 and very often above 0.7. This shows strong LRD between the frames. This characteristic can be exploited for the long-term prediction.

An important observation in Table 6.2 is that the Hurst parameter decreases when the QP value increases. This implies the LRD characteristic is stronger for higher-quality videos (i.e., lower values of QP) and also for videos with low aggregation levels. Thus, if we choose large values for M (in Section 6.3) and/or if the quality of the video is low, then the accuracy of the long-term prediction is expected to degrade. We will see in Section 6.4 that the results for the long-term prediction confirm this property.

Table 6.1: Hurst Parameter Estimates Using Different Methods for H.264 Movies with G12/B2

Movie	QP	Method	Level of Aggregation 1	2	3	12
Terminator 2	10	ASU Website	1.0019	1.0078	0.9544	0.8537
		Autocorrelation	0.9699	0.9697	0.9695	0.9682
		R/S Analysis	1.0173	1.0225	0.9480	0.8485
		Variance	0.8986	0.8444	0.7794	0.8065
		Periodogram	1.0540	1.1981	1.3600	1.4610
	28	ASU Website	0.9227	0.9615	0.9758	0.8758
		Autocorrelation	0.9608	0.9606	0.9605	0.9587
		R/S Analysis	0.9347	0.9783	0.9973	0.8424
		Variance	0.8983	0.8433	0.7533	0.6839
		Periodogram	0.8549	0.9963	1.1514	1.3875
	48	ASU Website	0.8456	0.9036	0.9394	0.8263
		Autocorrelation	0.9489	0.9543	0.9546	0.9528
		R/S Analysis	0.8650	0.9361	0.9730	0.8184
		Variance	0.8930	0.8301	0.6998	0.5948
		Periodogram	0.7319	0.8612	1.0025	1.3357
Horizon Talk Show	28	ASU Website	0.7559	0.7160	0.7542	0.7537
		Autocorrelation	0.9778	0.98	0.9805	0.9808
		R/S Analysis	0.6921	0.7262	0.7603	0.7879
		Variance	0.8882	0.8145	0.7223	0.6938
		Periodogram	0.7194	0.8351	0.9611	1.1885
From Mars to China	28	ASU Website	0.8953	1.1403	1.0853	0.8664
		Autocorrelation	0.9808	0.9817	0.9819	0.9813
		R/S Analysis	0.8454	0.8679	0.8773	0.8244
		Variance	0.9104	0.8713	0.8390	0.8437
		Periodogram	0.7777	0.9189	1.0533	1.3351

Table 6.2: R/S Hurst Parameter Estimates for Movies Encoded Using H.264 and MPEG-2

GOP Structure	Movie	Standard	QP	Level of Aggregation 1	2	3	12	24	48	96
G12/B2	Terminator 2	MEPG-4 AVC	10	1.0173	1.0225	0.9480	0.8485	0.8535	0.8093	0.8395
			22	0.9682	1.0018	0.9973	0.8405	0.8357	0.7791	0.7541
			28	0.9347	0.9783	0.9973	0.8424	0.8201	0.7535	0.7136
			34	0.9032	0.9559	0.9861	0.8213	0.7675	0.7218	0.6761
			48	0.8650	0.9361	0.9730	0.8184	0.7870	0.7971	0.7469
		MPEG2	10	0.9705	1.0080	1.0259	0.8709	0.8310	0.7807	0.7593
			15	0.9509	1.0008	1.0184	0.8520	0.8120	0.7819	0.7692
			20	0.9321	0.9876	1.0094	0.8401	0.8024	0.7819	0.7777
			25	0.9157	0.9750	1.0011	0.8380	0.8020	0.7840	0.7900
			30	0.9040	0.9623	0.9973	0.8331	0.8045	0.7817	0.7755
	Horizon	MPEG-4 AVC	28	0.6921	0.7262	0.7603	0.7879	0.7563	0.7280	0.6952
	Sony	MEPG-4 AVC	10	0.9772	1.0247	1.0247	0.9468	0.9311	0.8690	0.8171
			22	0.9582	1.0268	1.0522	0.9618	0.9629	0.9105	0.8798
			28	0.8864	0.9654	1.0102	0.9694	0.9809	0.9224	0.9064
			34	0.8411	0.9221	0.9802	0.9712	0.9746	0.9258	0.9071
			48	0.8081	0.8984	0.9884	0.9885	0.9615	0.9213	0.8857
		MPEG2	10	0.9396	1.0266	1.1133	0.9691	0.9498	0.8882	0.8597
			15	0.8948	0.9839	1.0810	0.9713	0.9523	0.8909	0.8585
		Part2	8	0.8930	0.9862	1.0769	0.9690	0.9475	0.8838	0.8353
			12	0.8778	0.9729	1.0687	0.9783	0.9704	0.8949	0.8452
	From Mars to China	MPEG-4 AVC	28	0.8454	0.8679	0.8773	0.8244	0.8180	0.8228	0.8240
	Multiplexed 1 : Terminator 2 + Sony with MPEG-4 AVC, QP=10			0.9861	1.0471	1.0133	0.8989	0.8732	0.8030	0.7921
	Multiplexed 2 : Terminator 2 + Sony with MPEG-4 AVC, QP=28			0.8800	0.9672	1.0440	0.9490	0.9202	0.8431	0.8529
	Multiplexed 3 : Terminator 2 + Sony + From Mars to China + Horizon Tak show with MPEG-4 AVC, QP=28			0.7326	0.7889	0.8375	0.8345	0.8201	0.8068	0.7961

Tables 6.2 and 6.3 compare the Hurst parameter estimate for movies coded with three different GOP structures: G12/B2, G16/B3, and G16/B1. G12/B2 comprises 12 frames with 2 consecutive B frames between each reference frame (i.e., I or P frames). Since a GOP contains only one I frame, G12/B2 comprises 1 I frame, 3 P frames, and 8 B frames. Similarly, G16/B3 contains 1 I frame, 3 P frames, and 12 B frames while G16/B1 contains 1 I frame, 7 P frames, and 8 B frames. Moreover, under low QP and aggregation levels, Table 6.3 shows that the H.264 videos consistently exhibit stronger LRD compared to videos encoded in MPEG-4 Part 2 with the same levels. This implies that the long-term prediction for H.264 videos can be more accurate than videos encoded using the older MPEG-4 Part 2 standard. In Table 6.2, the Hurst parameter for three multiplexed H.264

(G12/B2) streams is also computed. In this case, the values of the Hurst parameter decrease. Thus, the long-range characteristic is weaker for multiple video streams than for a single video. However, the Hurst parameter is still close to 0.8. Table 6.4 confirms this observation for multiplexed videos, when compared to the individual videos of Table 6.3.

Table 6.3: Hurst Parameter Values for Different Movies Encoded with H.264 and MPEG-4 Part 2 Using Different GOP Structures

GOP	Movie	Standard	QP	Level of Aggregation						
				1	2	3	12	24	48	96
G16/B3	Star Wars	H.264	16	0.9039	0.9299	0.9543	0.8817	0.8671	0.8740	0.8704
			24	0.8765	0.9084	0.9415	0.8934	0.8743	0.8750	0.8760
			28	0.8632	0.8988	0.9329	0.8954	0.8803	0.8819	0.8850
		Part2	16	0.8562	0.8818	0.8882	0.8693	0.8596	0.8649	0.8357
			24	0.8573	0.8829	0.8843	0.8749	0.8648	0.8649	0.8451
			28	0.8599	0.8836	0.8854	0.8730	0.8647	0.8614	0.8459
	NBC	H.264	16	0.9265	0.9413	0.9288	0.8814	0.8687	0.8668	0.8611
			24	0.8338	0.8530	0.8615	0.8591	0.8487	0.8394	0.8129
			28	0.7960	0.8191	0.8287	0.8451	0.8397	0.8409	0.8065
		Part2	16	0.8132	0.8309	0.8360	0.8446	0.8184	0.8473	0.8376
			24	0.8095	0.8241	0.8306	0.8362	0.8137	0.8417	0.8334
			28	0.9008	0.8839	0.9119	0.9169	0.9189	0.9043	0.8782
	Silence of the Lambs	H.264	16	0.9347	0.9598	0.9866	0.9319	0.9268	0.9180	0.9206
			24	0.9008	0.9358	0.9679	0.9236	0.9155	0.8935	0.8958
			28	0.8839	0.9206	0.9592	0.9231	0.9144	0.8850	0.8853
		Part2	16	0.9119	0.9304	0.9515	0.9089	0.9181	0.9472	0.8676
			24	0.9169	0.9305	0.9538	0.9133	0.9262	0.9532	0.8670
			28	0.9189	0.9296	0.9504	0.9097	0.9158	0.9294	0.8602
G16/B1	Star Wars	H.264	16	0.9043	0.9319	0.9509	0.8884	0.8790	0.8802	0.8601
			24	0.8782	0.9121	0.9359	0.8941	0.8815	0.8737	0.8642
			28	0.8662	0.9036	0.9283	0.8965	0.8833	0.8779	0.8692
		Part2	16	0.8641	0.8938	0.9126	0.8843	0.8730	0.8689	0.8370
			24	0.8675	0.8948	0.9125	0.8883	0.8776	0.8693	0.8393
			28	0.8675	0.8948	0.9125	0.8883	0.8776	0.8693	0.8393
	NBC	H.264	16	0.9513	0.9247	0.9453	0.8862	0.8698	0.8747	0.8608
			24	0.8404	0.8569	0.8701	0.8710	0.8576	0.8521	0.8234
			28	0.8047	0.8246	0.8362	0.8543	0.8477	0.8426	0.8098
		Part2	16	0.8055	0.8246	0.8254	0.8424	0.8230	0.8477	0.8324
			24	0.8049	0.8214	0.8243	0.8409	0.8215	0.8460	0.8378
			28	0.8082	0.8226	0.8273	0.8421	0.8202	0.8445	0.8426
	Silence of the Lambs	H.264	16	0.9329	0.9570	0.9967	0.9414	0.9385	0.9312	0.9372
			24	0.8987	0.9319	0.9728	0.9281	0.9228	0.8985	0.9054
			28	0.8824	0.9180	0.9621	0.9257	0.9216	0.8877	0.8925
		Part2	16	0.9131	0.9334	0.9679	0.9114	0.9148	0.9051	0.9089
			24	0.9191	0.9331	0.9689	0.9114	0.9145	0.8936	0.8842
			28	0.9170	0.9321	0.9679	0.9162	0.9219	0.9086	0.9046

Table 6.4: Hurst Parameter Values for Multiplexed H.264 Movies

GOP	QP	Level of Aggregation						
		1	2	3	12	24	48	96
G16/B3	16	0.8944	0.9294	0.9259	0.9040	0.9008	0.8823	0.8677
	24	0.8142	0.8487	0.8669	0.8803	0.8755	0.8538	0.8458
	28	0.7883	0.8249	0.8428	0.8755	0.8707	0.8504	0.8346
G16/B1	16	0.9191	0.9191	0.9430	0.9110	0.9093	0.8893	0.8706
	24	0.8195	0.8532	0.8705	0.8856	0.8812	0.8609	0.8508
	28	0.7957	0.8292	0.8463	0.8777	0.8733	0.8519	0.8382

6.3 Model Formulation

As illustrated in Figure 6.4, we wish to employ a linear algorithm to predict the size $x(t + \delta)$ of a future frame using a block size of M prior video frames and a total of L video frames. Prior frames are separated from the frame to be predicted by δ frames. As we shall see, the algorithm can be extended to predict the size of a group of pictures (GOP). In this case, we are now dealing with a group of frames and not individual frames. M determines the complexity of the model. A bigger M results in a more complex model. On the other hand, if M is too small, the accuracy of the prediction may be compromised. Clearly, a small δ corresponds to short-term prediction, whereas a larger δ corresponds to long-term prediction. The challenge is to find the best coefficients $g(i)$ in (1.1) that will minimize the error $E_{\delta,M}$ defined in (6.5). Recall that the short-term prediction algorithm presented in the previous chapter is highly accurate due to the close correlation between successive frame sizes. In this case, the autocorrelation function can be used to determine whether short-term prediction will give accurate results. The autocorrelation function can also be used to choose the best M and δ.

However, even if the Hurst parameter shows strong LRD, the correlation between the frames decreases when the lag increases, as illustrated in Figure 6.1. A reasonable value δ is less than 1,200 frames.

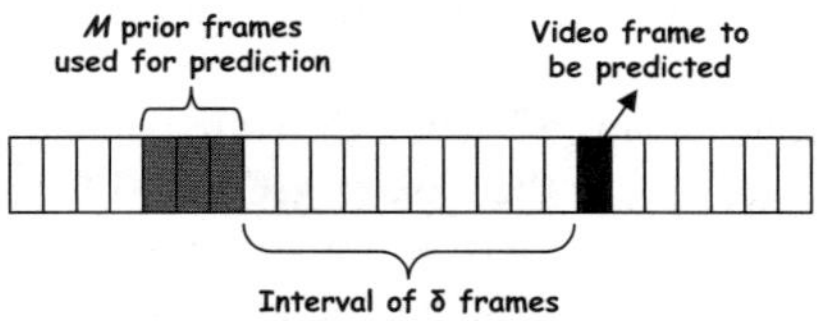

Figure 6.4: Model of the long-term prediction.

We choose a linear prediction model as presented in (6.4).

$$x(t+\delta) = \sum_{i=1}^{M} g(i)x(t+i-M-1) \tag{6.4}$$

In this equation, $x(t + \delta)$ corresponds to the predicted video frame in Figure 6.4. $x(t - M)$ to $x(t - 1)$ corresponds to the video frames used in the prediction. This equation is only valid for values of t that satisfy $1 < t + \delta < L$ and $1 < t - M < L$ at the same time. The inequalities can be combined as: $1 + M < t < L - \delta$. As an example, for a G12/B2 GOP structure of IBBPBBPBBPBB, the variable L corresponds to the number of GOPs of the video if we are predicting the size of the I frame or the GOP. Similarly, L is three times the number of GOPs if we predict the size of the P frames and 8 times the number of GOPs for the B frames. Now, we need to define the error $E_{\delta,M}$ and minimize this error, in order to determine the coefficients $g(i)$. $E_{\delta,M}$ is defined in (6.5). Once the optimal coefficients are known, we use (6.4) to construct the predicted vector. Note that if we set $\delta = 1$ and $M = 2$, this yields the special case of short-term prediction.

$$\begin{aligned} E_{\delta,M} &= \sum_{t=1+M}^{L-\delta} \varepsilon_{\delta,M}^2(t+\delta) \\ \varepsilon_{\delta,M}(t+\delta) &= x(t+\delta) - \hat{x}(t+\delta) \end{aligned} \tag{6.5}$$

To minimize $E_{\delta,M}$, its derivative to zero. The coefficients are determined by:

$$[g(1) \quad g(2) \quad \dots \quad g(M)]^T = R_{xx}^{-1} R_x \tag{6.6}$$

where

$$\forall\, i, j = 1, \dots, M: \begin{cases} R_{xx}(i,j) = \displaystyle\sum_{t=j}^{L-\delta+j-M-1} x(t)x(t-i-M-\delta-1) \\ R_x(i) = \displaystyle\sum_{t=1+M+\delta}^{L} x(t)x(t+i-M-\delta-1) \end{cases}$$

We are now in a position to evaluate the accuracy of the integrated prediction algorithm. Defining the relative percentage error (RPE) as the ratio in the difference between the predicted size of the frame and its actual size to the actual size of the frame, the performance of the prediction algorithm for various quality levels is shown in the following figures. Clearly the prediction becomes more accurate as the quality of the video increases (corresponding to lower values of QP). This is expected since the variability of the frame sizes tends to be higher for lower quality videos [5].

An interesting trend emerges when δ is increased to higher values. As illustrated in Figures 6.5 to 6.8, in the case of GOP and I frame size prediction, the prediction improves (i.e., smaller RPE) for small values of M and QP. Note that a small value for M (such as 1 or 3) reduces the complexity and buffer requirements of the prediction algorithm, which in turn improves the response time of the prediction. The choice of $M = 3$ achieves slightly better accuracy. With large δ (greater than 1,000 frames), the prediction becomes more accurate and is independent of M and QP. A similar trend appears for the P frame size prediction except that the prediction improves for both large and small δ. The B frame size prediction is only accurate for small values of δ. For large δ, the inaccuracy in the B frame size prediction can be attributed to fact that there are typically more B frames in a GOP and since B frames contain only temporal information, they are therefore more susceptible to changes than spatial information. However, since the B frame is the least important among all the frame types (and can be dropped in most cases), its long-term prediction accuracy is less critical.

In general, the GOP size prediction is better than the I frame size prediction, which is better than the P frame size prediction. This is consistent with the values of the coefficient of variability (CoV) for the three frame types. The CoV is the ratio of the standard deviation of the frame sizes to the mean of the frame sizes. The CoV is higher for the B frame than the P frame, for the P frame than the I frame, and for the I frame than the GOP.

To summarize, the long-term prediction model is accurate for the P and I frames, and the GOP in most cases. For the B frame prediction, the model is accurate if the CoV is low and the Hurst parameter is high. This is typically the case for the high-quality movies (i.e., movies encoded with a low QP values).

6.4 Impact of Video Quality and Video Coding Standard

This section compares the results of the long-term prediction for the two standards and takes into account the quality of the movie. The video traces of the two standards with different QP values are obtained from [5]. We employ the *Terminator 2* movie. The plots of the RPE are presented on Figure 6.9. The values of the QP for the H.264 standard are QP = 10, 22, 28, 34, and 48. The values of the QP for the MPEG-2 standards are QP = 10, 15, 20, 25, and 30. As can be seen from the figure, with the exception of the B frames, the long-term size prediction for the I and P frames, and the GOP generally decreases as δ increases. We now analyze the influence of QP for each standard. Then, we compare the two standards for movies encoded with the same QP (i.e., movies with the same resolution). Finally, a global comparison of H.264 and MPEG-2 is made.

6.4.1 Impact of Different QP Values

The prediction becomes more accurate as the quality of the video increases (corresponding to lower values of QP). This is expected since the variability of the frame sizes tends to be higher for lower quality videos. This is consistent with the high values of the Hurst parameter obtained in Table 6.2. Since the Hurst parameter decreases when the quality decreases, the long-range characteristic of the video degrades when the video is encoded with a high QP. As a consequence, the long-term prediction is less accurate when the QP increases. For the case of MPEG-2 encoded videos, the Hurst parameter shows a similar trend but the prediction accuracy for all frame types and QP values lie within an RPE of about 5%. This suggests that the frame and GOP size variation in MPEG-2 is less severe than H.264, which leads to roughly similar RPE in the prediction accuracy. This is reasonable given that H.264 exhibits a far more superior compression efficiency than MPEG-2, which can result in large variations in the frame sizes. The results suggest that for H.264, the prediction accuracy can be seriously affected by the QP value whereas for MPEG-2, the prediction accuracy is not highly dependent on the QP value. For sufficiently low values of QP (e.g., QP = 10), the H.264 prediction can be far more accurate than the MPEG-2 prediction.

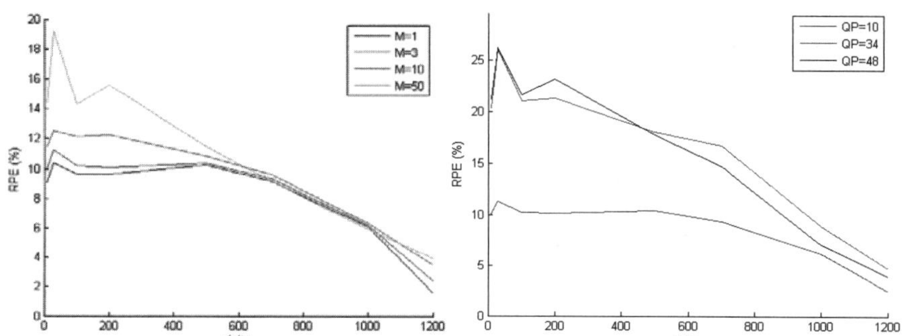

(a) QP = 10 (b) $M = 3$

Figure 6.5: Size prediction for GOP (*Terminator 2* H.264 video).

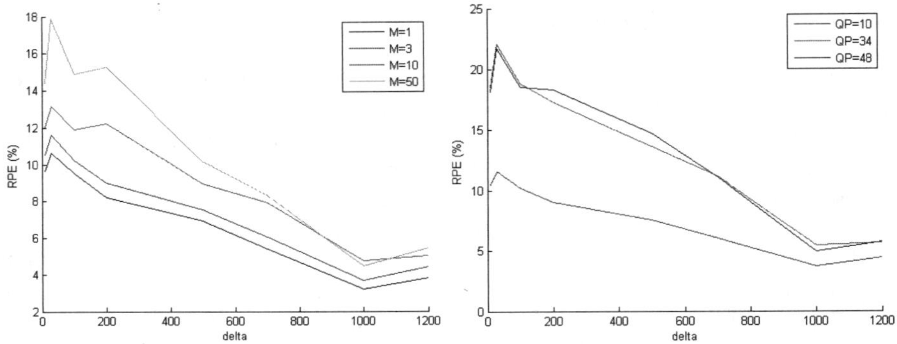

(a) QP = 10 (b) $M = 3$

Figure 6.6: Size prediction for I frame (*Terminator 2* H.264 video).

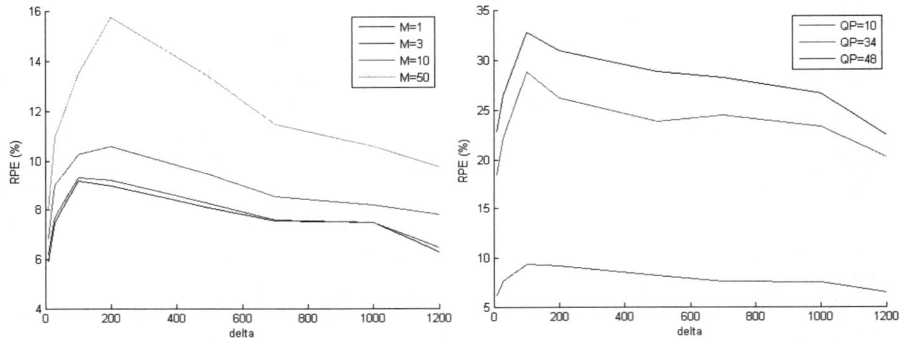

(a) QP = 10 (b) $M = 3$

Figure 6.7: Size prediction for P frame (*Terminator 2* H.264 video).

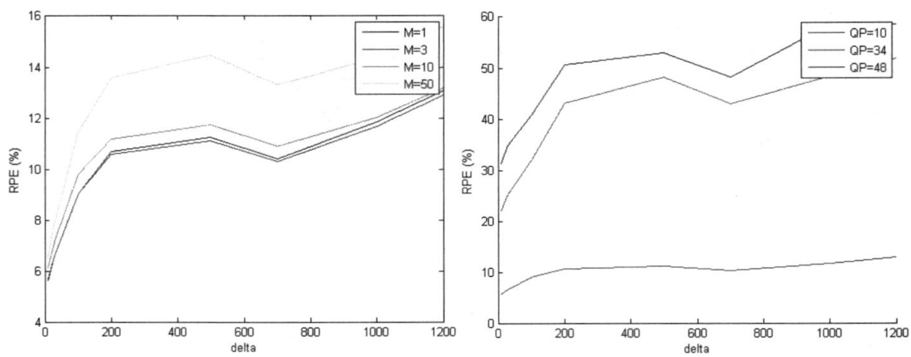

(a) QP = 10 (b) $M = 3$

Figure 6.8: Size prediction for B frame (*Terminator 2* H.264 video).

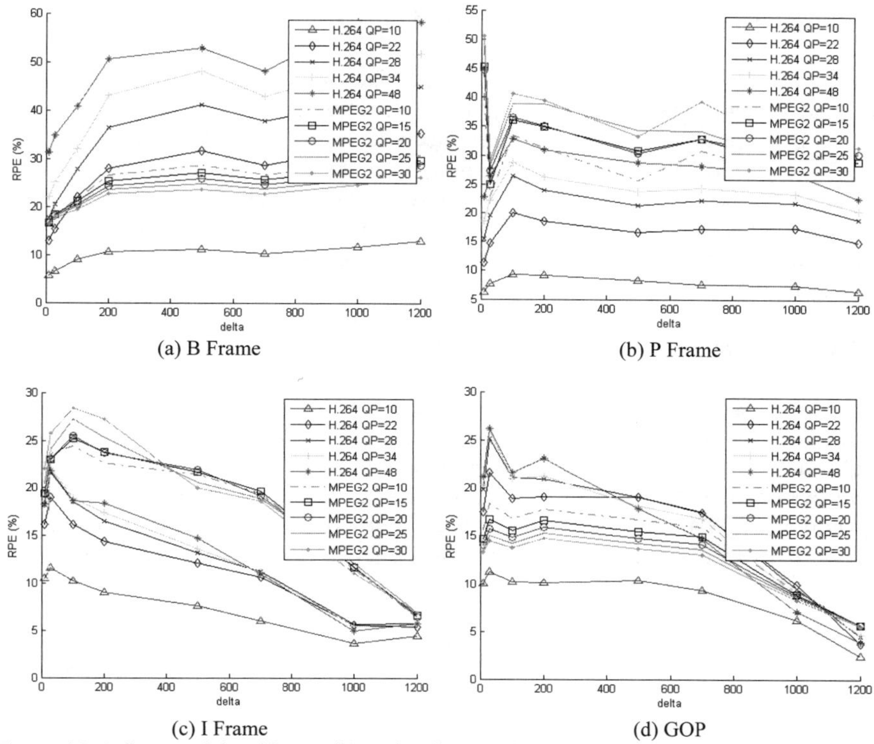

Figure 6.9: Influence of the video quality and coding standard on the RPE for *Terminator 2* encoded using H.264 and MPEG-2 with G12/B2 and $M = 3$.

6.4.2 Impact of Using the Same QP Value

Two videos coded with the same QP value present approximately the same quality. For videos coded with a small QP (QP = 10), the prediction for H.264 is more accurate than MPEG-2 in all cases (I, P, B frame and GOP size prediction). However, if we compare videos encoded with QP values higher than 10, the prediction is more accurate for the P and I frames encoded in H.264 and for the B frames and the GOP encoded in MPEG-2. We observe that when comparing H.264 video with QP = 22 and the same video encoded with MPEG-2 using QP = 20 and 25. A similar observation can be made for the H.264 video with QP = 28 and the same video encoded with MPEG-2 video using QP = 25 and 30.

Videos with approximately the same file size use approximately the same amount of bandwidth when they are transmitted. We compare the video encoded using H.264 with QP = 28 to the same video encoded using MPEG-2 with QP = 20. Their sizes are approximately 160 Mbytes. We observe that the B frame and the GOP size predictions are more accurate for the video encoded with MPEG-2 with QP = 20, and that the I and P frame size predictions are more accurate for the video encoded with H.264 with QP = 28.

6.4.3 Global Comparison of MPEG-2 and H.264

For high QP values, the prediction of the size of the I and P frames is more accurate when the video is coded with H.264. However, the prediction of the size of the B frames and the GOP is more accurate when the video is encoded with MPEG-2. For low QP values, the prediction accuracy for the H.264 video is more accurate for the I, P, B frame and GOP size prediction than MPEG-2 in all cases. With a good choice of δ (between 50 and 1,000 frames), the RPE is below 11% for QP = 10. This implies that the long-term prediction accuracy is excellent for high quality H.264 videos.

6.4.4 Impact of Multiplexing H.264 Videos

Figure 6.10 shows the RPE for the multiplexed videos with different values of QP. *Multiplexed 1* is the sum of the video traces of *Terminator 2* and *Sony* encoded with QP = 10. *Multiplexed 2* is the sum of the video traces of *Terminator 2* and *Sony* encoded with QP = 28. *Multiplexed 3* is the sum of the video traces of *Terminator 2*, *Sony*, *From Mars to China*, and *Horizon Talk* show encoded with QP = 28. In all cases, the total number of frames for the multiplexed videos is the same as the individual videos.

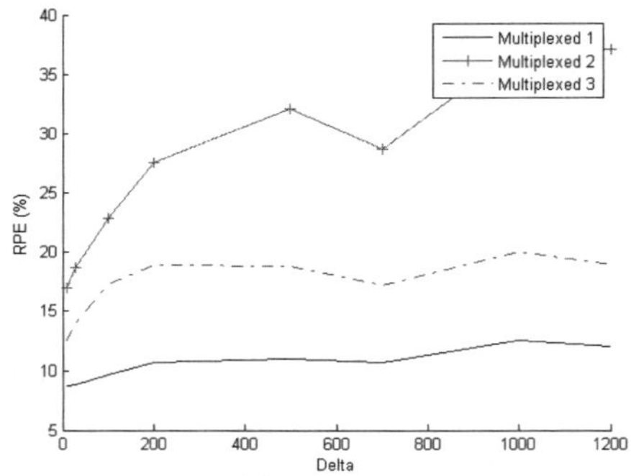

(a) B Frame

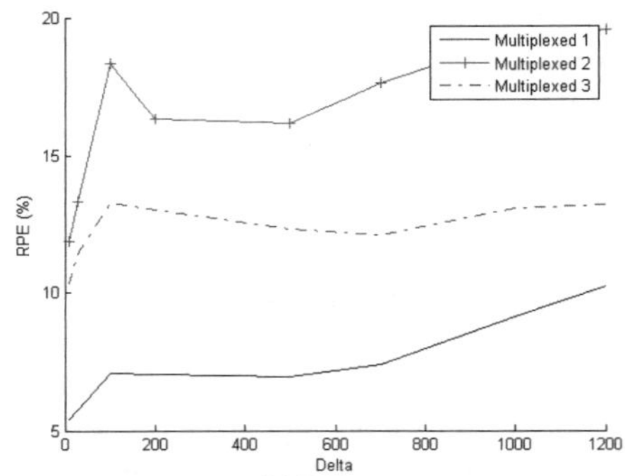

(b) P Frame

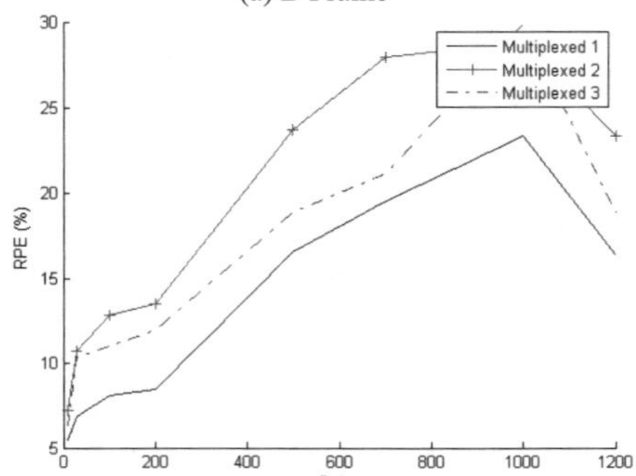

(c) I Frame

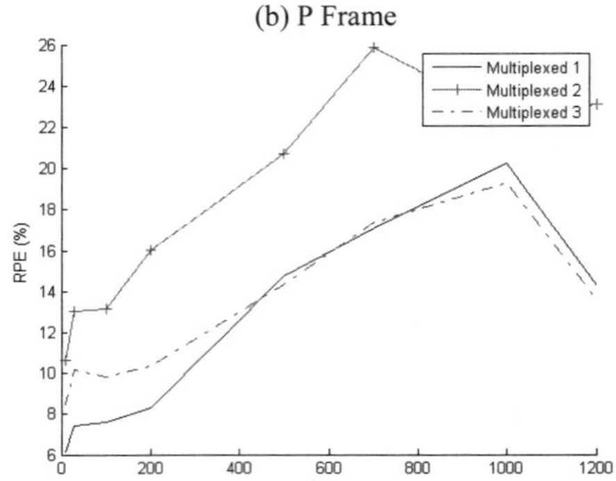

(d) GOP

Figure 6.10: RPE for multiplexed movies encoded using H.264 with G12/B2.

Table 6.5 shows that as more VBR videos are multiplexed together, the value of the Hurst parameter decreases for low aggregation levels. However, the CoV also decreases. The CoV reduction is interesting because it shows that a larger number of aggregated H.264 videos actually tends to "smooth" and this is due to the weaker LRD. Note that the standard deviation for 4 multiplexed videos at QP = 28 (*Multiplexed 3*) is higher than for 2 multiplexed videos (*Multiplexed 2*) even though the CoV is lower for 4 videos. The long-term prediction is more accurate when the CoV is lower and this can be confirmed by a higher accuracy of *Multiplexed 3* over *Multiplexed 2* (4 multiplexed movies versus 2 multiplexed movies, all with same QP values). In addition to the number of movies, the QP value also affects the prediction accuracy for multiplexed videos. This is evident by comparing *Multiplexed 1* and *Multiplexed 2* in Figure 6.10. Both curves employ the same movies (*Terminator 2* and *Sony*) but different QP values (10 versus 28). Like before, the prediction accuracy degrades for higher values of QP. Thus, the long-term prediction accuracy improves as the number of multiplexed H.264 videos and the video quality increases. This provides a positive incentive for multiplexing more high-quality videos encoded using H.264. We shall revisit the multiplexing of VBR H.264 streams in Chapter 9.

6.5 Conclusions

To summarize, a linear algorithm can be used as an integrated mechanism for predicting the size of a single or group of H.264 video frames in the short-term and long-term. In particular, long-term prediction for the size of the I frame and a GOP is highly accurate and has the benefit of being independent of the quality of the H.264 video and the size of the GOP. In general, the accuracy of the prediction is excellent for H.264 movies encoded with a small QP value. We have applied the model to MPEG-2 movies as well. It has been shown that in most cases, the prediction works better for the H.264 movies than for the MPEG-2 movies. This provides an incentive for using the H.264 standard to transport videos with high-quality resolution such as HD videos. The utility of the prediction algorithm may also be applicable for multiplexed videos although the LRD characteristic tends to be weaker.

Table 6.5: CoV for VBR H.264 Movies with G12/B2

Movie	Standard	QP	GOP		I-frames		P-frames		B-frames	
			CoV	Std. Dev.	CoV	Std. Dev.	CoV	Std. Dev.	CoV	Std. Dev.
Terminator 2	MPEG-4 AVC	10	0.3074	3.52E+06	0.2710	4.15E+05	0.2913	3.68E+05	0.3698	2.83E+05
		28	0.5442	4.82E+05	0.4374	8.96E+05	0.5602	6.84E+04	0.9159	3.55E+04
	MEPG-2	10	0.4493	5.80E+05	0.5291	9.28E+04	0.5524	8.87E+04	0.6769	5.37E+04
		15	0.4159	4.15E+05	0.5362	7.58E+04	0.5718	7.41E+04	0.6484	3.80E+04
Sony	MPEG-4 AVC	10	0.4102	3.74E+06	0.4478	9.18E+04	0.4253	4.82E+05	0.5607	2.61E+05
		28	0.5472	5.38E+05	0.5622	2.43E+05	0.7302	9.43E+04	1.1888	2.42E+04
	MEPG-2	10	0.4722	5.86E+05	0.7111	2.15E+05	0.8772	1.28E+05	0.6441	4.03E+04
		15	0.4194	3.67E+05	0.7147	1.65E+05	0.9849	9.93E+04	0.6188	2.65E+04
From Mars to China	MPEG-4 AVC	28	0.5722	1.11E+06	0.5194	3.84E+05	0.6641	1.66E+05	0.9365	5.27E+04
Horizon Talk show	MPEG-4 AVC	28	0.3429	2.11E+05	0.2796	9.37E+04	0.5239	3.31E+04	0.8872	9.93E+03
Multiplexed 1 : Terminator 2 + Sony with MPEG-4 AVC, QP=10			0.5472	5.34E+06	0.5622	9.76E+05	0.7302	6.16E+05	1.1888	3.96E+05
Multiplexed 2 : Terminator 2 + Sony with MPEG-4 AVC, QP=28			0.3926	7.34E+05	0.3867	2.48E+05	0.4612	1.16E+05	0.7204	4.27E+04
Multiplexed 3 : Terminator 2 + Sony + From Mars to China + Horizon Talk show with MPEG-4 AVC, QP=28			0.2678	1.03E+06	0.3264	5.08E+05	0.3312	1.59E+05	0.4556	4.80E+04

References

[1] J. Beran, et. al., "Long-Range Dependence in Variable Bit Rate Video Traffic," *IEEE Transactions on Communications*, Vol. 43, No. 2/3/4, February/March/April 1995.
[2] J. C. Cano and P. Manzoni, "On the Use and Calculation of the Hurst Parameter with MPEG Video Data Traffic," *Euromicro Conference*, 2000.
[3] R. Clegg, "A practical guide to measuring the Hurst Parameter," *International Journal of Simulation: Systems, Science & Technology*, October 2006.
[4] N. Cackov, Z. Lu, M. Bogdanov, and L. Trajkovic, "Wavelet-Based Estimation of Long-Range Dependence in MPEG Video Traces," *IEEE ISCAS*, May 2005.
[5] Video trace resource, http://trace.eas.asu.edu/hd/index.html.
[6] R. Adler, R. Feldman, M. Taqqu, *A Practical Guide to Heavy Tails*, Birkhauser, Boston, 1998.
[7] M. Garrett and W. Willinger, "Analysis, Modeling, and Generation of Similar Traffic VBR Video Traffic," *Proceedings of ACM SigCOMM*, August 1994, pp. 269–280.
[8] M. E. Crovella and A. Bestavros, "Self-Similarity in World Wide Web Traffic: Evidence and Possible Causes," *Proceedings of ACM Sigmetrics*, May 1996.
[9] R. Gusella, "Characterizing the Variability of Arrival Processes with Indexes of Dispersion," *IEEE JSAC*, Vol. 9, No. 2, February 1991, pp. 1374–96.
[10] W. E. Leland, M. S. Taqqu, W. Willinger, and D. V. Wilson, "On the Self-Similar Nature of Ethernet Traffic," *IEEE/ACM Transactions in Networking*, Vol. 2, No. 1, February 1994, pp. 1–15.
[11] S. Deng, "Empirical Model of WWW Document Arrivals at Access Link," *Proceedings of ICC*, 1996.
[12] D. P. Heyman, A. Tabatabai, and T. Lakshman, "Statistical Analysis and Simulation Study of Video Teleconference Traffic in ATM Networks," *IEEE Transactions Circuits and Systems for Video Technology*, Vol. 2, No. 1, March, 1992, pp. 49–59.

Appendix: Traffic Modeling

A.1 Data Traffic

The Poisson model is a classic model used to model data packet arrivals. However, it is not realistic for video traffic modeling since it does not capture any correlation between consecutive packet arrivals [6–12]. In addition, burstiness for the Poisson process will smooth out over a long period, unlike LRD video traffic, which is bursty over different time scales.

A.2 Web Traffic

Web traffic can be modeled as on-off sources 11] with durations modeled as Weibull distribution. Furthermore, packet arrivals in the on periods are generated according to an exponential distribution. The Weibull distribution of random variable X is given as:

$$P(X \le x) = F(x) = 1 - e^{-\left(\frac{x}{\beta}\right)^{\alpha}}, \forall x \ge 0$$

where $\alpha > 0$ and $\beta > 0$ are real and are called the shape and scale parameters, respectively. As α decreases, the Weibull distribution becomes more and more heavy tailed. When $\alpha = 1$, the Weibull distribution is reduced to the exponential distribution. Since packet arrivals during the ON periods are exponentially distributed, this implies that when $\alpha = 1$, the ON/OFF process reduces to a 2-state Markov modulated Poisson process (MMPP) or more precisely, to an interrupted Poisson process (IPP).

A.3 Voice Traffic

This is typically modeled as ON/OFF sources with the durations of ON and OFF periods exponentially distributed. This model is only relevant when silent suppression is employed. Otherwise, a CBR model is used where fixed-length voice packets are generated periodically.

A.4 Video Traffic

More realistic to use real traces or employ a discrete autoregressive DAR model [9]. This is a discrete Markov chain with N states and transition probability matrix P given by:

$$P = \rho \mathrm{I} + (1 - \rho)Q$$

where

ρ = autocorrelation coefficient
I = identity matrix

Each row of Q consists of the negative-binomial probabilities $(f_0, f_1, \ldots, f_k, F_k)$

$$F_K = \sum_{k>K} f_k$$

where K = peak rate per video frame.

The parameters of DAR are computed to fit with real video sequence.

A.5 Heavy Distributions

Heavy distributions of network traffic lead to undesirable effects like long-range dependence and self-similar traffic. Such traffic have high variance, thus higher buffer demands than Poisson traffic. However, high variance may be useful. For a blocking system (e.g., telephone exchange, Internet access servers), the blocking probability decreases with increased variance of the holding time. It is shown in [8, 11] that traffic transferred using HTTP, FTP, and e-mail are heavy tailed.

A.6 Stationary Traffic Models

Short, long-range (for decoupling), and arbitrary distributions are needed.

A.6.1 Short Range

Markov and regression models are characterized by an autocorrelation structure that decays exponentially. The Markov model is widely used to model voice sources because of a small number of states. Regression models are simple to generate but queuing makes them intractable.

A.6.2 Long Range

Fractional autoregressive integrated moving average (ARIMA) and fractional Brownian models are characterized by an autocorrelation structure that decays a slower rate. ARIMA employs Gaussian processes. Fractional Brownian motion has one parameter controlling autocorrelation function but no flexibility in modeling short-range dependence.

A.6.3 Arbitrary Distributions

DAR and TES processes have arbitrary distributions.

Exercises

6.1. VC-1 does not employ a fixed GOP structure. When a scene change occurs on a B frame, the size of the frame can be large. VC-1 employs BI frames, which are B frames containing only intracoded MBs. The BI frames reduce the overheads for the syntax elements. In addition, when a frame is similar to its reference, VC-1 uses a skipped frame. At the decoder, the reference frame replaces the skipped frame. Can the linear prediction algorithm be applied to VC-1?

6.2. Contrast the pros and cons of exploiting LRD for long-term bandwidth prediction versus removing this dependency for more efficient multiplexing of compressed videos. Can these contradictory objectives be reconciled using the quantization parameter?

6.3. In Table 6.5, explain why the difference in standard deviation for B frames is the lowest for two and four multiplexed videos when compared to I and P frames. Note that in contrast, the difference in the CoV is the highest for B frames.

6.4. Will randomizing the packet stream (i.e., packet interleaving) of a coded video weaken the LRD? Is this equivalent to error resilience as described in Section 3.6.2? How about randomizing the order of the coded video frames, employing encryption, and bit interleaving? By performing randomization at the source (bit-, packet-, or frame-wise), what are the consequences on the video playback at the receiver? Evaluate the maximum bit, packet, and frame interleaving depth (i.e., the maximum number of bits, packets, or frames to be used for each randomization time period) in order to maintain proper playback at the desired frame rate.

6.5. Assuming the same coding efficiency for a video coded in MPEG-2 and H.264, which video will produce a higher Hurst parameter value? Will you revise your answer if the H.264 video has a file size that is two times smaller than the MPEG-2 video? Explain why the Hurst parameter value tends to be lower for CBR videos than VBR videos. Note that the frame sizes for CBR videos are almost the same to maintain a constant coded bit rate, although they may change occasionally due to the complexity of the scene.

Chapter 7

Lossless FMO Removal for H.264 Videos

This chapter presents a new lossless scheme to remove the FMO structure from H.264-encoded videos to allow playback on any H.264 decoder/player. Extending this transcoding method to non-FMO-encoded videos with multiple slices per frame can help reduce their size. The proposed scheme works in the compressed domain and does not require extended computations or memory storage. Moreover, it is capable of modifying the slice structure when slices are lost during the transmission and is frame-based, which makes it suitable for live video streaming applications. We first describe the FMO removal process at the receiver and then demonstrate, via actual implementation, the trade-offs in terms of the relative overhead for decoding different FMO types under varying degrees of packet losses. We also present a novel method that accurately predicts the average overheads. The scheme also allows the reduction of the sizes of videos encoded with several slices. Our results show that FMO type 1 incurs the highest relative overheads and these overheads should be managed carefully, especially in a network environment with high losses.

7.1 Introduction

One of the key features of H.264 is slice coding, which is the ability to segment a video frame into slices. A slice comprises an integer set of MBs coded in raster scan order, which can be modified when an error-resilient method such as FMO is used. The slice structure of a frame is flexible and can be changed during the encoding of each new frame. Slices are designed to be independently decodable meaning that a slice does not require other slices from the same frame to be decoded. The use of slices is an effective error resilience tool that prevents the propagation of errors due to losses to other parts of the video frame. However, due to interprediction (i.e., prediction from other frames), the loss of a slice can still impact the VQ of a received video.

FMO allows the assignment of MBs to slices in non-raster scan order, thus spreading possible losses across the frame in a manner similar to bit interleaving. FMO allows EC to operate more effectively since it can now conceal smaller areas of a degraded frame. Thus, FMO improves the overall quality of the received video and is suitable for transmission over lossy channel. It has been proved to be effective with loss rates up to 10% for video conferencing applications [1]. However, FMO requires an overhead and this creates a trade-off between error resilience performance and coding efficiency.

Like many error resilience tools, FMO is only available in the baseline and extended H.264 profiles. This means that not all H.264 encoders and decoders have the capability of handling a bitstream with FMO. For instance, many popular H.264 players such as Quicktime and Flash cannot decode FMO-encoded videos. This problem has been addressed at the *transmitter* in [2, 3], where lossless transcoding techniques are employed to modify the slice structure and FMO is introduced separately after encoding the video content but before streaming.

In this chapter, we introduce a lossless scheme at the *receiver* that allows the display of FMO-encoded videos by H.264 players that are non-FMO-compliant. By lossless, we mean that the visual quality of the decoded videos remains the same as if no FMO removal had occurred. This presents an attractive alternative to transcoding techniques involving requantization which leads to higher computational costs and also degrades video quality due to re-encoding. Unlike methods reported in [2, 3] the proposed scheme has the capability of modifying the slice structure when slices are lost during the transmission and enables playback on non-FMO-compliant H.264 players. We assess, via actual implementation, the relative overhead of our technique, and propose a new model for predicting the overhead. We also show how our scheme allows the reduction of the size of videos encoded with several slices when they are received. The proposed scheme is applied to each frame before decoding as shown in Figure 7.1. For all experiments, the JM 15.0 reference software was used.

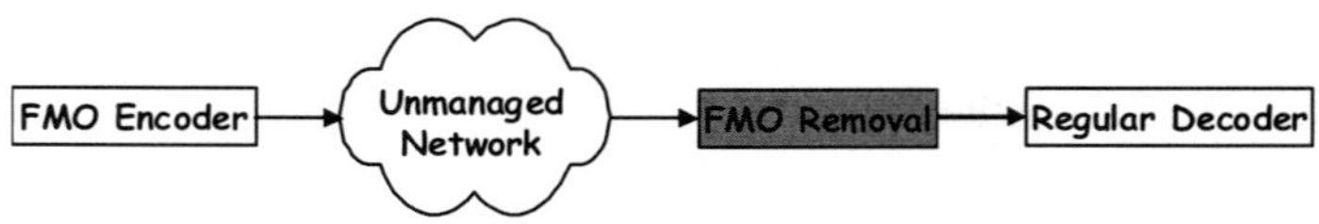

Figure 7.1: FMO removal.

The remainder of this chapter is organized as follows. The technique and its restrictions are described in Section 7.2. The performances in terms of visual quality are evaluated and compared to re-encoding techniques in Section 7.3. In Section 7.4, we show how the size of videos encoded with several slices and transmitted over lossless channels can be reduced. Finally, we present the experimental results obtained with different FMO types and under various packet loss rates, and propose a model for predicting the overheads induced by our scheme when FMO type 1 is used in Section 7.5.

7.2 FMO Removal

We now describe the overall process of decoding an H.264-encoded video. First, the beginning of the bitstream is parsed to extract the information related to the entire stream, or at least to a sequence of frames between two IDR frames, coded into the sequence parameter sets (SPS). Next, the picture parameter sets (PPS) containing information related to all the slices belonging to a single frame are read. Most of the information related to FMO is encoded into the PPS. Note that in general, resending the PPS with each coded frame is not required as long as the information contained in the PPS is still valid, which is often the case. As an alternative, a set of SPS and PPS can be sent in an out-of-band channel before transmitting the remainder of the video and each slice header contains a reference to one of the previously transmitted PPSs. Once the parameter sets are received, the video is decoded frame by frame. A frame comprises a sequence of slices, each slice being decodable independently. The decoding process for one slice is simple. The slice header (SH) contains the number of the first MB in the slice, which is decoded first. The remainder of the slice is then decoded MB by MB, the information of the previous MBs being used to decode their neighbors present in the same slice. The key difference when FMO is used is that the next MB to be decoded may not be in raster scan order. In this case, a few bits are allocated to the encoding of the extra information needed to find the next MB in the PPS and in the SH depending on the FMO type used. The amount and type of information needed also vary with the FMO type used, as shown in Figure 7.2.

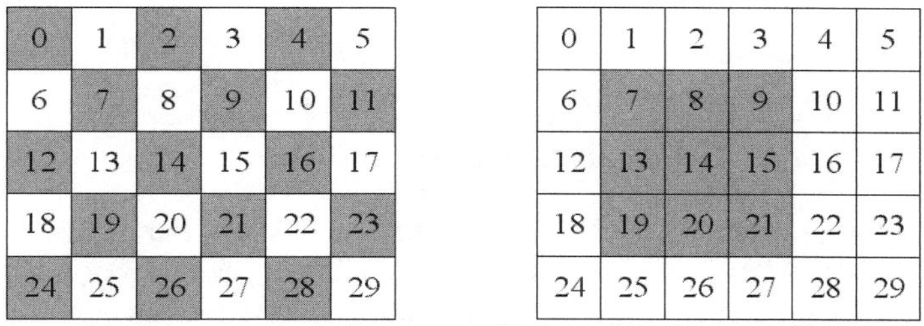

(a) A possible slice map with FMO type 1 (b) A possible slice map with FMO type 2

Figure 7.2: (a) The only extra information needed is the type of FMO used, which is coded in the PPS. (b) Additional information such as the numbers of the top left and bottom right MBs of the foreground slices are specified in the PPS.

Our scheme rearranges the slice structure to form a new slice map where all the MBs in a slice are successive MBs in raster scan order. In this way, the FMO structure is removed. A flowchart of the algorithm is presented in Figure 7.3. The FMO-related information is removed from the PPSs and the remaining information is written to the output file. When a frame arrives, the entropy decoding step extracts the information related to all received MBs (belonging to the received slices). Once this step is completed, a new slice map is created. The frame is scanned in raster scan order; as long as no MB is missing, MBs are added to the same slice. If one MB is missing and some received MBs have not been processed yet, the frame is further scanned until the next received MB is found. Because the previous MB in raster scan order is missing, this MB constitutes the beginning of a new slice. The same process is applied until all received MBs belong to a slice. Once the last received MB has been found, the frame is sent to the output and the scheme is applied to the following frame. Since this process is frame-based and can be activated as soon as an entire frame has been received, it does not require the entire video to be downloaded, which makes it suitable for real-time video streaming applications. In addition, since the computations are performed in the compressed domain and the raw video is not reconstructed, this reduces the storage requirements.

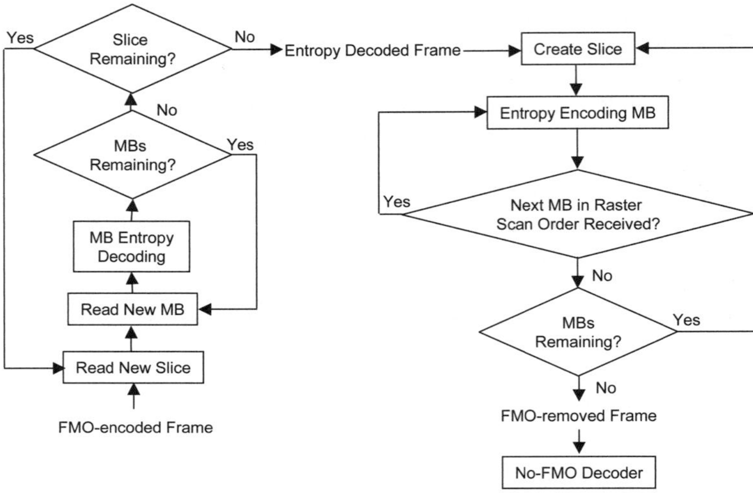

Figure 7.3: Algorithm flowchart.

When a new MB is coded within the same slice, the information from previously coded MBs is used to improve the coding efficiency. For interpredicted MBs (i.e., MBs predicted from other frames), this allows a reduction in the length of the syntax elements. Removing this dependency and coding an MB and its neighbors in different slices is always possible. On the other hand, for intracoded MBs (i.e., MBs predicted from previously decoded neighbors in the same frame), prediction is not allowed across slice boundary. Thus, assigning an intracoded MB to a slice and its predictors to another is impossible. For example, in Figure 7.4(a) MB 20 is predicted from MBs 13, 14, 15, and 19 and they all belong to the gray slice. If the white slice is lost, three new slices will be created by the proposed scheme as shown in Figure 7.4(b) and MBs 13, 14, 15, and 20 will belong to two different slices, preventing intraprediction. Thus, intrapredicted blocks prevent the use of the proposed scheme. As a result, it cannot be applied to slices containing intra-predicted blocks, including I slices. This means that I slices should be protected by other means, for example, redundant slices. Our implementation can be applied to P and B slices with intraprediction disabled.

Interpredicted MBs can be subdivided into smaller blocks. There are several partition sizes for interpredicted MBs as described in Chapter 3. In the 8×8 sub-MB, each 8×8 block can be further subdivided into up to four 4×4 blocks. In all cases, each block is assigned a motion vector (MV) pointing to an area in the reference frame, which is used for prediction. Once the MVs have been computed, the residual information for luma and chroma components is computed. However, in order to improve the coding efficiency, all of this information is not coded directly into the bitstream. First, motion compensation is applied, that is, the horizontal and vertical components of the MVs of each block are predicted from the neighboring blocks and the difference between the actual MV and its prediction is coded into the bitstream. Next, if any luma or AC-chroma residual information is present, the number of nonzero coefficients for each 4×4 luma or AC-chroma block is predicted from the neighboring blocks. Based on the values obtained, on the actual number of nonzero coefficients, and on the number of trailing ones of the block, a codeword is chosen and coded into the bitstream. The remaining information for the luma and AC-chroma blocks is then coded into the bitstream. Note that unlike the AC-chroma residuals, the DC-chroma residuals of each MB are independently decodable. When the MV predictions are exact and there is no residual for one MB, no information is coded into the bitstream other than the fact that this MB is skipped.

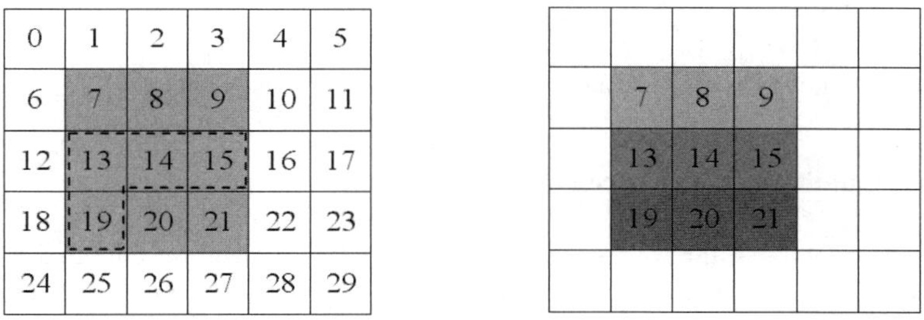

(a) FMO-encoded frame (b) FMO-removed frame after loss of the white slice

Figure 7.4: One example where intraprediction is allowed. The different gray levels correspond to different slices.

During the entropy decoding step, the MVs, the number of nonzero coefficients, and the number of trailing ones are computed and stored for all MBs. Then the new slice map is created. Figure 7.5 shows that with the new slice map, some MBs can use new neighbors for prediction purposes: it is necessary to re-encode each MB into the new bitstream according to the new slice map. For skipped MBs in the FMO-encoded video, their new predicted MVs under the new slice map are compared to their actual motion data and, if they remain the same, the MB is skipped in the FMO-removed video, otherwise its new MV compensations are coded. Note that new MBs may be skipped under the new slice map. Each MB without residual information and predicted with a 16×16 partition is tested and, if the new MV predictions correspond to its actual motion data, the MB is coded as a skipped MB.

In the previous sections, we described how the proposed scheme can be used to remove the FMO structure. The method is also useful for videos encoded without FMO but with several slices. Some encoders do not have the capability of encoding with FMO. Nevertheless, the slice map can still be modified via the number of slices. For example, in Internet transport, [1] recommends encoding videos with slices of less than 1,500 bytes to avoid fragmenting the packet containing the slice. However, using several slices per frame induces overheads. Assuming a video has been encoded with several slices to improve transmission reliability, our scheme allows modifications of the slice structure by grouping all the MBs in one unique slice when no losses occurred. This reduces the size of the video. Even though the video has already been transmitted reducing its size may still be useful when it has to be stored before playback, which is typically the case in applications such as VoD.

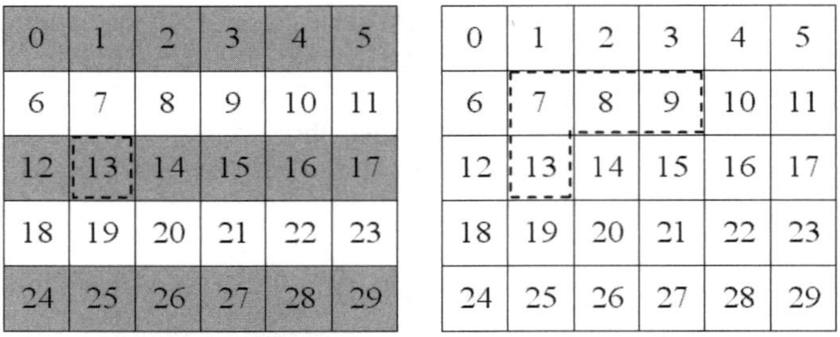

(a) FMO-encoded frame (b) FMO-removed frame

Figure 7.5: An example of interleaved FMO.

7.3 Visual Quality Performance Evaluation

In this section, we use the *Foreman* sequence in QCIF resolution encoded with a QP value of 20. The first frame is an I frame encoded without FMO and the remainder of the video is only composed of P frames encoded with FMO type 1 with two slices. The experiment was carried out on other videos with different resolutions and similar results were obtained.

Our main concern when designing the proposed scheme is to keep the same visual quality between the FMO-coded video and the FMO-removed video. In the previous section, we explained the process of removing the FMO structure from a received video. No information is lost during the processing of the algorithm. Hence, when there are no losses, the algorithm does not compromise visual quality, as demonstrated in Figure 7.6(a).

In a lossy environment, an additional error concealment step is needed at the decoder. However, applying the FMO-removal process before the decoding does not affect its efficiency, as shown in Figure 7.6(b). When a frame is received, all received slices are decoded before the missing MBs are concealed. When the proposed scheme is applied, the status of the frame before error concealment is the same as if FMO had not been removed. Therefore, the scheme will not affect the error concealment performance.

We compare our results to the performance obtained by decoding the FMO-encoded video and then re-encoding it without FMO. The setup of the experiment is presented in Figure 7.7. For the re-encoding, a QP value of 0 is chosen, which corresponds to the best quality. Although this process has a higher computational cost and is done at the expense of the compression efficiency, Table 7.1 indicates that the average Y-PSNRs of the reconstructed videos are slightly smaller.

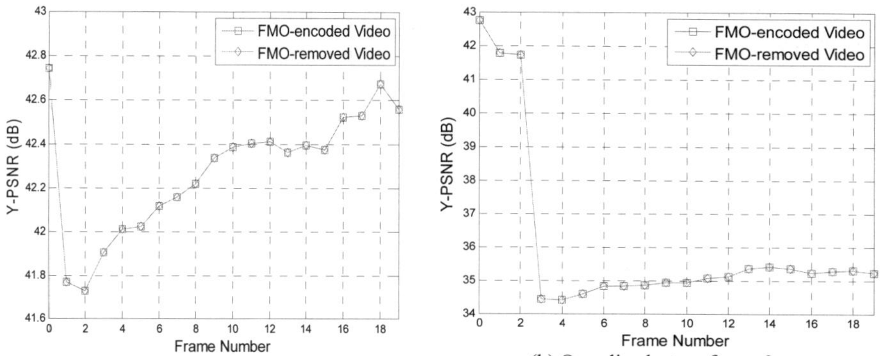

(a) No Loss occurred (b) One slice lost on frame 3

Figure 7.6: Y-PSNR for the 20 first frames of the *Foreman* QCIF sequence.

7.4 Using Multiple Slices

In this section we show the reduction in size achievable for non-FMO-encoded videos with several slices per frame. To assess the possible gains, an experiment is carried out. The first 300 frames of the *Foreman* and *News* QCIF videos are encoded with varying numbers of slices per frame and with different QP values (20 and 30). Next, the proposed scheme is applied to each encoded video so that only one slice per frame is present at the output (no losses occurred). Figure 7.8 presents the sizes of the different files before and after applying our scheme. For one slice per frame, the slice structures are the same and the two files are identical. As expected, with several slices per frame, the output video is always smaller but the savings vary with the videos and the QP values.

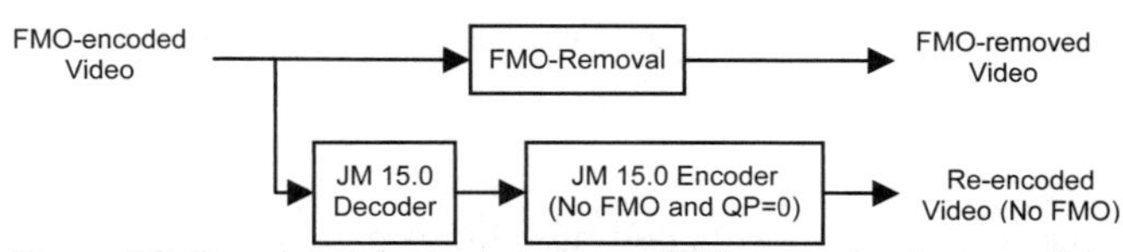

Figure 7.7: Experimental setup to compare FMO removal and re-encoding.

Table 7.1: Comparison between FMO-Removed Videos and Re-encoded Videos

	Video	Size (bytes)	Average Y-PSNR (dB)
No loss	FMO-removed	812,683	42.7313
	Re-encoded	2,951,161	42.7241
5 % random losses	FMO-removed	792,388	28.6476
	Re-encoded	2,713,557	28.6469

The reduction in file size does not only depend on removing slice-related information but is also due to improvements in the coding efficiency, especially for the choice of the codewords for luma and chroma residual information. This is the reason why the gains are higher for lower QP values. Next, the gains increase with the number of slices because more MBs had a reduced number of neighbors and more slice-related information can be removed when more slices were used.

Finally, removing the slice structure does not allow recovering the size of the video directly encoded with one slice. This is due to the dependency of the MB partitioning on the slice structure (see next section for more details).

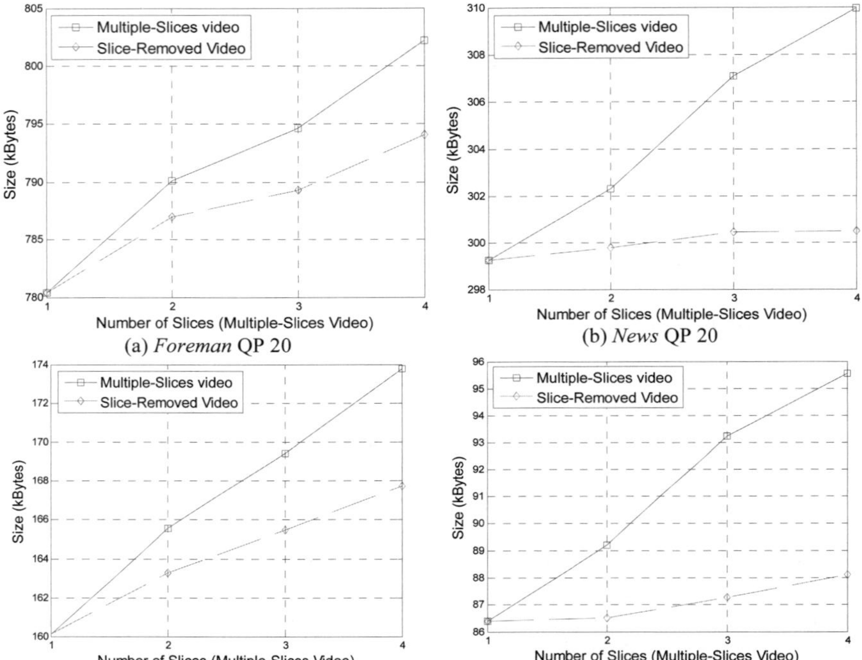

(a) *Foreman* QP 20 (b) *News* QP 20 (c) *Foreman* QP 30 (d) *News* QP 30

Figure 7.8: Size of the videos when varying the number of slices.

7.5 Overheads

In the previous section, we explained how the proposed scheme may reduce the size of the received videos. In a lossy environment, the loss of a slice may result in the creation of new slices requiring extra bits to code the headers. This is not the only reason for a change in the video size: in some cases, breaking the slice structure may degrade the efficiency of entropy coding. In this section, we assess the relative overheads incurred by our scheme for various classical FMO types and we propose a model to predict the overhead for FMO type 1. For a description of the different FMO types, the reader can refer to [5]. Although the presented experiments are performed on a fixed FMO slice mapping, the scheme can be applied to videos where the slice mapping changes with each frame.

In this subsection, we consider the transport of videos over a lossy network subject to uniform random losses. We assume that I slices and parameter sets are never lost. We vary the packet loss rate between 0% and 10% and compute the relative overheads between the FMO-encoded video and the corresponding FMO-removed video. The experiments are carried out on the same videos as in the previous section. The first frame of each sequence is an I frame encoded without FMO and all other frames are FMO-encoded P frames. In all cases each packet transports only one slice. Figure 7.9 presents the average results over 10 traces.

As expected, the relative overheads strongly depend on the original slice map. When there are no losses, only one slice per frame is required whatever the original slice map. When the proposed scheme is applied after slice losses the slice map in the FMO-removed frames depends on the original slice structure.

Even in loss free environment (0% packet loss), the relative overheads differ and are difficult to predict. In most cases, they remain negative but they can turn positive as shown in Figure 7.9(c) for FMO type 1 and Interleaved. In this particular case, the bits saved by removing slice-related information are overcome by the loss in coding efficiency. This is due to the process of selection of the prediction mode in the encoder [4]. MB partition modes are chosen to minimize the number of bits required to code the MV differences. Because the MV predictors depend on the MVs of the available neighbors, the selected partition mode depends on the slice map used by the encoder. After the FMO removal process, the slice map is changed and the MB partition modes may not be optimal anymore; this may result in a positive overhead. Note that this phenomenon may happen with any slice map but it is significant only when the number of neighbors used for prediction is reduced, for example, for FMO type 1 and interleaved, where no more than one neighbor can be used for MV prediction.

On the other hand, the number of nonzero coefficients and trailing ones do not depend on the partition mode: the higher the number of neighbors, the smaller the codewords. When an entire frame is correctly received, the FMO-removed frame has only one slice and the number of available neighbors is maximized. This means that, on average, bit savings are guaranteed for codeword coding. By decreasing the QP value, the number of bits required to code the MV differences remains the same but the number of residual coefficients increases. This increase results in more coded codewords or in longer codewords, both effects inducing higher savings when increasing the number of neighbors. When the QP value is sufficiently small, the savings on codewords overcome the possible worsening of the MV prediction under the new slice map. This is why all overheads for no loss turn negative for a QP value of 20.

Another observation is the increase in relative overhead when the packet loss rate increases. For FMO types 1 and 5, and interleaved, this is due to the increase in the number of slices as shown in Figure 7.10. This results in an increase in the relative overhead for two reasons. First, Figure 7.10 shows that creating new slices may result in reducing the number of neighbors for some MBs and thus, the coding efficiency. Second, each new slice requires a new slice header and additional bits (see following subsection). This overhead is almost constant and does not depend on the QP value. This explains why the relative overheads observed for low QP values remain smaller: the absolute values of the overheads are of the same order but the size of the videos with low QP values is bigger. Thus, the relative overhead is higher. The same applies for the *News* video, which shows little motion and is small compared to the *Foreman* video (306,601 bytes versus 814,070 bytes for QP = 20 and FMO type 1): its relative overheads are higher.

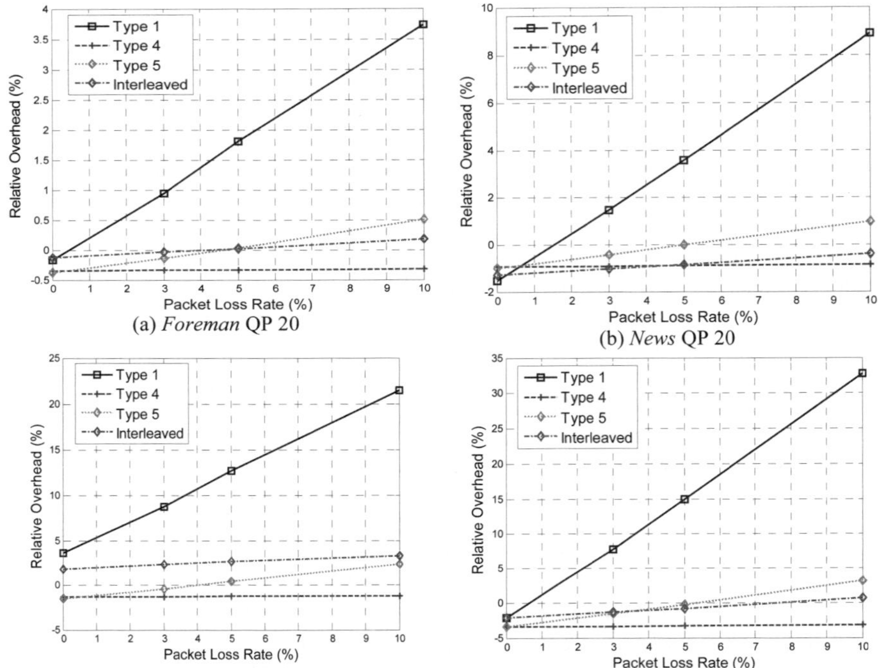

(a) *Foreman* QP 20 (b) *News* QP 20

(c) *Foreman* QP 30 (d) *News* QP 30

Figure 7.9: Relative overheads for different packet loss rates.

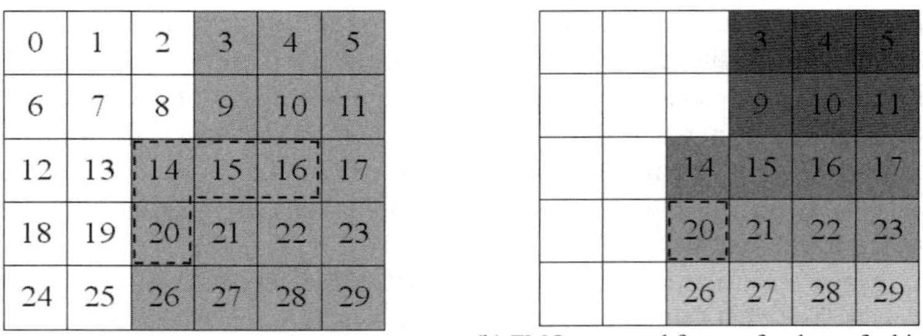

(a) Type 5 FMO-encoded frame (b) FMO-removed frame after loss of white slice

Figure 7.10: (a) MB 21 has four neighbors. (b) Five new slices are created and only one neighbor of MB 21 can be used for prediction. The different gray levels correspond to different slices.

The use of FMO type 1 usually results in the lowest coding efficiency. When applying our scheme at the receiver, the increase in relative overhead becomes more significant. This is expected since when a slice is lost, consecutive MBs in raster scan order are not present in the frame. As a result, for each received MB, a new slice has to be created and no neighboring MBs can be used for improving the coding efficiency. This is the worst possible situation. Having a model for the overhead allows prediction of the storage space for the receiving device.

The general expression for the overhead is:

$$a = \sum_i (\tilde{f}_i - f_i) \tag{7.1}$$

where $\tilde{f}_i$ is the size of the ith frame of the FMO-removed video, and f_i is the size of the ith frame of the FMO-encoded video. Taking the expected value on both sides:

$$E[a] = E\left[\sum_i (\tilde{f}_i - f_i)\right] = \sum_i E\left[(\tilde{f}_i - f_i)\right] \tag{7.2}$$

Using the packet loss rate probability p, we derive the following expression:

$$E\left[(\tilde{f}_i - f_i)\right] = p(1-p)S1_i + p(1-p)S2_i + (1-p)^2 a_i^0 \tag{7.3}$$

where $S1_i$ and $S2_i$ are the overheads induced by the loss of one slice when the other is received and a_i^0 is the overhead when both slices are received. If the degradation of coding efficiency due to the unavailability of neighbors is negligible compared to the size of the new headers and other slice-related extra bits, we can assume that $S1_i$ and $S2_i$ do not depend on the frame content. In other words, $S1_i$ and $S2_i$ are almost the same for all frames and we can drop the subscripts. Equation (7.2) becomes:

$$\begin{aligned} E[a] &\approx np(1-p)(S1+S2) + (1-p)^2 \sum_i a_i^0 \\ E[a] &\approx np(1-p)(S1+S2) + (1-p)^2 a^0 \end{aligned} \tag{7.4}$$

where n is the number of frames from which FMO type 1 has been removed and a^0 is the overhead for the entire video when no loss has occurred. a^0 is known before transmitting the video. $S1$ and $S2$ can be predicted as follows. In one frame, all new slice headers are the same except for the bits coding the number of the first MB of the slice. However, because the slice structure is predetermined with FMO type 1, when one slice is lost we know where the remaining MBs are: we can predict exactly the number of additional bits. Next, a 3-bytes-long flag is added before each new slice for parsing at the decoder. Finally, we have to take into account the fact that each slice must be coded in an integer number of bytes. When the information bits do not sum up to an integer number of bytes, some stuffing bits are added at the end of the slice. With two slices per frame, these extra bits produce a negligible overhead. However, this no longer holds when the number of slices is half the number of MBs per frame, which is the case when one of the two slices is lost for FMO type 1. We model the number of stuffing bits as a uniform random variable: one byte is added with a probability of 0.5 for each new slice. Table 7.2 and Figure 7.11 depict the accuracy of the prediction model. As can be seen from our experiments, the model is in general very accurate, with a relative percentage error of less than 6% for packet loss rates up to 10%.

Table 7.2: Average Actual and Predicted Overheads

Video	***Foreman* QCIF**		***News* QCIF**	
QP	20	30	20	30
		$p = 3\%$		
Predicted (bytes)	7606	14918	4542	7092
Actual (bytes)	7445	14734	4295	6894
Relative error	2.17%	1.25%	5.75%	2.87%
		$p = 5\%$		
Predicted (bytes)	13294	20308	10355	12801
Actual (bytes)	13981	20947	10391	12909
Relative error	4.91%	3.05%	0.35%	0.84%
		$p = 10\%$		
Predicted (bytes)	26437	32732	23799	25994
Actual (bytes)	27383	33640	24545	26882
Relative error	3.45%	2.70%	3.04%	3.30%

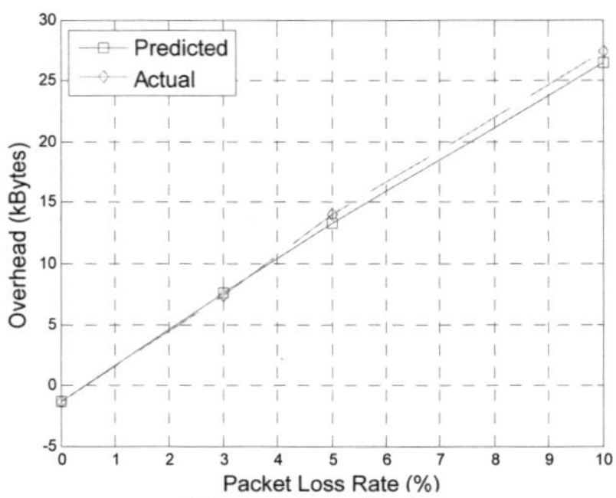

(a) *Foreman* QP 20

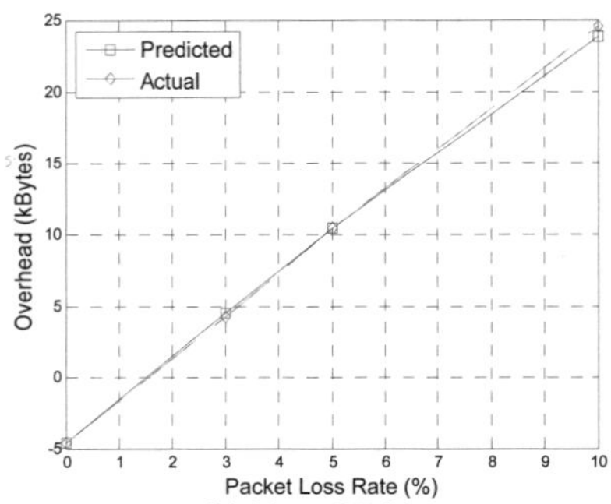

(b) *News* QP 20

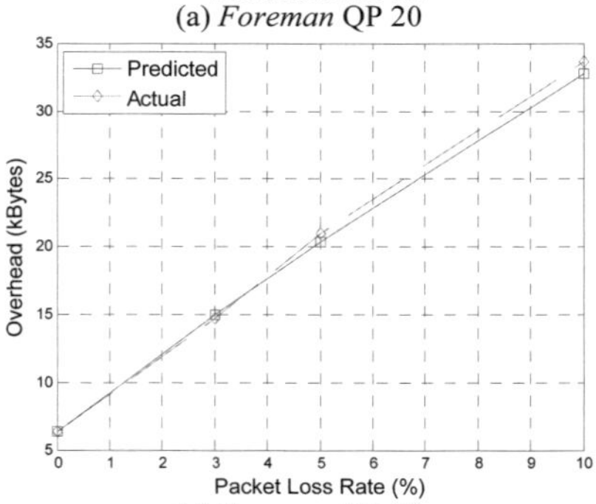

(c) *Foreman* QP 30

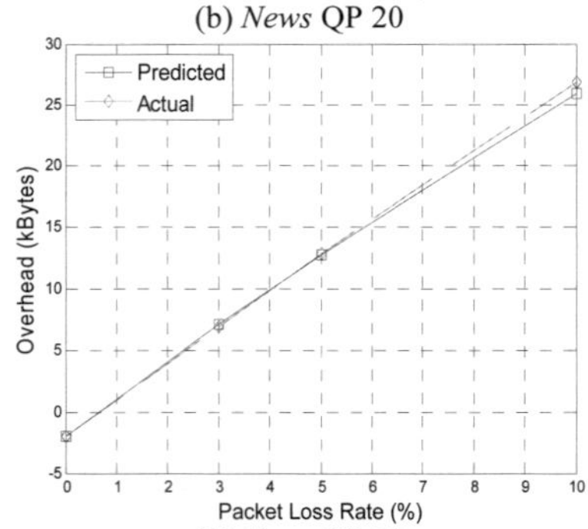

(d) *News* QP 30

Figure 7.11: Actual average overheads and their predicted values for FMO type 1 under different packet loss rates.

7.6 Conclusions

We have presented a new lossless scheme that is capable of modifying the slice structure when slices are lost during the transmission and enables playback of non-

FMO compliant H.264 players. Some overheads may be induced in lossy environments. We have also shown that FMO type 1 incurs the highest overheads and proposed a simple model to predict the average overheads.

References

[1] S. Wenger, "H.264/AVC Over IP," *IEEE Transactions on Circuits and Systems for Video Technology*, Vol. 13, No. 7, pp. 645–656, July 2003.

[2] M. Naccari, G. Bressan, M. Tagliasacchi, F. Pereira, and S. Turbaro, "Unequal Error Protection Based on Flexible Macroblock Ordering for Robust H.264/AVC Video Transcoding," *Picture Coding Symposium*, November 2007.

[3] W. Tan, E. Setton, and J. Apostolopoulos, "Lossless FMO and Slice Structure Modification for Compressed H.264 Video," *IEEE International Conference on Image Processing*, Vol. 4, pp. 285–288, December 2007.

[4] "Joint Model Reference Encoding Methods and Decoding Concealment Methods," JVT-I049d0, San Diego, CA, USA, September 2003.

[5] P. Lambert, W. De Neve, Y. Dhondt, and R. Van de Walle, "Flexible Macroblock Ordering in H.264/AVC," *Journal of Visual Communication and Image Representation*, Vol. 17, pp. 358–375, 2006.

Chapter 8

Error Concealment Methods for Improving Video Quality

This chapter presents some error concealment (EC) schemes for videos encoded using the H.264 and VC-1 codecs. These schemes enhance video quality in online Internet transmission without relying on feedback mechanisms. We will show that the peak signal-to-noise ratio (PSNR) improvement can be substantial: 17 to 24 dB for a variety of HD videos under a loss rate of 20% and 25%, and 18 dB for a SD video under a loss rate of 50%. Although the schemes are not dependent on the use of error resilience methods such as flexible macroblock ordering (FMO), the best performance is usually achieved when FMO is employed. We will compare the performance of these schemes and evaluate the overheads associated with FMO.

8.1 Introduction

H.264 and VC-1 are two powerful codecs that can support efficient HD delivery because they provide higher compression ratios than legacy codecs such as MPEG-2. However, these encoders may increase the bit rate variability for HD videos, which can result in higher packet losses when peak transmission rates are suddenly demanded. This is especially true when variable bit rate (or VBR) encoding is used. This encoding mode is becoming popular in online Internet video transmission because it can maintain constant video quality for each encoded frame. Note that the occurrence of packet losses need not be caused by a lossy network protocol such as the user datagram protocol (UDP). Even if the network transport is loss-free, a sudden burst in the number of received packets may result in buffer overflow (and packet losses) at the client device. Packets arriving late may also be discarded. One of the ways to mitigate the impact of packet losses is to employ EC in conjunction with error-resilience methods. With such a technique, video artifacts associated with packet losses can be concealed quickly without relying on time-consuming channel feedback, and the overall video quality (VQ) is improved.

Flexible macroblock ordering (FMO) is an effective error-resilience method that partitions the frame into several slice groups (SGs), thus allowing the restructuring and reordering of macroblocks (MBs) in the frames. MBs are no longer assigned to slices in raster scan order. Bits associated with adjoining MBs are scattered more randomly throughout the bitstream, thereby improving error immunity and reducing the probability of a packet loss affecting a large region of the frame. For example, SGs can be constructed in such a way that if one SG is not available, each missing MB can be surrounded by MBs of other SGs. This in turn enhances EC by ensuring that neighboring MBs will be available for prediction of a missing MB. Each MB can be assigned freely to a specific SG using an MBA map, up to 8 SGs in one frame. Within an SG, the MBs are coded in scan order. When only one SG is activated, FMO is deactivated. FMO can be considered as an example of multiple description coding. Each SG represents a "description" and is independently decoded. Examples of the interlaced and dispersed SG mappings are shown in Figure 8.1.

While error-resilience such as FMO is not explicitly specified in VC-1, it can be modified to allow a limited form of error resilience. For example, even though the VC-1 standard allows only one SG (see Chapter 3), several slices can be implemented but the length of each slice is restricted to a single row of MBs in the frame. This is less flexible than in H.264 where several SGs can be used and in addition, the length of the slices can be less than a row of MBs (granularly can be a single MB if desired). Thus, error resilience can be incorporated into VC-1 as an extension using a restricted form of H.264's FMO type 0.

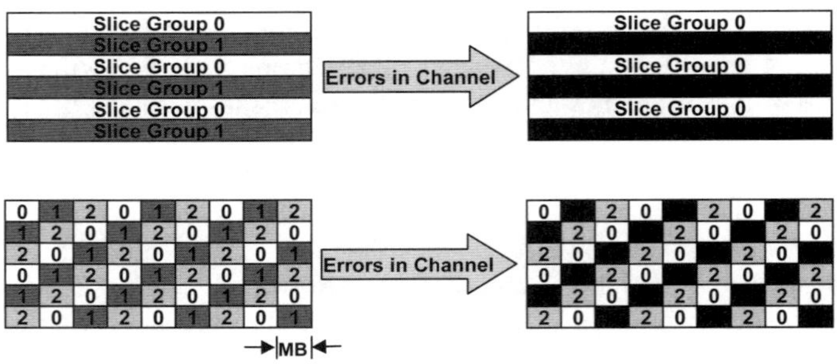

Figure 8.1: Spreading errors using multiple slice groups.

8.2 Error Concealment for HD Videos

In our experiments, all HD videos are encoded using only P frames except for frames on the scene change, which are encoded with an I frame. A quantization parameter value of 25 was selected for 720p videos and a value of 20 for 1080p videos. No B frames were used. Each video contains 300 frames and is encoded twice: one with FMO type 1 (2 SGs) and one without FMO (1 SG). The case for FMO type 1 corresponds to H.264 whereas the no-FMO case corresponds to VC-1. For simplicity, we use the term "no-FMO" although the implementation is a restricted form of FMO type 0 as explained in Section 8.1. Thus, the results without FMO correspond to the best possible EC for VC-1 videos with one slice corresponding to a row of MBs. For the 720p video, slices of 80 MBs correspond to one row of MBs when FMO. For the 1080p video, slices of 120 MBs correspond to one row of MBs.

The losses are generated as follows:

- For the 720p and 1080p videos, two frames (frames 99 and 199) have been corrupted, one on an abrupt scene change and another in the middle of a scene.
- In each corrupted frame of the 720p video, 9 (out of 45) slices (or 720 out of 3600 MBs) have been removed randomly. This corresponds to a loss of 20%. The slices at the same position in the bitstream have been removed for the FMO and no-FMO frames. Note that since the 720p video comprises 1280 × 720 pixels, with 16 × 16 MBs, this gives 3600 MBs and 720/16 or 45 slices per frame.
- In each corrupted frame of the 1080p video, 17 (out of 68) slices (or 2040 out of 8160 MBs) have been removed randomly. This corresponds to a loss of 25%. The slices at the same position in the bitstream have been removed from the FMO and no-FMO frames. Since a 1080p video frame comprises 1920 × 1080 pixels with 16 × 16 MBs and 1080 is not divisible by 16, 8 pixels are added by copying the 8 upper rows of each frame. This gives 8,160 MBs and 1,088/16 or 68 slices per frame. After decoding, each frame is cropped to get back the original resolution.

Note that the loss rate of 25% requires a contiguous number of missing packets within a 1080p HD video frame, and are only likely to occur in heavily congested networks. For instance, a 1920 × 1080 HD frame correspond to 2,073,600 (about 2 million) pixels. Each 4:2:0 pixel is represented by a 12-bit value, which implies 2,488,3200 bits make up an uncompressed 1080p frame. With a compression efficiency of 0.05 bits/pixel (or a compression efficiency of 240), this gives 103,680 bits per compressed 1080p frame or roughly 71 maximum-length (1,460 bytes) TCP packets. A 25% loss rate therefore corresponds to 17 lost packets. Similarly, a 720p frame corresponds to roughly 31 maximum-length TCP packets and a loss rate of 20% is equivalent to about 6 lost packets.

The EC methods are applied as follows:

- For intracoded I frames with scene change, the error in the frame is concealed by the weighted average of the luma and chroma values of the pixels surrounding the missing MB. By default, only the correctly received pixels are used. In some instances when there are consecutive rows of missing MBs, the pixels from the concealed MBs may be used for concealing other neighboring MBs.
- For intracoded I frames with no scene change, the MV of the 8 × 8 block is predicted by searching a matching area in the previous frame. Subdividing the 16 × 16 MB into four 8 × 8 blocks makes the concealment more efficient. Note that there is a need to compute MVs for the I frames which is not needed for P frames. Even if the I frame is not on a scene change, computing the MV is still needed because it is not available in I frames

- For the P frame, each missing MB is concealed by "guessing" its motion vector (MV). We first compute all the MVs of the surrounding blocks, their average, their median, and the zero MV. We then choose the MV minimizing the difference of the luma pixels at the edge of the MB to be concealed. In other words, we compute the error difference of the MVs at the MB boundaries and select the smallest error. In some instances when there are consecutive rows of missing MBs, the same process is employed but the MVs for the concealed neighboring MBs may be used.

The order in which the MBs are concealed can be done in two ways. They can be concealed column-wise starting from the leftmost column and moving to the rightmost column (i.e., the last column to be concealed is the rightmost column). Another way is to conceal by alternating between the leftmost column and the rightmost column (i.e., the last column to be concealed is the center column). The complexity of the two methods is identical. However, the alternating column method may give better results on the average because the center of the frame is usually more difficult to conceal. Depending on the video content, this method may allow the easiest MBs to be concealed first and then use them to conceal the more difficult MBs located in the center of the frame. We have employed the alternating column method in our implementation.

8.2.1 Results

As shown in Figures 8.2 to 8.7, in all cases (with or without scene change, 1080p or 720p), FMO achieves better performance for the concealment of the corrupted frame, and this can measured in terms of the improvement in peak-signal-to-noise (PSNR) or subjective visual quality. The PSNR values are measured in decibels (dB) and the improvement in VQ can be substantial (17 to 24 dB for a variety of HD videos under a packet loss rate of 20% and 25%).

As expected, the concealment of the I frame (corresponding to an abrupt scene change) is not as good as the concealment of the P frame in all cases because the frame to conceal does not resemble the previous frame. In this case, we are restricted to the use of interpolation for the scene change and this method may not always perform well. We note that in all cases, the concealed frames are of better quality (i.e., higher PSNR) when FMO is used. However, the impact of error propagation on subsequent frames may differ, especially when the error occurs on a scene change. For example, in the 720p video with an error on the scene change, the artifacts created by imperfections of the EC in the corrected FMO frame propagate more than the no-FMO case. It arises because the missing MBs are not at the same spatial location in the FMO and no-FMO frames, and this is a result of the randomness of the error. In most situations however, FMO improves the VQ for all frames, including frames following the concealed frames. For example, in the 1080p video, the FMO case exhibits less error propagation than the no-FMO case after the same scene change, a reverse phenomenon to the 720p video. This is because the losses are different in the 720p and 1080p videos and this changes the error propagation characteristics.

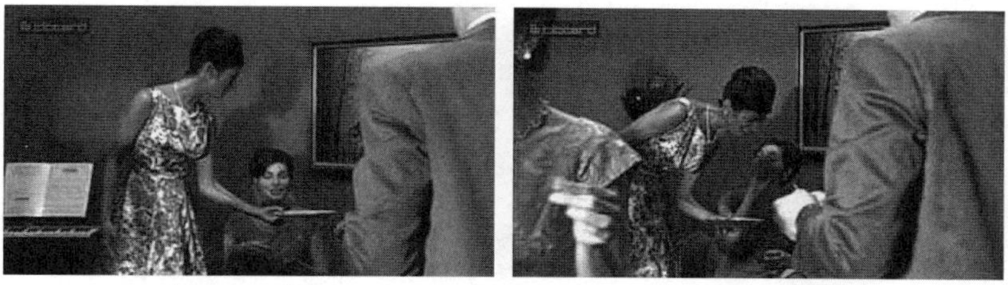

(a) Scene Change (b) No Scene Change

Figure 8.2: Original 720p video frames, with and without scene change.

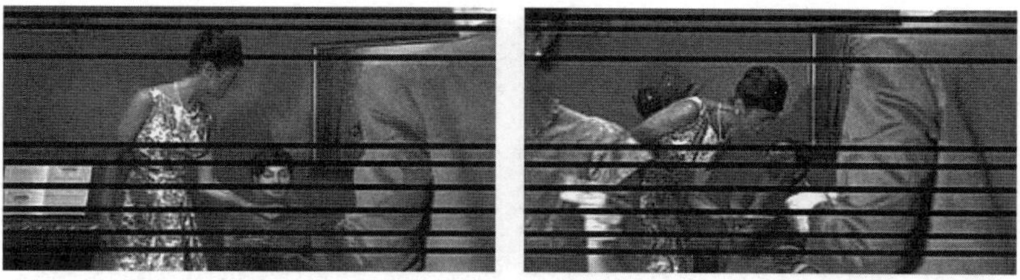

(a) No FMO Scene Change PSNR 17.66 dB (b) No FMO No Scene Change PSNR 17.83 dB

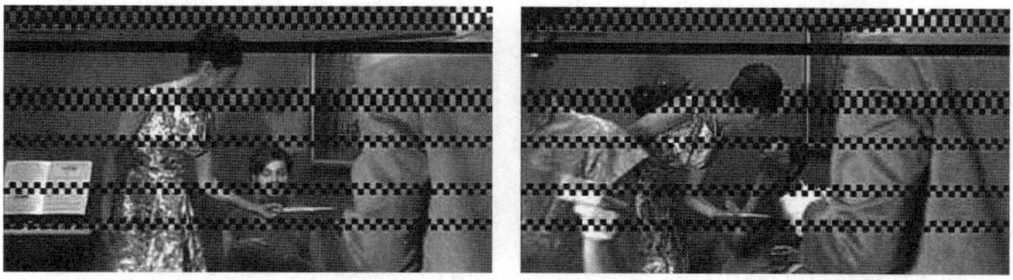

(c) FMO Scene Change PSNR 17.78 dB (d) FMO No Scene Change PSNR 17.94 dB

Figure 8.3: 720p video frames with losses, with and without FMO.

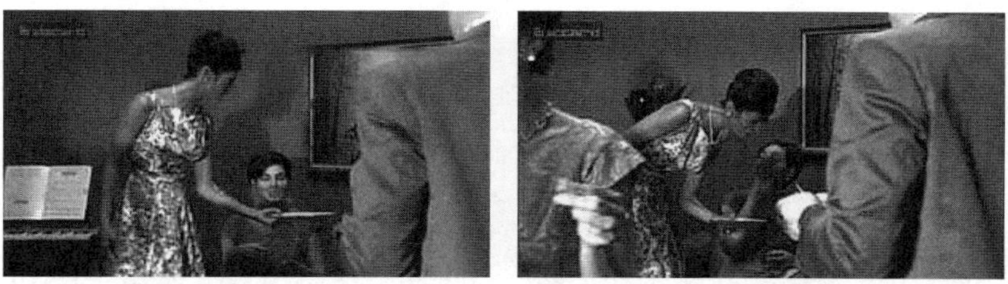

(a) No FMO Scene Change PSNR 32.18 dB (b) No FMO No Scene Change PSNR 37.84 dB

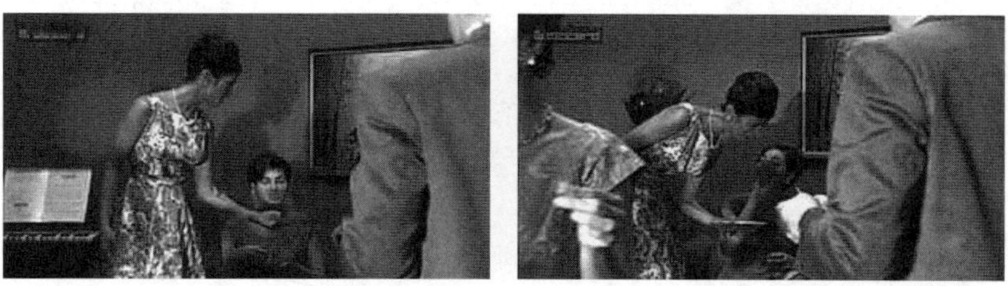

(c) FMO Scene Change PSNR 34.52 dB (d) FMO No Scene Change PSNR 41.82 dB

Figure 8.4: 720p video frames with error concealment, with and without FMO.

(a) Scene Change (b) No Scene Change

Figure 8.5: Original 1080p video frames, with and without scene change.

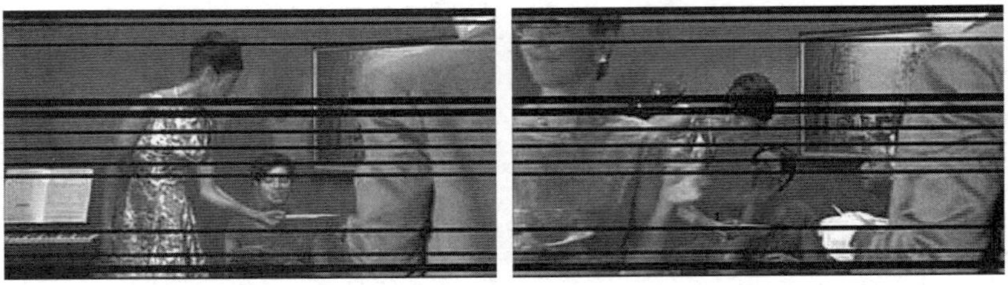

(a) No FMO Scene Change 16.44 dB (b) No FMO No Scene Change 16.46 dB

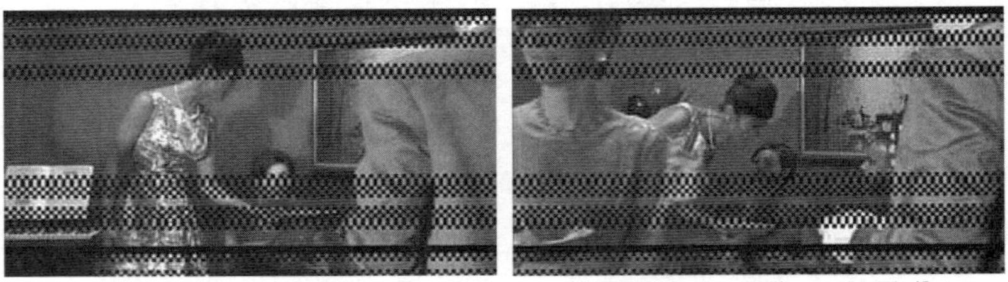

(c) FMO Scene Change 16.67 dB (d) FMO No Scene Change 16.71 dB

Figure 8.6: 1080p video frames with losses, with and without FMO.

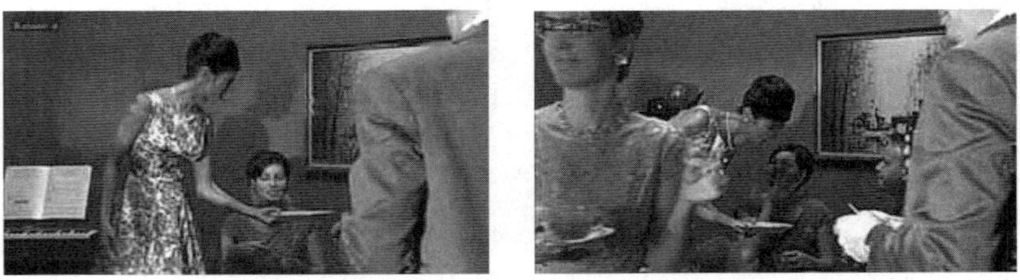

(a) No FMO Scene Change PSNR 33.34 dB (b) No FMO No Scene Change PSNR 36.67 dB

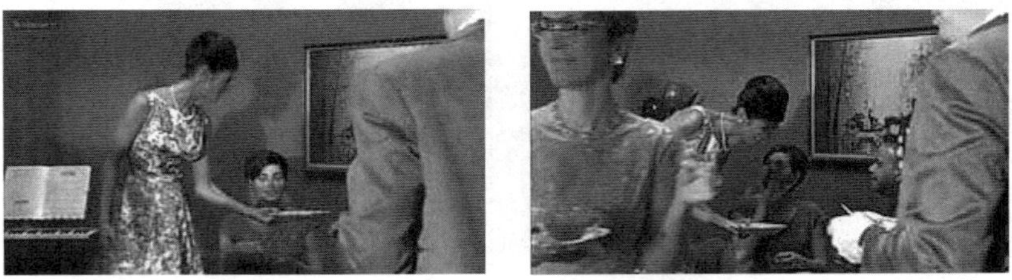

(c) FMO Scene Change PSNR 34.95 dB (d) FMO No Scene Change PSNR 40.54 dB

Figure 8.7: 1080p video frames with error concealment, with and without FMO.

8.2.2 FMO Overheads

As shown in Table 8.1, the FMO file sizes are all greater than the no-FMO file sizes. As explained in Chapter 4, this increased overhead is due to the reduction in coding efficiency when several SGs, and thus slices, are used, and the extra slice headers and picture parameter sets. However, these overheads can be offset with reduced or even removal of FEC overheads at the PHY layer and intentional removal of unimportant frames (e.g., B frames) at the transmitting source (which further conserves bandwidth). Thus, potential savings in bandwidth are possible with the added advantage of improved VQ and the ability to deal with packet losses (where FEC may not be effective).

Table 8.1: File Sizes, with and without FMO

File	No FMO	With FMO	Overhead
720p, QP 25	4.198 MB	4.583 MB	9.17 %
1080p, QP 20	14.944 MB	16.241 MB	8.68 %

8.3 Error Concealment for SD Videos

We apply EC to a 198-frame 480p SD video where 675 MBs (out of 1,350 MBs) are removed from frames 1, 109, and 165. Each MB contains 16 × 16 pixels. With 720 × 480 pixels, this gives 1,350 MBs per frame. Thus, the corrupted frames correspond to a 50% error. The figure on the left in Figure 8.8 shows error frame 165 overlapped on a previous frame (frame 164). As shown in the figure on the right, the concealment was excellent, achieving an 18-dB improvement in VQ. A sharp reader will however, notice some EC imperfections of the plane's tail, but with a frame rate of 30 Hz, such artifacts become unobservable and may not be perceptible.

Figure 8.8: Error concealment when applied to a 480p SD video.

8.4 Temporal Error Concealment

Temporal EC can be applied to I frames not corresponding to scene changes and P and B frames. I frames are often used even at instances when there are no scene changes to prevent error propagation and can be made to recur periodically by defining a fixed GOP. In this case, the content of the I frame may be highly correlated with the previous frame. Thus, the use of temporal EC may be more efficient than spatial EC. However, because I frames are entirely intrapredicted, the bitstream does not directly provide the motion data needed to perform EC. We describe a temporal EC scheme for I frames that performs effectively even when consecutive MBs in raster scan order are missing and can therefore be applied without FMO. We focus on EC for partial frame loss, which may arise when multiple slices are employed and when high quality video is used. The algorithm employs 8 × 8 blocks and optimizes the MV search by using the motion data of the previous frame.

8.4.1 Algorithm

The concealment is performed when all received MBs of the frame have been decoded. A status map showing the current state of the MB is created. The state can correspond to correctly received, concealed, or lost. When an MB is concealed, the status map is updated. Figure 8.9 shows a possible status map where the concealment is performed column-wise.

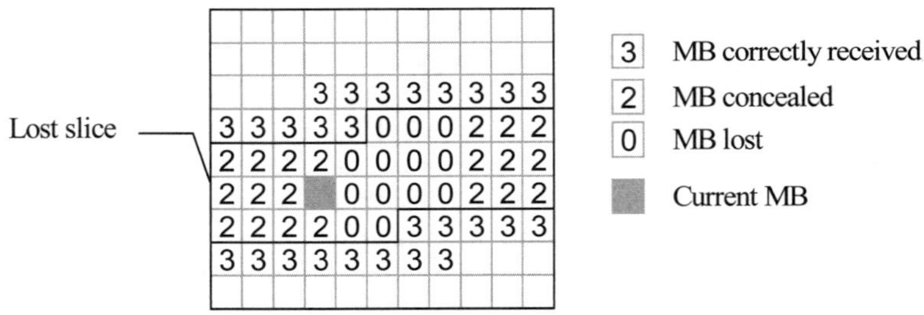

Figure 8.9: MB status map.

The 16 × 16 MB to be concealed and its surrounding MBs are split into 8 × 8 blocks. The MV of an 8 x 8 block is predicted by searching a matching area in the previous frame. Suppose the MV of block A is to be computed. If the motion is smooth and continuous across the frames, the matching block is probably close to the location of A displaced by the MV of the block at the same location in the previous frame. If this MV is not available, it is considered null. Next, the search is performed on a small area around this location. The correct block A' minimizes ε_A as follows.

$$\varepsilon_A = \sum_i (A[i] - C[i])^2 \tag{8.1}$$

where $A[i]$ is the ith luma component of block A and $C[i]$ is the 8 × 8 candidate block. Eventually, the MV for A corresponds to the translation from A' to A.

When all MVs are obtained, the correct MV needed for EC is determined by trial and error (Figure 8.10). If two MVs have been computed in the previous step, four candidate MVs are considered: the null MV, the two computed MVs, and the average of the two MVs. If only one MV has been computed, then it is compared to the null MV. The correct MV minimizes the edge distortion with the received or already concealed surrounding blocks. This distortion is measured as the SAD of the luma samples on the edges of the blocks. This process of electing the concealing MV reduces the production of visible blocky effects.

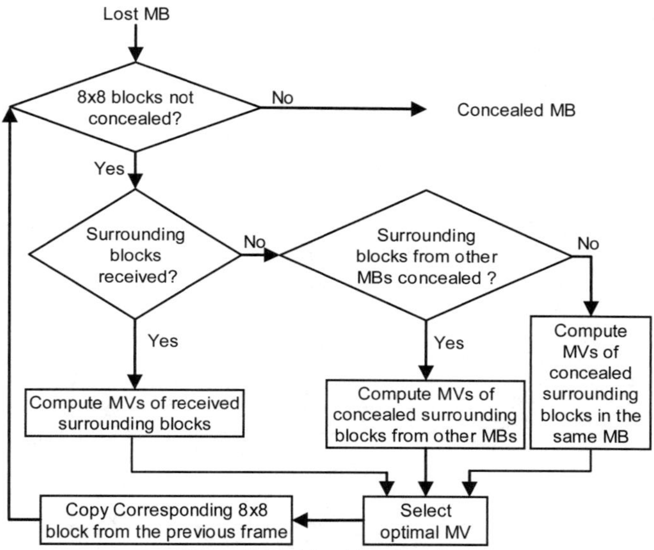

Figure 8.10: Algorithm flowchart.

8.4.2 Performance Evaluation

We present the results obtained for the *Foreman* and *Bus* CIF videos and for the *Shields* and *Parkrun* 1080i HD videos. All videos are encoded with a QP value of 20. Note that the *Bus* and *Parkrun* videos contain fairly high motion. Each video is encoded with and without FMO type 1 using 22 MB slices for the CIF videos and 160 MB slices for the HD videos. An I frame is introduced every 10 frames need not correspond to a scene change. The other frames are P frames. The I frames are subject to uniform random packet losses with probabilities of 10% and 20%. The P frames are not corrupted but may be affected by errors propagating from the corrupted I frames.

The proposed method is compared to the following EC methods:

- JM 15.0 EC using interpolation;
- MB copy EC (i.e., if an MB is lost, it is replaced by the MB at the same location in the previous frame);
- MV copy algorithm (if a block is lost, the MV corresponding to the same block in the previous frame is used).

A simplified version of the proposed algorithm (reduced EC) compares only the samples at the edges of the blocks and not all luma samples of the 8x8 blocks in (8.2). This reduces the computation by more than 50%. The results are presented in Table 8.4 and in Figures 8.11 and 8.12. As can be seen, the proposed EC outperforms the other methods in terms of Y-PSNR. Moreover, the visual artifacts are also reduced and the subjective visual quality is improved. The best performance is obtained for FMO-encoded videos but the results for non-FMO-encoded video are also acceptable. The simplified algorithm performs better than the other techniques for most videos and in all cases when FMO is used. Note that the performance of the MB and MV copy methods depends on the motion content. For the *Bus* video (fast moving), MV Copy outperforms MB Copy while it is the opposite on the *Foreman* video. However, in both cases, the proposed EC gives better results. As expected, the JM EC performs better when FMO is used.

Table 8.2: Average Y-PSNR after Error Concealment in dB (Number of Frames in Parentheses)

EC Method	***Bus*** **(150)**		***Foreman*** **(300)**		***Parkrun*** **(100)**		***Shields*** **(100)**	
	No FMO	**FMO**	**No FMO**	**FMO**	**No FMO**	**FMO**	**No FMO**	**FMO**
10% loss								
JM	26.45	30.21	29.26	31.55	27.48	29.45	30.31	33.06
MB Copy	28.33	29.00	33.01	35.30	29.93	29.08	33.54	32.99
MV Copy	29.27	29.93	31.80	33.55	31.28	30.08	34.58	34.36
Reduced	29.76	31.95	33.86	36.01	30.41	30.23	35.17	36.47
Proposed	32.87	34.28	34.54	37.09	31.75	32.77	36.63	38.55
20% loss								
JM	24.40	24.61	26.71	27.93	24.17	26.78	27.50	31.10
MB Copy	25.46	25.49	31.00	30.41	26.49	25.89	30.35	30.14
MV Copy	26.27	26.24	30.87	29.70	27.71	27.19	31.55	31.68
Reduced	26.77	27.50	30.94	31.64	27.01	27.21	31.93	34.33
Proposed	29.79	29.86	31.74	33.20	28.49	30.25	33.46	37.18

8.5 Conclusions

In most situations, EC is more effective for HD videos encoded with FMO than for videos encoded without FMO. Nevertheless this benefit should be balanced with the additional overheads associated with FMO encoding. A key challenge is in developing error-resilient schemes that can overcome consecutive frame errors, thereby enabling the transport of HD video over the Internet even under severe network congestion. We also describe an EC method for I frames that is shown to be superior to the method adopted in the JM reference software and to other classical EC algorithms. Future research will focus on finding ways to dynamically adapt the size of the search area for MVs using the local motion data and to automatically select the type of EC (spatial or temporal) depending on whether the I frame corresponds to a scene change or not. Besides eliminating transmission overheads, flickering due to the reconstruction of lost slices can be avoided if the reliance on error resilience can be removed. As such, efforts have been focused on this important research area as well as the application of EC methods to 3D-HD videos.

(a) No EC (17.30 dB) (b) JM EC (25.07 dB) (c) MB Copy (23.10 dB)

(d) MV copy EC (23.99 dB) (e) Reduced EC (26.07 dB) (f) Proposed EC (29.96 dB)

Figure 8.11: Y-PSNR of a no-FMO *Bus* video frame with artifact in the fence.

(a) No EC (11.61 dB) (b) JM EC (25.60 dB) (c) Copy EC (32.01 dB)

(d) MV Copy EC (29.65 dB) (e) Reduced EC (33.41 dB) (f) Proposed EC (33.93 dB)

Figure 8.12: Y-PSNR of FMO *Foreman* video frame.

Exercises

8.1. Suppose each bit of video information can be manipulated at a rate of 1 GHz in an Android system. For a 1080p video (30 Hz, 4:4:4 color format, 8 bits/sample) with 16×16 MBs, compute the time needed to perform FMO type 1 error resilience. Assuming no additional bit overheads associated with FMO and a coding efficiency of 0.25 bit/sample, compute the time needed to transmit a 1080p video frame using maximum-length RTP packets on a 6 Mbps error-free link.

Based on the delays computed thus far, calculate the minimum amount of receiver buffer space to store the video frames to avoid buffer starvation during playback. Repeat the above with a 720p video containing 8 × 8 blocks.

8.2. Can EC convert an interlaced H.264 video to progressive? If so, describe the EC methods that can be applied. Since error resilience is implicit in this case, list the compatible FMO type(s). Can this idea be extended to interlaced VC-1 videos? Can a 1080i 30 Hz video be converted to 1080p 60 Hz? Does the spatial correlation of adjacent lines of a field increase or decrease for parts of the field containing moving objects? How can the spatial correlation be exploited in EC?

8.3. Can multiple 3D views be generated from spatial EC? From a coding efficiency standpoint, which is better: coding the video as a single frame or as a pair of fields? Note that H.264 allows the coding decision to be made at the vertically-adjacent MB-pair level (called MB adaptive frame-field or MBAFF) or at single frame level (called picture adaptive frame-field or PicAFF).

8.4. In general, quick recovery of a damaged image region can be achieved when image regions are coded in the intra mode (i.e., without reference to previously coded frames). Suppose FMO type 1 is chosen as the error resilience method. Explain how certain MBs in the FMO frame can be intracoded to improve error resilience and reduce decoding delay.

Chapter 9

Video Traffic Smoothing and Multiplexing

This chapter presents methods to improve the efficiency and performance of 2D and 3D HD video streaming over the Internet. We first present the basics of traffic smoothing for transmitting compressed video traffic. Smoothing essentially transforms the peak bit rate of the video stream to the average rate. It will be shown that such an algorithm reduces bit rate variability and offers high buffer utilization, and can also minimize packet losses due to buffer overflows at the receiving device. Traditionally, the approach of a video server has been to send the contents of the entire video stream as quickly as possible, which we shall refer to as rapid downloading. We will show that rapid downloading does not offer optimal bandwidth usage for sending VBR HD videos, and in addition, may result in a high number of packet losses over the Internet when peak rates are demanded by the video stream. A better solution is to adopt progressive downloading where the stream is sent in segmented blocks. With VBR videos, particularly 1080p HD videos, this approach provides improved buffer and bandwidth utilization. We will also show that by properly smoothing and shaping the traffic to reduce the peak rates, this will help minimize packet losses due to buffer overflow at the receiver. With compound TCP and efficient multiplexing, further performance gains can be achieved.

9.1 Introduction

Traditional Web servers primarily work with Hypertext Markup Language (HTML), Cascading Style Sheets (CSS), Hypertext Preprocessor (PHP), Scalable Vector Graphics (SVG), and related files. These file sizes are often in the range of a few Kbytes to occasionally a few Mbytes. Web servers typically send these files as quickly as network resources permit. While this approach works well for Web pages, it is not scalable to streaming larger audio and video files. This is especially so for HD and 3D videos. As an example, Figure 9.1 shows the high and variable encoded rates for the *Eye* 720p 3D VC-1 video. Keeping that in mind, the next approach is to send media at a constant rate. This will work for media files that are encoded at a constant bit rate (CBR) but will not perform optimally in the case of variable bit rate (VBR) videos. In this chapter, we explore alternative approaches that would allow better scalability for streaming media, particularly VBR media.

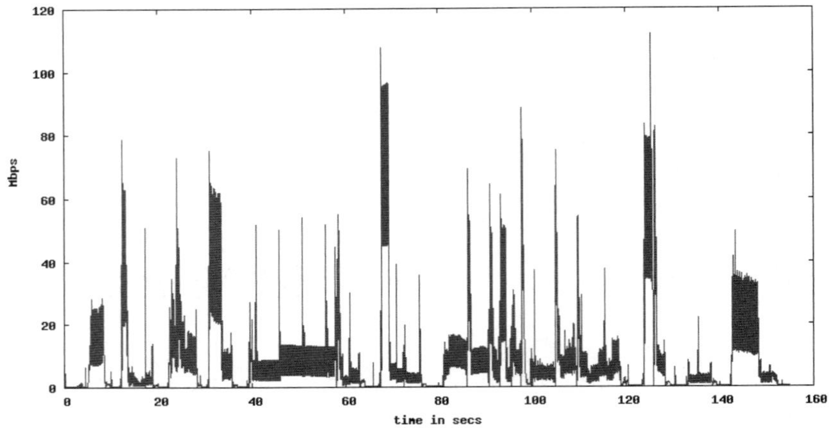

Figure 9.1: Variable bit rates for 3D 720p VC-1 video.

9.2 Basics of Video Smoothing

Our goal is to smooth live video and simultaneously transmit data under a predefined rate required for proper playback. With the widespread use of HD-capable devices, HD videos are favored over SD videos. Thus, we tested several schemes over HD-quality H.264 video traffic. H.264 is chosen because it has become the player of choice in many video streaming Web sites (primarily due to the popularity of Adobe's Flash player, which uses H.264). H.264 is also increasingly supported by desktop and Web-based media players. H.264 is more efficient than MPEG-2 in terms of higher compression gains and hence, higher bandwidth savings. However, this also leads to a higher bit rate variability compared to MPEG-2. This rate variability can sometimes be represented by the coefficient of variability (CoV) of the frame sizes, which is computed as follows. A low CoV will result in a lower range of bit rate requirements, which allows more efficient utilization and allocation of network bandwidth. For a video sequence consisting of M frames encoded with a given quantization level, if X_m ($m = 1, 2, \ldots, M$) denotes the sizes of the encoded video frames, then the CoV of the encoded video is defined as in (5.1). In this case, σ is the standard deviation and $\overline{X}$ is the mean of the frame sizes.

In VBR video smoothing, the startup latency, transmission delay, and buffer size of the client device (e.g., a set-top box or STB) all affect the peak rate of smoothed video transmission. The start-up latency is needed to reduce the peak rate of the initial segment since the first frame of the MPEG compressed video (an I frame) is much larger than the immediately subsequent B and P frames. The most critical factor affecting smoothing performance is the STB buffer size. Large STB buffers may be required since the VBR video can be very lengthy and bursty (both short-term and long-term). VBR video with multiple-time-scale variability cannot be smoothed completely without a very large STB buffer and long start-up latency.

With small STB buffers, smoothed video streams continue to exhibit long-term, slow-time rate variability, as will be demonstrated in Section 9.3. An optimal frame transmission rate is required for both online and offline video smoothing. A lower bound ensures continuous playback at the STB while an upper bound is dictated by the availability of video frames at the video source and the buffer space at the STB. With a proper rate controlled smoothing algorithm, packet losses due to buffer overflows at the STB can be virtually removed.

Online video smoothing is intended for live video transmission and therefore low time-complexity is important. A typical method involves the smoothing of small segments (time slots) of live video. Smoothed video streams can be sent in a piecewise-CBR fashion [1]. Offline smoothing for prerecorded video is ideal for video-on-demand systems. Some examples include work-ahead smoothing where only small buffers are considered (< 1 Mbyte) [2], Enhanced Piecewise Constant Rate Transmission and Transport (e-PCRTT) [3], and Efficient Changes and Variability Bandwidth Allocation (ECVBA) [4]. However, many of these smoothing algorithms cannot satisfy QoS and achieve high utilization at the same time. In addition, the performance of these schemes is evaluated for the older, standard-definition MPEG-1 and MPEG-2 encoded video traffic. Given the maximum bandwidth and limited STB buffer size, many of these smoothing algorithms cannot satisfy QoS and achieve high utilization at the same time. In addition, the performance of these schemes is evaluated for the older, standard-definition MPEG-1 and MPEG-2 encoded video traffic.

A proper rate must be selected. Define the network cost function (c) as:

$$c = R_{\max}(L_{\text{video}} + t_{\text{smooth}}) + B_p - B_b - B_c \tag{9.1}$$

where

R_{max} = maximum rate limit

L_{video} = length of video

t_{smooth} = total smoothing time

B_p = extra bits in pause time

B_b = bits reduced due to B frame deletion

B_c = STB buffer size

R_{max} is selected such that c is minimized without sacrificing video quality. The smoothing time must be minimized while maintaining high buffer utilization.

We further define:

$B(t)$: Maximum cumulative data that can be received by STB over time t (value depends on available STB buffer).

$D(t)$: Cumulative data that must be sent by server over time t (value depends on STB buffer utilization).

C_{max}: Maximum rate at which the server may transmit over a given interval $[a, b]$ without overflowing the STB buffer, starting from an initial buffer level q. C_{max} is the slope (rate) between points on the curves $B(t)$ and $D(t)$ when the STB buffer is full.

$$C_{\max}(t) = \frac{B(t) - [D(a) + q]}{t - a} \tag{9.2}$$

t_B: Latest time at which the STB buffer is full when server transmits at C_{max} over $[a, b]$ starting with an initial buffer level q.

C_{min}: Minimum rate at which the server may transmit over a given interval $[a, b]$ such that the STB buffer never starves, starting from an initial buffer level q. C_{min} is the slope (rate) between points on the curve $D(t)$ when STB buffer is empty.

$$C_{\min}(t) = \frac{D(t) - [D(a) + q]}{t - a} \tag{9.3}$$

t_D: Latest time at which the STB buffer is empty when server transmits at C_{min} over $[a, b]$ starting with initial buffer level q.

Figures 9.2 and 9.3 illustrate the impact of adjusting the different slopes for optimizing the rate limits for different values of C_{max} and C_{min}. t_e and $t_{e'}$ are used to test if the smoothed video segment is going beyond the threshold.

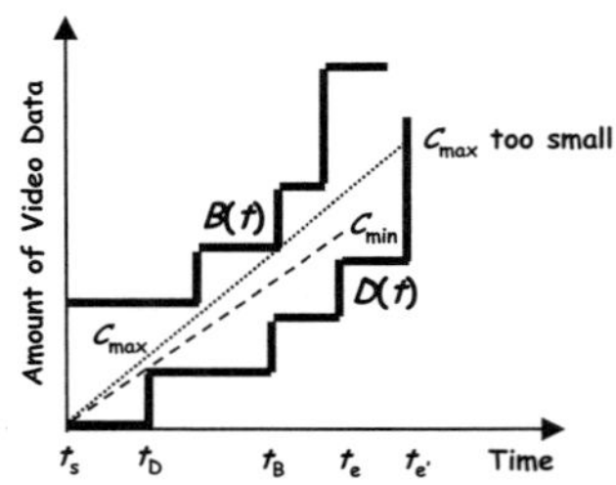

Figure 9.2: C_{max} causes buffer starvation at $t_{e'}$. Therefore, end the segment at t_B and start the next segment at a higher rate.

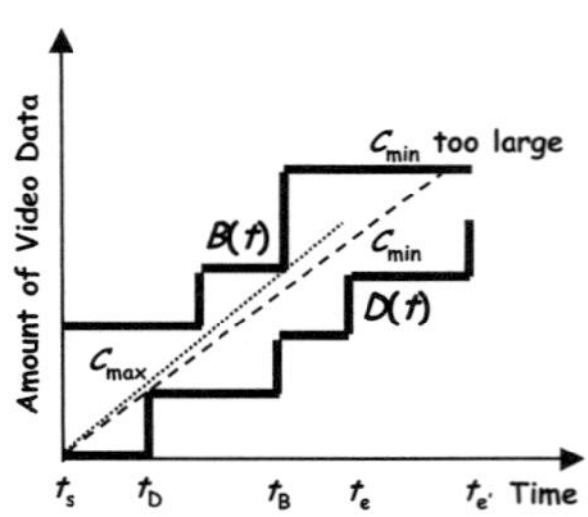

Figure 9.3: C_{min} causes buffer overflow at $t_{e'}$. Therefore, end the segment at t_D and start next segment at lower rate.

9.3 A Video Smoothing Algorithm

Figure 9.4 shows the flowchart of a rate-limited smoothing algorithm. We pause the video transmission whenever the video transmission rate (from the video source) exceeds the rate limit. We employ the trace of the H.264 encoded *Sony Demo* (HD) video of 10,000 frames (~5 minutes) from the Web site http://trace.eas.asu.edu. Similar results are also obtained with other video types. We first present the results for the case without smoothing (Figure 9.5). The CoV is 2.27 in this case. With piecewise smoothing (no rate control) of 500 frames per segment (Figure 9.6), the CoV becomes 0.43. We will demonstrate shortly that this reduction in CoV can be reduced further with rate limited smoothing. Note that piece-wise smoothing is similar in concept to the ABR streaming approach described in Chapter 1.

The network simulator NS2 (http://www.isi.edu/nsnam/ns) is used to evaluate the performance of the rate-controlled smoothing algorithm. We first simulated the algorithm for the MPEG-2 encoded HD video (*Sony Demo*) and then repeated the simulation on the same video that is encoded using H.264. Each video is encapsulated as UDP packets and the transmitting STB sends packets from a video source generating the frames of the trace at a rate of 30 Hz.

Table 9.1 shows the results obtained for the case of the MPEG-2 video with an STB buffer of 4 Mbyte. The pause time and buffer utilization decrease with an increase in rate limit. However, the CoV increases as the rate limit increases. The network cost as defined in (9.1) is minimized at a 2.4 Mbps rate limit.

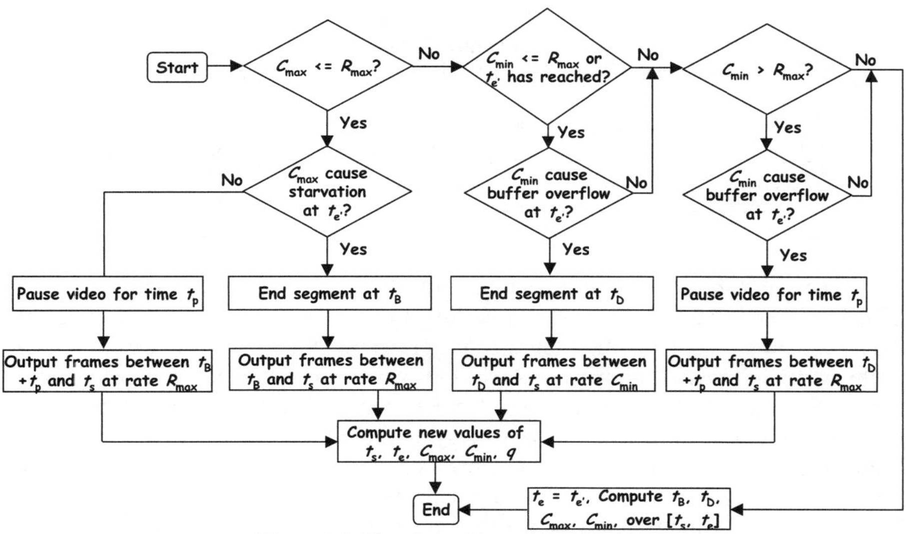

Figure 9.4: Flowchart of smoothing algorithm.

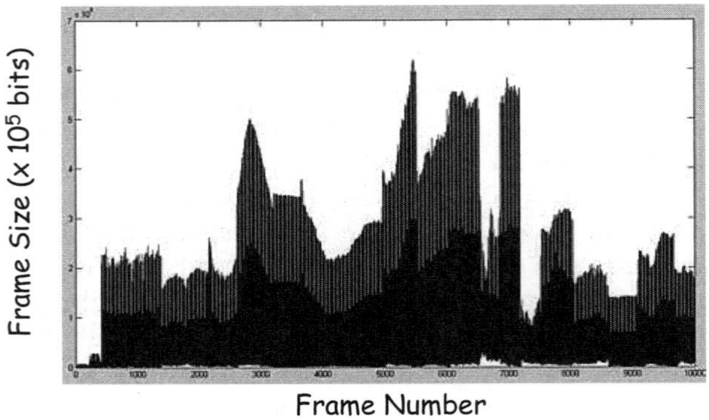

Figure 9.5: Original H.264 HD video trace (*Sony Demo*).

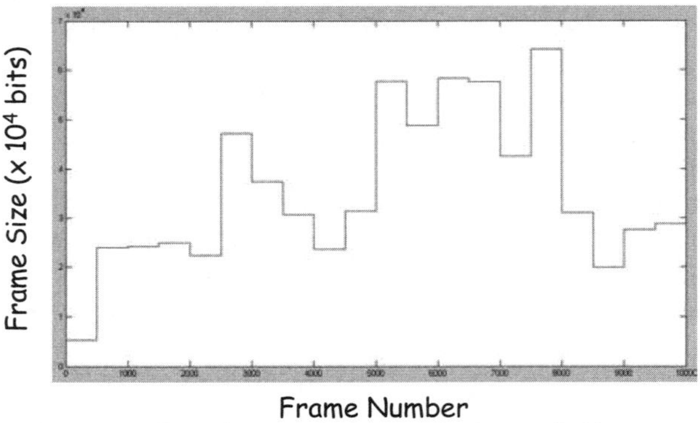

Figure 9.6: Piecewise-smoothed H.264 *Sony Demo* (HD) trace.

Table 9.1: MPEG-2 HD Video Smoothing with Rate Control (4-Mbyte STB Buffer)

Rate Limit (Mbps)	CoV	Pause Time (s)	Buffer Utilization (%)
1.8	0.0106	89.3	99.9
2.4	0.1239	18.6	90.0
3	0.1749	0	76.1

Table 9.2 shows the results obtained for the case of the H.264 video with an STB buffer of 1 Mbyte. The lower rate limits compared to the MPEG-2 video is due to the higher H.264 compression gains. However, the CoV is higher compared to the MPEG-2 video, although it is still lower compared to piecewise smoothing. The pause time and buffer utilization decrease with an increase in rate limit. This is because the probability that the transmission rate of the video exceeding the rate limit decreases. In this case, the network cost is minimized at a 0.9 Mbps rate limit. Table 9.3 shows the results obtained for the case of the H.264 video with an STB buffer of 2 Mbyte. The CoV is reduced compared to the 1-Mbyte buffer case. The network cost is minimized at a 0.9 Mbps rate limit. Table 9.4 shows the results obtained for the case of the H.264 video with an STB buffer of 4 Mbyte. The CoV is comparable to the 2-Mbyte buffer case. We observe a reduction in rate limit since the pause time reached 0 seconds when the rate limit was 1.2 Mbps.

With 1- and 2-Mbyte buffers, the pause time reached 0 seconds when the rate was 1.65 Mbps. This shows that with a higher buffer capacity at the STB, a lower rate limit can be set, which translates to lower bandwidth requirements for the video transmission. Table 9.4 also shows the highest buffer utilization. Furthermore, for a pause time of 0 seconds, the buffer utilization was higher when compared to the 1- and 2-MB buffers, *and* the MPEG-2 video. However, this case incurs a higher network cost; network cost is minimized at a 1.05 Mbps rate limit, 0.15 Mbps higher than the previous cases.

Table 9.2: H.264 HD Video Smoothing with Rate Control (1-Mbyte STB Buffer)

Rate Limit (Mbps)	CoV	Pause Time (s)	Buffer Utilization (%)
0.75	0.030634	139.83	99.63
0.9	0.071577	81.13	94.81
1.05	0.140538	50.06	87.85
1.2	0.205606	26.76	81.83
1.35	0.2646	8.7	76.58
1.5	0.2939	0.9	70.54
1.65	0.2992	0	64.31

Table 9.3: H.264 HD Video Smoothing with Rate Control (2-Mbyte STB Buffer)

Rate Limit (Mbps)	CoV	Pause Time (s)	Buffer Utilization (%)
0.75	0.0016	138.43	99
0.9	0.0646	80.96	94.8
1.05	0.1372	49.76	87.92
1.2	0.2034	26.35	81.93
1.35	0.2639	8.8	76
1.5	0.2906	0.86	70.55
1.65	0.2959	0	64.3

Table 9.4: H.264 HD Video Smoothing with Rate Control (4-Mbyte STB Buffer)

Rate Limit (Mbps)	CoV	Pause Time (s)	Buffer Utilization (%)
0.75	0.0017	138	99.93
0.9	0.0558	68.13	97.89
1.05	0.1061	20.06	95.32
1.2	0.1310	0	88.43

9.4 Live HD Video Streaming

The typical raw streaming model sends packets as quickly as possible. This is used for example when watching video clips from Fox Sports online video portal (http://msn.foxsports.com/video). TCP streaming takes advantage of TCP's rate control mechanism to define the quickest sending rate. To provide more fine-grained control of the video stream, one can define a "progressive" download mechanism. A progressive download can be activated in Flash servers and is widely used in YouTube video streaming with a higher burst at the beginning to reduce the startup delay for playback. Interestingly, Lala, an online music portal, also adopts a similar progressive download model for streaming music. We can compute the average streaming rate as:

$$S = \frac{F - H}{P} = \frac{A + V}{P} \tag{9.4}$$

where
A = Audio file size
F = File size
H = Container (e.g., MP4, TS, and so forth) header size
P = Playback time
S = Streaming rate
V = Video file size

This ensures a constant streaming rate and shows good performance for CBR videos. With VBR videos, this is more of a problem, since fast moving scenes (such as an action scene) will have more than the average number of frames. To counter this, one approach would be to queue frames for a few scenes in a buffer before playing them. However, determining a good buffer size is not easy beforehand, especially with 1080p HD videos with large frame sizes, and this will also increase the start playback delay.

To deal with the problem of delayed frames, we refine our approach, where we stream a file at the same frame rate that the video is meant to be played at. Typical frame rates are 25 Hz and 30 Hz. Instead of sending a file based on file or frame sizes, we now send a file based on the number of frames in order to maintain the natural frame rate at the receiver. While it is possible to employ this for variable frame rate (VFR) videos, we will primarily consider fixed frame rate videos. We take advantage of this frame rate to send frames. We aggregate all the packets to be played over a small time interval, and send the combined stream to the client. The individual frame sizes are unlikely to be similar, thus, there is a need to ensure that outgoing streams are in the same range of network packet sizes, at the expense of sending a different number of frames.

To aim for a range of sending rates, we employ the following algorithm:

1. Decide on range of streaming rates before transmission.
2. Accumulate frames at a rate equal or slightly greater than the frame rate.
3. If sending these packets exceeds or trails the limits of the range, add another packet or remove the last packet added, respectively.
4. Repeat step 3 until we are within the range or overshoot the range.
5. If the frame sizes are such that adding or removing a packet would overshoot or undershoot the range, we choose a value closer to the mean of the limits.

The above algorithm approximates achieving a fixed streaming rate while still sending entire frames in a timely fashion. We tested the algorithm by sending a fixed number of frames every fixed time interval to maintain a constant frame rate at the receiver. We set up a Windows 7 server and all 1080p videos are streamed using the MP4 container. We wrote our own software to perform streaming. We used the Win32 API to utilize Windows' TCP/IP protocol suite.

Streaming was performed over TCP using the HTTP protocol. To read the MP4 files, we used the libav* libraries from m-player, which are part of the FFmpeg software package (http://ffmpeg.org). We implemented our scheduling algorithm as described in Sections 9.4.1 to 9.4.3. We noted sending times and rates for streaming media using the raw sending, average bit rate streaming model, and the per-frame streaming model. We employ the 1080p computer-animated video, *Big Buck Bunny* (*BBB*), for streaming. The server was connected to the Internet over a 802.11g wireless network operating at 54 Mbps. We use the following assumptions in modeling our software:

- Once a user has started playing a streaming file (either from the beginning or somewhere in the middle), he or she will not pause or skip to a later segment in the video (i.e., no trick modes).
- The HD video files stored at the server contain the header, an audio stream, and a video stream but no subtitles, menus, or other textual content.

We also compare the performance of the native Windows TCP with the Compound TCP (CTCP) enhancement [5]. CTCP is the next generation TCP/IP stack that optimizes the sender-side throughput. Together with receive window autotuning, CTCP can increase link utilization and improve performance for connections with large bandwidth-delay products. It is therefore optimized for TCP connections with large receive window sizes. It aggressively increases the amount of data sent at a time, yet ensures that its behavior does not negatively impact other TCP connections. CTCP introduces a delay-based component in addition to the standard TCP loss component to efficiently detect and use spare bandwidth that may be available on a high speed and long delay network. In addition, the delay-based component detects the onset of congestion early and gracefully reduces the sending rate. The component ensures fairness, even in the presence of differing RTTs. The loss-based component, on the other hand, ensures there is effective response to losses in the network while in the absence of losses, it keeps the throughput of CTCP lower bounded by TCP Reno. Thus, CTCP is not timid, nor induces more self-induced packet losses than the standard TCP flow.

9.4.1 Raw Streaming

In this case, the server does not control the streaming rate. The operating system's native TCP algorithm streams the video across the Internet as quickly as possible. Under ideal network conditions, the file transfer can be transmitted very quickly. In our case, the 105-Mbyte, 10-minute *BBB* video takes little more than a minute to stream completely (Figure 9.7). We note the average streaming rate is about 14 Mbps. We shall use this figure to benchmark our other streaming models.

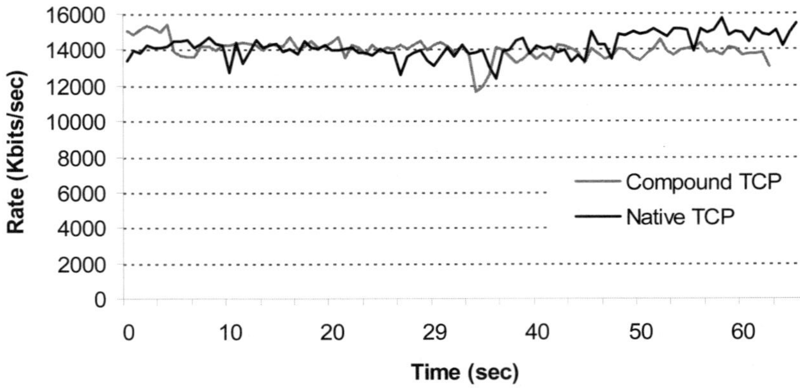

Figure 9.7: Raw streaming rates.

9.4.2 Progressive Streaming

With progressive streaming, a finer level of control over the streaming rate can be achieved (Figure 9.8). For smooth, continuous playback at the receiver end, we require an average streaming rate of about 180 Kbytes/sec for the BBB video (using 9.5). Variation from the expected rate is small compared to the other streaming models. However, this is not a dynamic rate, due to possible packet losses and retransmissions. In addition, sometimes the sender will have to buffer incoming packets while at other times, will have to wait for the remainder of a video frame to arrive.

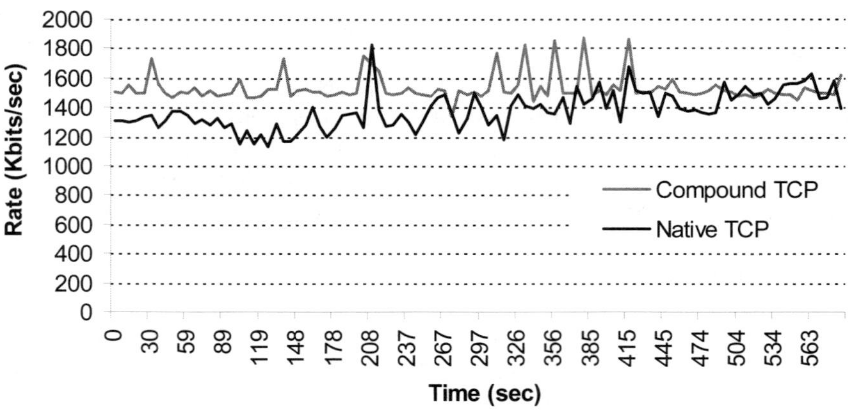

Figure 9.8: Progressive streaming rates.

9.4.3 Frame Smoothed Streaming

With frame-level smoothing, frames are sent at a rate equal to the frame rate of the video (Figure 9.9). We interleave the audio stream along with the video frames.

We observe that the streaming rate oscillates about our earlier calculated rate of 180 Kbytes/sec. We notice a greater variation in the streaming rate here, when compared to progressive streaming, but it is still considerably lower than the raw streaming. The higher variation is due to the fact that the frames sent correspond to the actual frame sizes. Larger frame sizes equate to greater streaming rates. The lower rates indicate a smaller frame size was transmitted. CTCP takes a longer time to complete the sending of the file than traditional TCP. However, the measured CTCP's performance shows less bit rate variation than regular TCP. CTCP starts out more aggressive than traditional TCP, and gradually slows down, whereas with native TCP, the starting rate is smaller (due to native TCP's slow start congestion algorithm). By comparing Tables 9.5 and 9.6, it is clear that CTCP results in a lower standard deviation in the bit rates for all streaming methods and achieves better overall TCP performance when streaming HD videos.

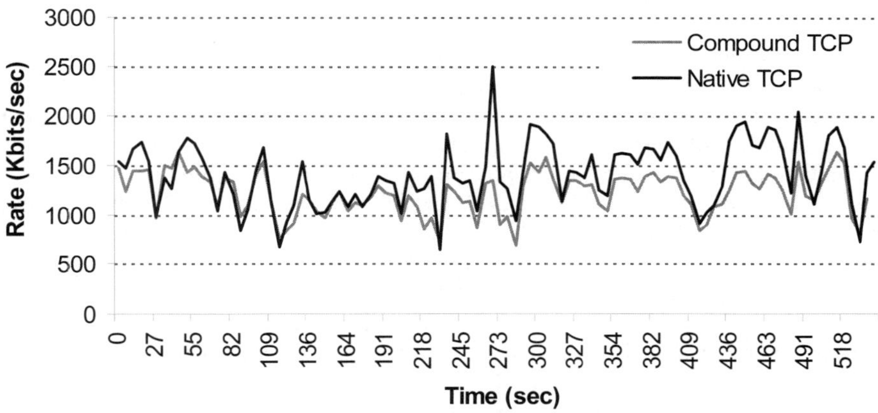

Figure 9.9: Frame smoothed streaming rates.

Table 9.5: Mean and Standard Deviation for Streaming Modes under Native TCP

Streaming Mode	Mean (Kbps)	Standard Deviation (Kbps)
Raw TCP	14189.95	623.45
Progressive	1379.23	124.53
Frame smoothed	1408.50	327.64

Table 9.6: Mean and Standard Deviation for Streaming Modes under CTCP

Streaming Mode	Mean (Kbps)	Standard Deviation (Kbps)
Raw TCP	13987.79	576.89
Progressive	1528.41	90.23
Frame smoothed	1221.74	215.63

9.5 Impact of Player's Buffer Size

Figure 9.10 shows the impact of the player's buffer size on the TCP streaming throughput for the 72.2s FCL 720p H.264 video over a Wi-Fi wireless LAN (54

Mbps). A larger buffer of 8 Mbytes allows more information to be stored, hence the transmission can be completed at an earlier time than the case with a 1-Mbyte buffer. However, the peak rates have to be managed. Note that the advertised receive window size may throttle the throughput to a smaller value when the buffer has received sufficient information. This is because the sender will take the minimum of the congestion window (which attempts to avoid congestion) and the receive window, when deciding on the appropriate window size to use. When some of the frames are played out, more buffer space is released, and the receive window will increase again. Figure 9.11 shows the impact of shaping the streaming throughput using a rate limit of 1.5 Mbps. As expected, the video file is received later than the unshaped case but the player with a larger buffer size achieves more efficient bandwidth utilization and receives the video earlier than the player with a smaller buffer size.

9.6 Impact of Transport Protocols

The performance of traditional TCP and TCP with SACK over a 54 Mbps WLAN is shown in Figures 9.12 to 9.14. The player's buffer size was set to 600 Kbyte. The theoretical average rate is computed by taking the video file size and dividing by the duration of the video. The initial rate is usually high because the player attempts to buffer as much data as possible for playback (including the first frame, a large I frame), and thus advertises a large receive window size. Subsequently, the advertised window gets throttled to a more reasonable value. However, in the case of the *Avatar* 720p 3D movie, the initial high rate can be attributed to the nature of the 3D video content (two 720p frames instead of one) and the player. Table 9.7 shows that TCP with SACK reduces the average streaming rate. Recall from Chapter 2 that TCP with SACK employs aggregated or block ACKs, which reduces the overheads of sending individual ACKs and enables multiple packet losses to be handled more efficiently.

Table 9.7: Overhead Comparison for TCP (Average Rates in Mbps)

	72.2s *FCL* 720p	**72.2s *FCL* 1080p**	**208s *Avatar* 720p 3D**
TCP with SACK	1.147	1.956	1.502
TCP without SACK	1.182	1.959	1.517
Theoretical	1.079	1.838	1.391

Figures 9.15 and 9.16 show the streaming of the *Avatar* 3D trailer using UDP. In general, UDP incurs less overheads than TCP since it is a unidirectional protocol. Unlike TCP, there is no congestion flow control in UDP, hence the raw UDP sending rates can be very high (a few orders of magnitude higher than TCP) and the video gets transferred in a very short duration. This may result in unacceptably high loss rates due to network congestion and receiver buffer overflow. These losses cannot be recovered in native UDP. A solution to manage this problem is to send the video at the frame rate that the video is meant to be played back (i.e., 30 Hz). The resulting sending rates are shown in Figure 9.15. Alternatively, progressive streaming can be employed to shape the rate variation (Figure 9.16).

In both cases, the entire video gets transferred according to its duration (208s). For progressive streaming, an average rate of 1.434 Mbps is achieved, thus showing that the UDP overheads are lower compared to TCP (Table 9.7). Figure 9.17 shows the UDP rates (computed every 1s) versus the theoretical rates computed by taking the size of each video frame and dividing by the natural frame interval (33 ms). Since audio, video, and UDP overheads are included in the UDP rates, the UDP rate is higher (1.434 Mbps versus 1.259 Mbps average rate). As can be seen, there is a close correlation between the natural frame rate of the video and the actual UDP rates. Note that the peak to average rate of the video is high.

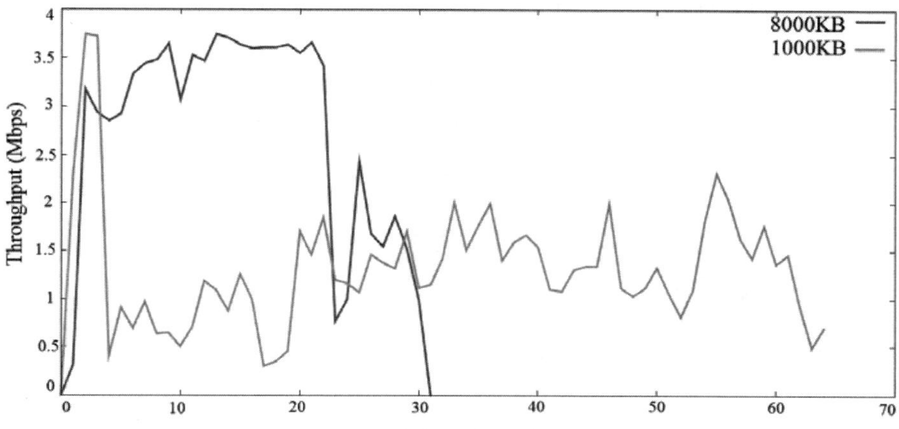

Figure 9.10: Impact of player buffer size on TCP streaming throughput.

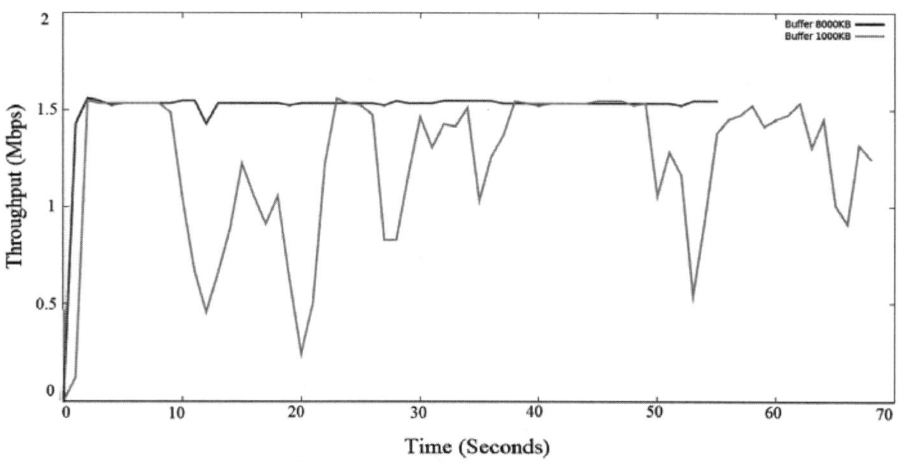

Figure 9.11: Impact of buffer space and shaping on TCP streaming throughput.

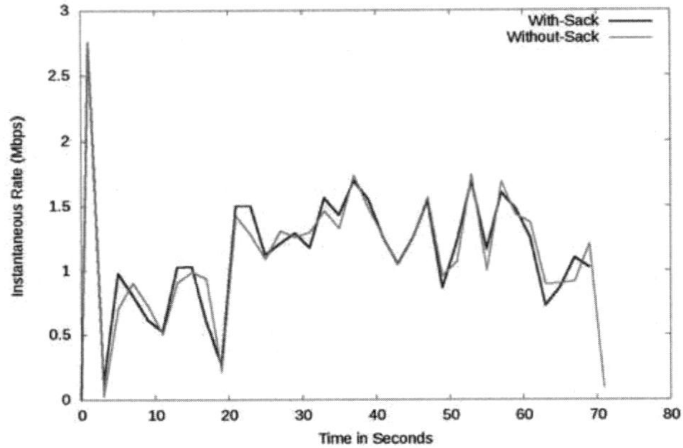

Figure 9.12: Instantaneous TCP rates for streaming the FCL 720p H.264 video.

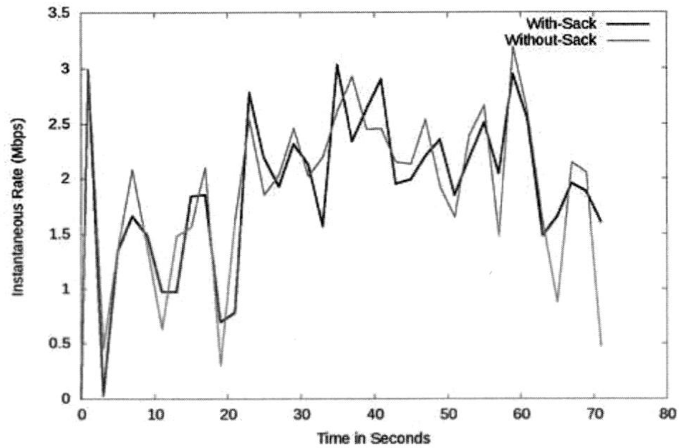

Figure 9.13: Instantaneous TCP rates for streaming the FCL 1080p H.264 video.

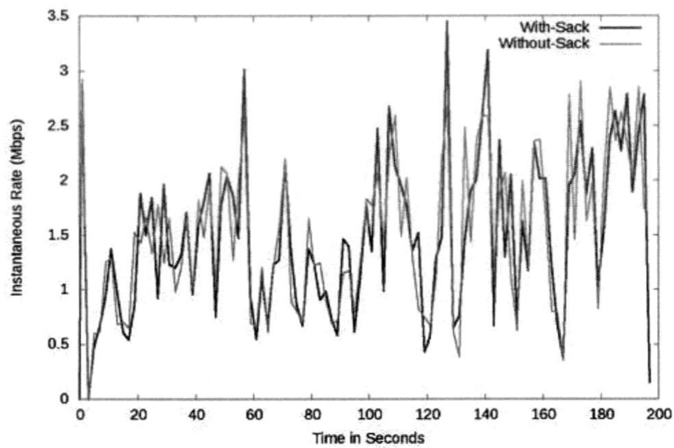

Figure 9.14: Instantaneous TCP rates for streaming the *Avatar* 720p H.264 3D video.

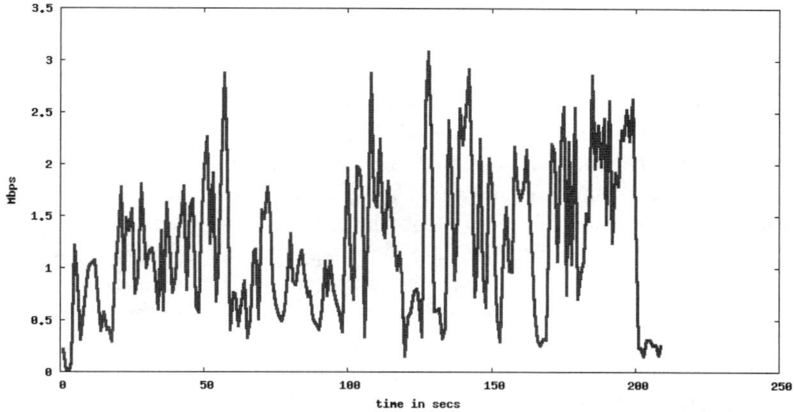

Figure 9.15: Instantaneous UDP rates for streaming the *Avatar* 720p H.264 3D video at 30 Hz.

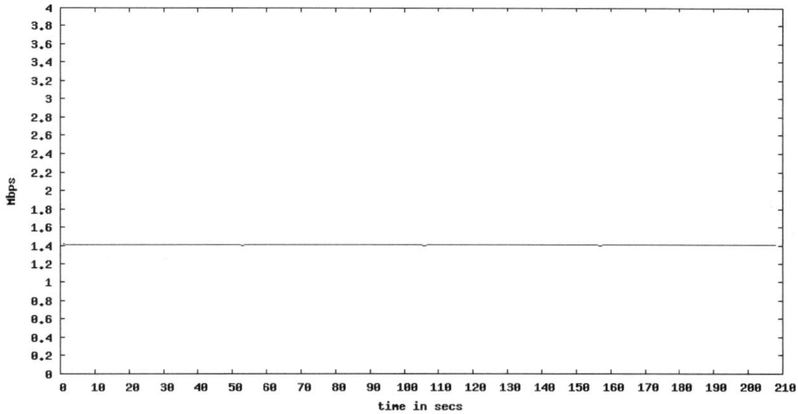

Figure 9.16: Shaped UDP rates for streaming the *Avatar* 720p H.264 3D video.

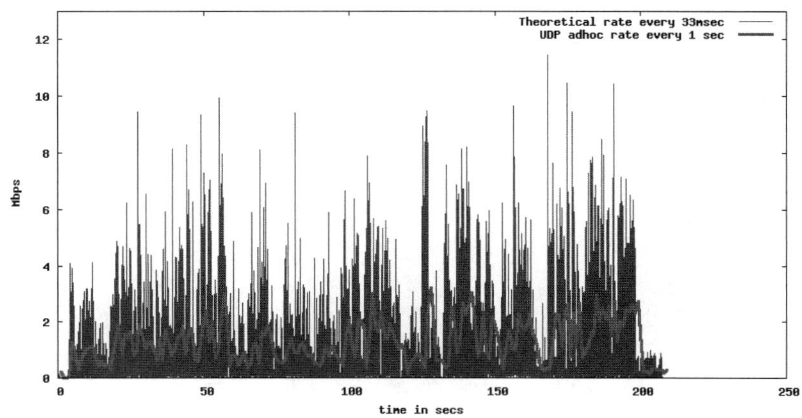

Figure 9.17: UDP rates measured every 1s versus the theoretical video rate measured every 33 ms.

9.7 Peak to Average Rate (PAR)

The PAR normalizes the actual variation of the VBR video rates. This is computed by selecting the instantaneous rate of the compressed video and dividing by the average rate within a predefined interval. For example, if the video frame rate is 25 Hz, then the frame interval is 40 ms. The size of the frame divided by the frame interval gives the instantaneous rate. The choice of the predefined interval is important. If the selected duration is too short, then it may not accurately reflect the actual variation of the rates as the video is played back. In the limiting case, the predefined interval is the same as the instantaneous interval, giving a PAR of 1. On the other hand, if the predefined interval is chosen to be the duration of the entire video, and the instantaneous rate for each frame interval is averaged over this period, this gives the most accurate PAR because it correlates perfectly with the instantaneous video rates. We will call this the long-term average PAR.

The impact of the choice of the predefined interval is illustrated in Figures 9.18 to 9.26, which use the *Dark Knight* 720p H.264 movie trailer. The 100-ms, 1s, and 10s moving average PARs are always greater than 1 and unlike the long-term average PAR, do not correlate well with the instantaneous video rates. The static average PAR is computed by first dividing the entire video into equal segments (e.g., 10s, 20s, 30s). The PAR is then computed by taking the ratio of the instantaneous rate and the average rate in these segments. As can be seen, a longer segment (e.g., 40s) results in a static average PAR that approaches the long-term average PAR. Although this is usually the case for most videos, it is highly dependent on the video content. Note that in Figure 9.18, the peak rate for the 720p VBR H.264 video is nearly 28 Mbps. Shaping the traffic to a lower rate requires additional buffering delay that has to be accommodated. As shown in Figure 9.27, choosing a low shaping threshold (fixed rate limit) can incur significant delay. However, moderate shaping results in minimal delay.

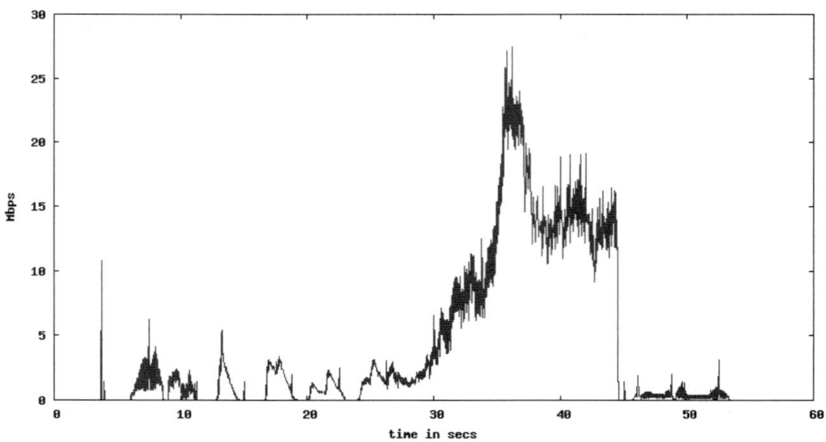

Figure 9.18: Instantaneous rates for a 720p H.264 video.

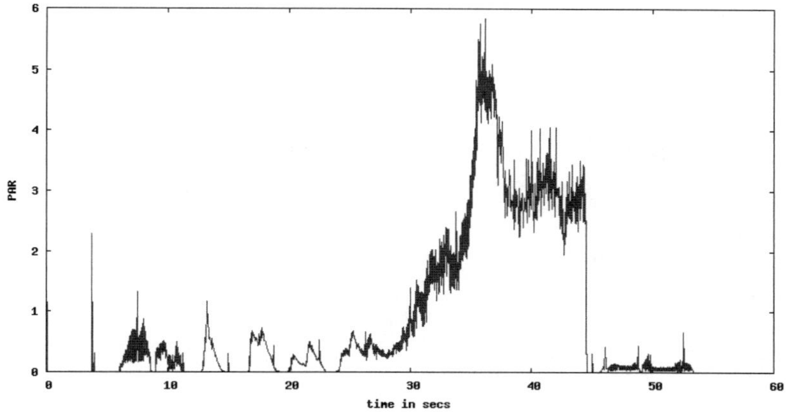

Figure 9.19: Long-term average PAR for a 720p H.264 video.

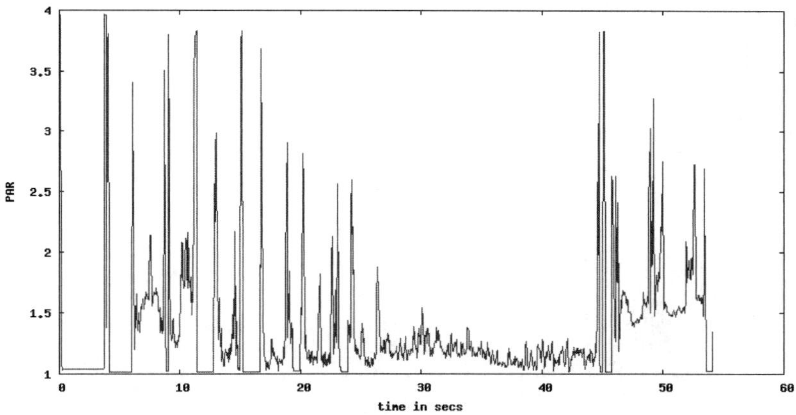

Figure 9.20: 100-ms moving average PAR for a 720p H.264 video.

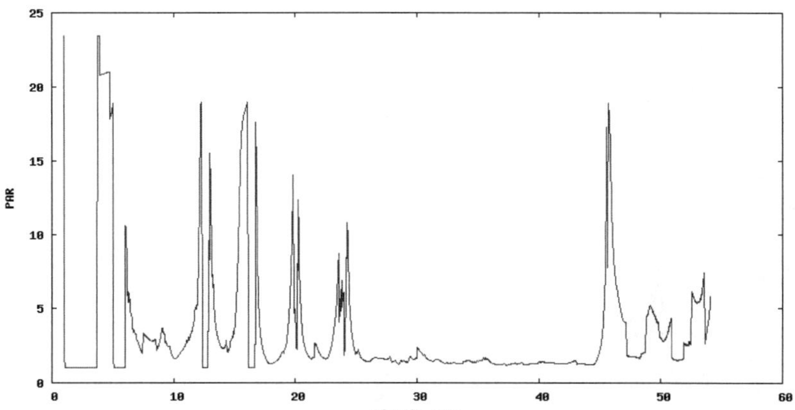

Figure 9.21: 1s moving average PAR for a 720p H.264 video.

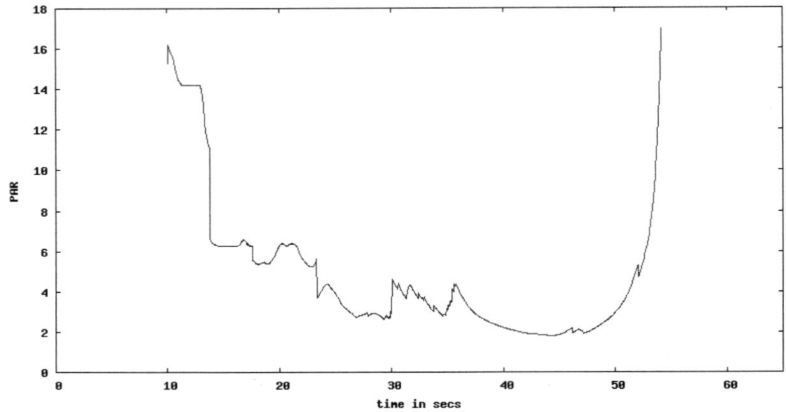

Figure 9.22: 10s moving average PAR for a 720p H.264 video.

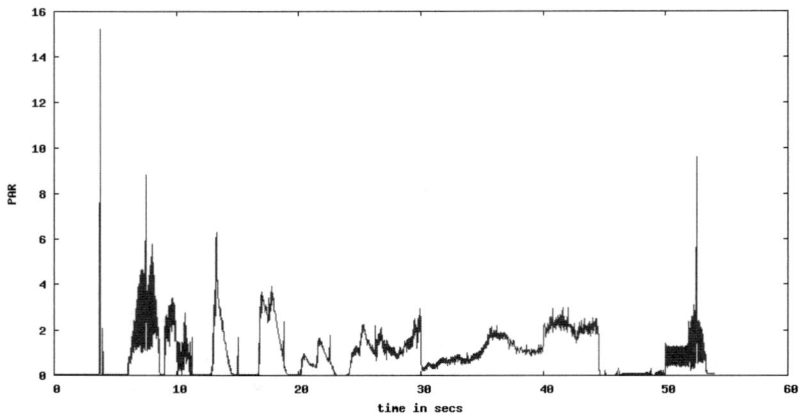

Figure 9.23: 10s static average PAR for a 720p H.264 video.

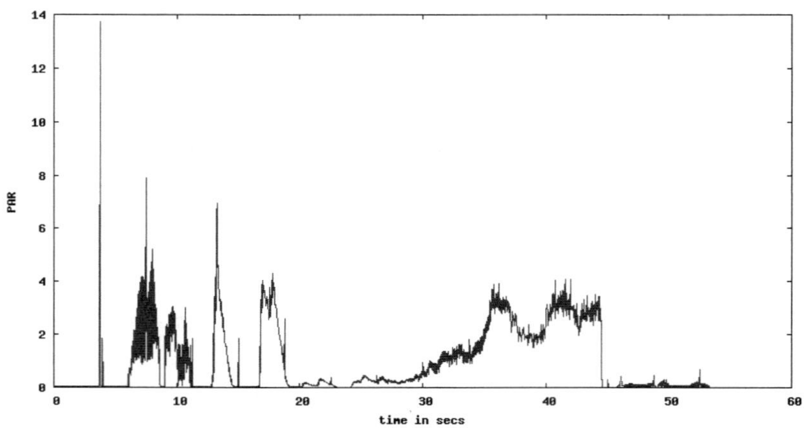

Figure 9.24: 20s static average PAR for a 720p H.264 video.

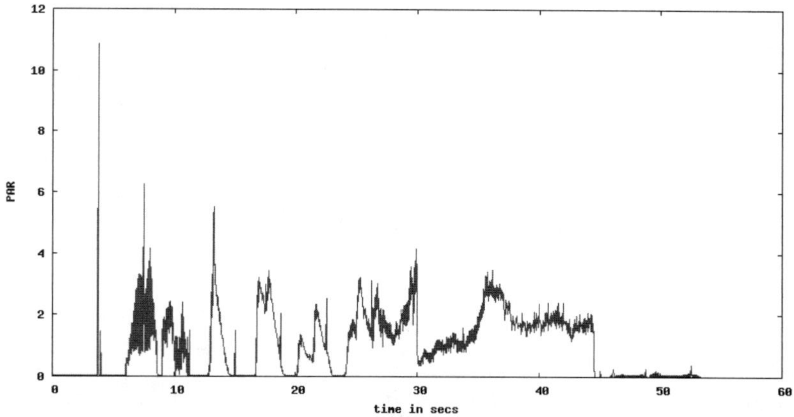

Figure 9.25: 30s static average PAR for a 720p H.264 video.

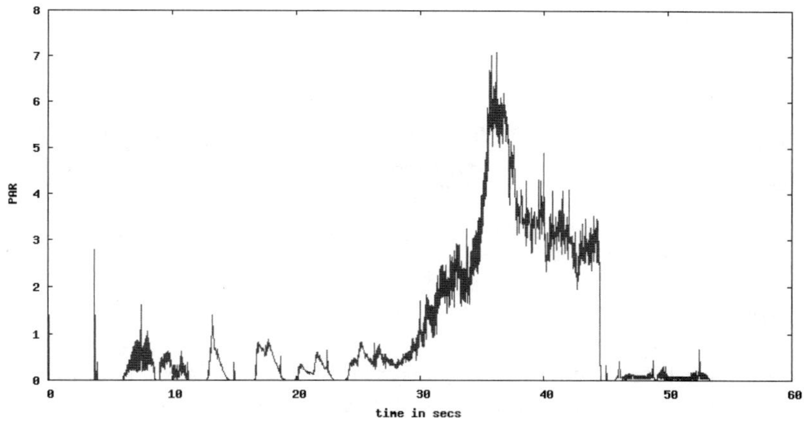

Figure 9.26: 30s static average PAR for a 720p H.264 video.

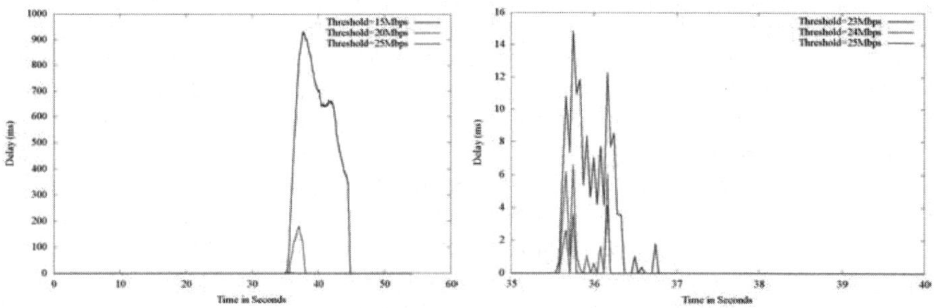

Figure 9.27: Shaping thresholds versus buffering delay.

The variation of the long-term average PAR is very high. A PAR of 1 is desirable because it achieves perfect utilization of the link bandwidth. A PAR value below 1 implies underutilization whereas a PAR greater than 1 requires more buffering or bandwidth to accommodate the peak rates of the VBR video. By appropriately shaping the video traffic, a PAR close to 1 can be achieved. Figure 9.28 shows the streaming of a *Vampire* 720p H.264 video with a fixed shaping threshold. A PAR close to 1 is achieved. Since the video was streamed in real-time using TCP and the duration of the video was not known beforehand, a running average method was used to compute the PAR. The running average is computed by averaging the instantaneous rates since the beginning of the video play back. Smoothing the video stream will further reduce the PAR variation.

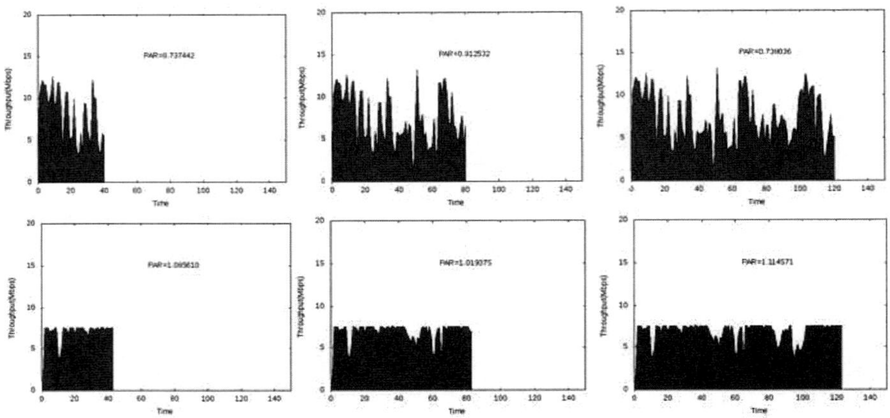

Figure 9.28: Evolution of the running average PAR for live TCP streaming of an unshaped (top row) and shaped (bottom row) 720p H.264 video.

9.8 Multiplexing of Composite VBR Videos

So far we have focused on a single coded video stream. It is common for a video headend or server to deal with multiple streams. Statistical multiplexing is a real-time decision based on the number of clients currently subscribing to the video channel or movie. When applied to composite VBR compressed videos at the headend or server, it can exploit the inherent variations in the instantaneous bit rates and increase the number of video streams within a fixed channel bandwidth while keeping the picture quality constant. For example, if one stream is demanding high bit rate, it is likely that other streams have capacity to spare. Thus, a large number of aggregated streams may tend to "smooth" to a normal distribution (based on the central limit theorem). Unlike per stream buffer-based smoothing and traffic shaping, statistical multiplexing introduces minimal delay.

The data rates for a MPEG stream can vary quite dramatically depending on the video content. For example, the data rate for a high-action sports program can be several times higher than a news reader because the picture is still most of the time. As shown in Figure 9.29, the video resolution also plays an important part in the frame size distribution (and hence the data rates). The 480p Dell video contains very fast scene changes and the peak of the frame sizes occurs in the 170 Kbyte range. The 1440 × 1080 *Terminator 2* trailer is fast action and the peak of the frame sizes occurs in the 340 Kbyte range. We compare these videos with the slower motion 1080p FCL movie (only five scene changes) where the peak of the frame sizes occurs in the 480-Kbyte range. This video exhibits the broadest range of frame sizes. Thus, the effect of video content (e.g., fast motion versus slow motion) may only be relevant for videos with the same resolution. However, there is no strong correlation between the frame size distribution and the video duration: *Dell* (75.3s), *Terminator 2* (125s), *FCL* (72.2s).

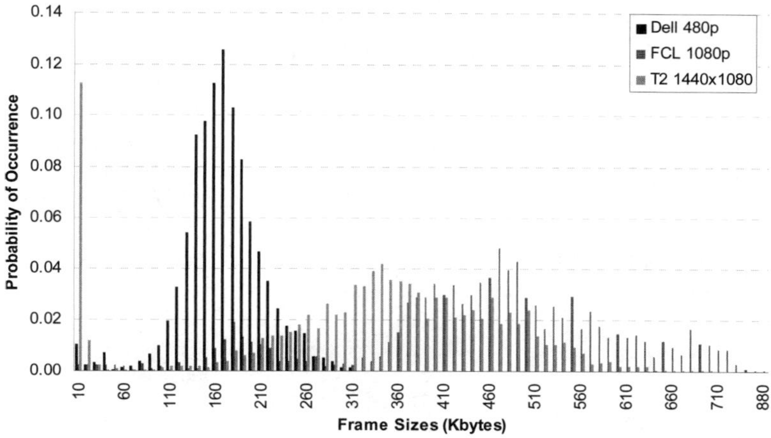

Figure 9.29: Frame size variability of H.264 videos.

In a typical multiplexing setup, each encoder assesses the video quality and complexity and outputs the compressed information to statistical multiplexer. The statistical multiplexer acts as bit rate arbiter and informs each encoder the available bit rate over a time period. An example of the bandwidth efficiencies that can be achieved using CBR and VBR 720p video multiplexing is shown in Figure 9.30. With channel bonding, higher efficiencies tend to be possible but not guaranteed. This is because with more streams, the standard deviation of the overall instantaneous rate becomes higher (as we will see shortly). To attain the same shaping delay as the lower number of streams, some bandwidth efficiency must be sacrificed. Alternatively, more aggressive shaping can be employed but this results in higher delays, which in turn, increases channel change latency.

Many coded videos exhibit long-range dependency (LRD) or long-tail traffic characteristic. This can be proved by computing the Hurst parameter, which is shown to be consistently greater than 0.5 and close to 1 in Table 9.8. An aggregation level of a means that the sizes of a consecutive video frames are averaged. A Hurst parameter close to 1 indicates very strong long-term dependency, and in general, this appears to be the case for videos encoded in high quality (i.e., low QP values). Because of this dependency, video traffic tends towards clustering and the overall bandwidth requirements become less predictable as the number of streams increases (which is in contrast to Poisson or completely random distributions that become smoother as the aggregation volume increases). To illustrate this phenomenon, we multiplex 19 and 38 720p H.264 movie trailers with different (uncorrelated) content but with the same frame rate (24 Hz). These videos are encoded with low QP values (9 to 20). As shown in Table 9.9 and in Figures 9.31 and 9.32, the standard deviation of the composite rates is increased almost two fold when the number of multiplexed videos is doubled, thereby proving the increased variability for a higher number of multiplexed streams. Thus, for multiplexed video streams, the buffers need to be larger to accommodate more extreme traffic-burst scenarios and traffic shaping may be needed. However, the standard deviation of the long-term average PAR reduces. This is because the composite average rate for 38 streams is larger than 19 streams, and this in turn, masks the effect of the overall variation. The average rate per stream remains about the same for the 19 and 38 streams—4.7 Mbps. The composite average rate for 38 streams is 178.555 Mbps. Using a shaping threshold of 193 Mbps (Figure 9.33) results in a peak delay of less than 1s for the composite stream. This gives an average rate of 5.1 Mbps per stream, far lower than the 28 Mbps peak rate in Figure 9.18. Similarly, for 19 streams (Figure 9.34), a threshold of 101 Mbps results in a delay comparable to 38 streams (i.e., less than 1s), giving an average rate of 5.3 Mbps per stream.

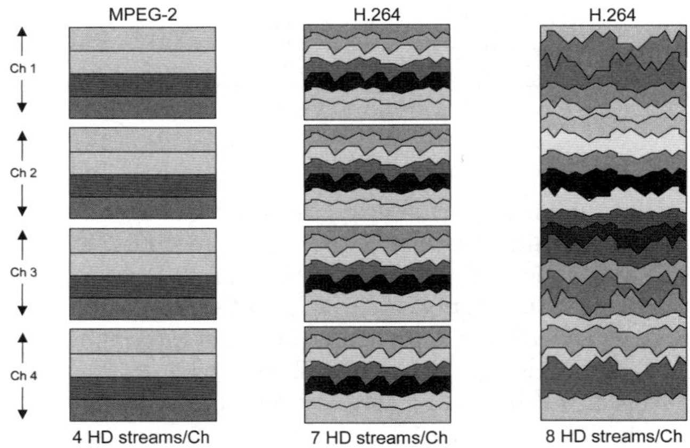

Figure 9.30: Efficiencies of 720p video multiplexing with and without channel bonding.

Table 9.8: Hurst Parameter Values for *Terminator 2* HD H.264-Encoded Video

	Aggregation Level						
Quality	1	2	3	12	24	48	96
QP=10	1.0173	1.0225	0.948	0.8485	0.8535	0.8093	0.8395
QP=28	0.9347	0.9783	0.9973	0.8424	0.8201	0.7535	0.7136
QP=34	0.9032	0.9559	0.9861	0.8213	0.7675	0.7218	0.6761
QP=48	0.8650	0.9361	0.9730	0.8184	0.7870	0.7971	0.7469

Table 9.9: Standard Deviation for Instantaneous Rates and PAR of Multiplexed Videos

Number of Streams	**Instantaneous Rate (Mbps)**		**Long-Term Average PAR**	
	0 to 52s	**10 to 52s**	**0 to 52s**	**10 to 52s**
19 streams	28.7967	18.5273	0.300595	0.193398
38 streams	50.3912	28.9135	0.250105	0.143505

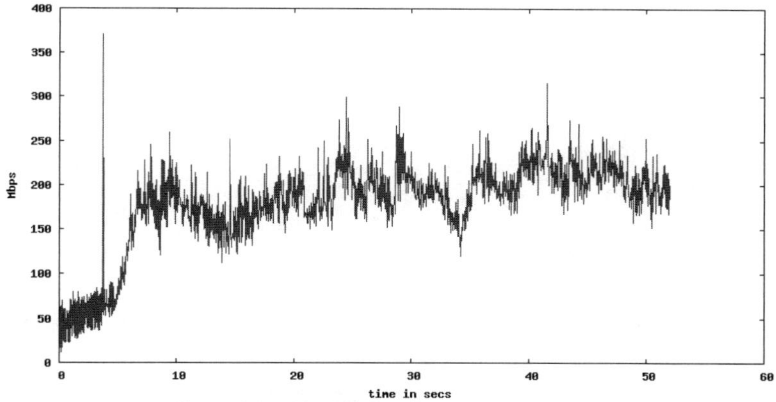

Figure 9.31: 38 multiplexed 720p H.264 videos.

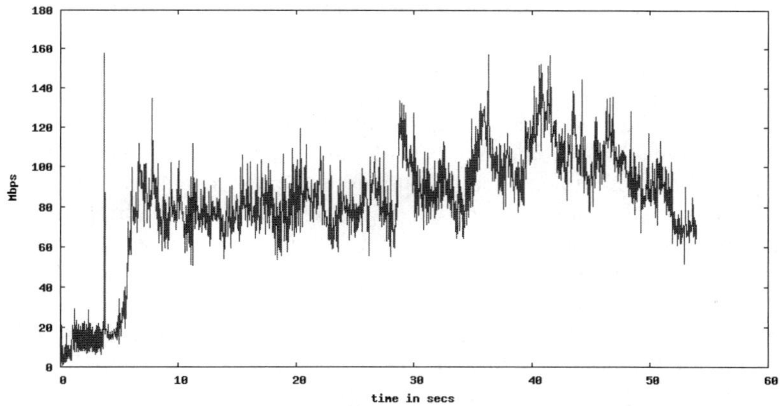

Figure 9.32: 19 multiplexed 720p H.264 videos.

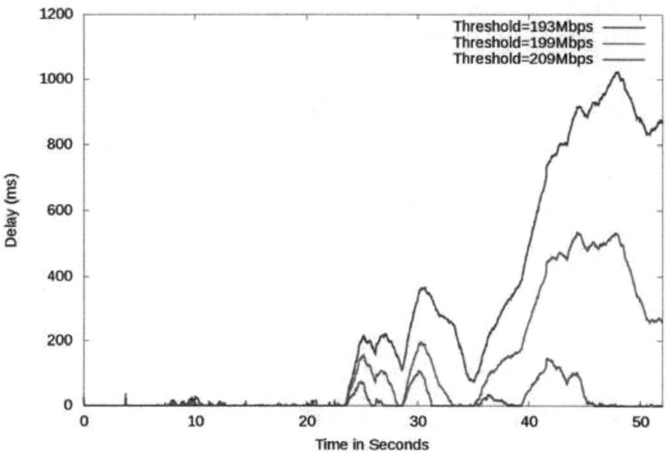

Figure 9.33: Shaping threshold and delay tradeoff for 38 multiplexed 720p H.264 videos.

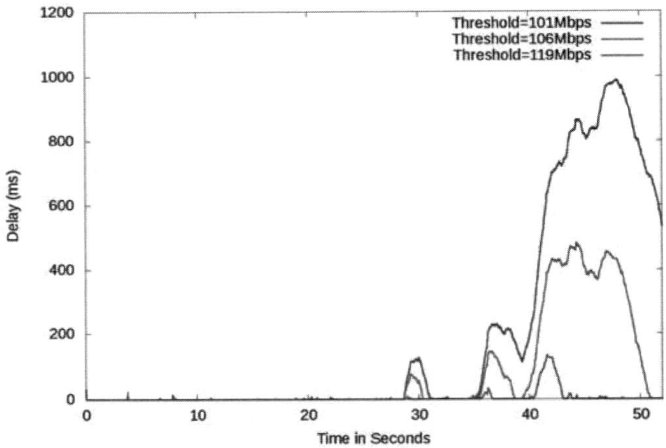

Figure 9.34: Shaping threshold and delay tradeoff for 19 multiplexed 720p H.264 videos.

To demonstrate the potential gains of statistical multiplexing a time-displaced video, we multiplex the same H.264-encoded video (*Foreman*, CIF resolution) via a sequence of client requests submitted at different times. Hence, during some interval of transmission, the number of concurrent videos sent to the clients overlap, and the multiplexing algorithm is activated to achieve the bandwidth gains. The statistics are summarized in Table 9.10. The multiplexing gain is substantial at higher transmission rates and with a larger number of videos. As an example, for the lossless case, the average bandwidth per video is 12 Mbps/10 or 1.2 Mbps when 10 videos are multiplexed, as opposed to 1.5 Mbps when 4 or 5 videos are multiplexed. The loss percentage indicates the proportion of video data (out of the total) that arrived late for playback, and hence could not be displayed. The startup latency is the time required between the start of transmission and the beginning of playback at the client side. As the bandwidth available increases, the loss decreases and the startup latency also reduces to a minimum. This is expected since more bandwidth can now be utilized for prefetching the video data.

Table 9.10: Statistical Multiplexing of H.264 *Foreman* Videos (352 × 288 CIF).

Number of Multiplexed Videos	Total Bandwidth	Startup Latency (Frame Intervals)	%Loss	Number of Paused Playback Occurrence
4	1.5 Mbps	6	45	28/300
	3 Mbps	4	28	16/300
	4.5 Mbps	3	8	5/300
	6 Mbps	2	0	0/300
5	1.5 Mbps	6	55	35/300
	3 Mbps	4	30	17/300
	4.5 Mbps	3	8	5/300
	6 Mbps	2	2	1/300
	7.5 Mbps	2	0	0/300
10	6 Mbps	3	20	13/300
	9 Mbps	2	2	1/300
	12 Mbps	1	0	0/300

Many 3D trailers are VC-1 encoded. The characteristics of some 3D videos are listed in Table 9.11. Each video frame has a twin to create a stereoscopic effect (one for the left eye, the other for the right eye) and encoded side by side, either horizontally (e.g., Katana) or vertically (e.g., *Heidelberg*, *Knights Quest*, *The Eye*, *Mouldpenny*). We wish to compare the performance of multiplexing five 2D HD H.264 videos versus five 3D HD VC-1 videos. The 2D HD 24-Hz movie trailers were selected from Table 3.7 (*The Dark Knight*, *Speed Racer*, *Terminator*, *Transformers*, *Oceans 13*). The PARs of the multiplexed VBR videos for the first 52s are shown in Figures 9.35 to 9.38. As can be seen, the instantaneous rate and PAR variation for the 3D videos is considerably higher than the 2D videos, even though the compression efficiency for H.264 and VC-1 is roughly the same.

Table 9.11: 3D HD VC-1 Video Characteristics (Total Bit Rate = 33.311 Mbps)

Video	Frame Rate	Resolution	Bit Rate	Duration
Heidelberg (720p)	25 Hz	1280 × 1440	6.108 Mbps	365s
Knights Quest (576p)	25 Hz	1024 × 1152	8.045 Mbps	104s
The Eye (720p)	25 Hz	1280 × 1440	9.224 Mbps	154s
Katana (720p)	25 Hz	1920 × 720	3.939 Mbps	156s
Mouldpenny (720p)	25 Hz	1280 × 1440	5.995 Mbps	289s

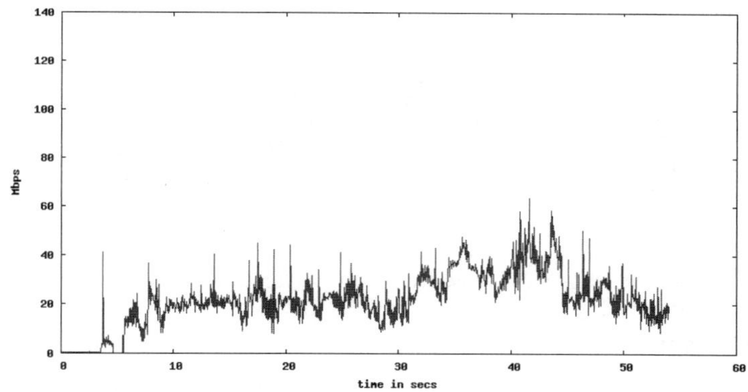

Figure 9.35: Instantaneous rates for five multiplexed 2D HD H.264 videos (average = 22.16 Mbps).

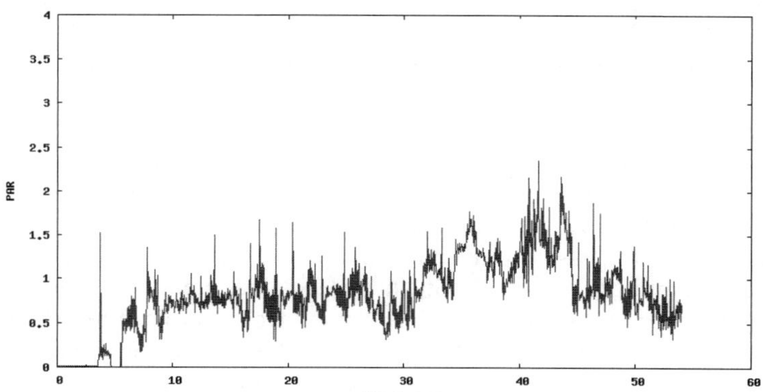

Figure 9.36: Long-term average PAR for 5 multiplexed 2D HD H.264 videos.

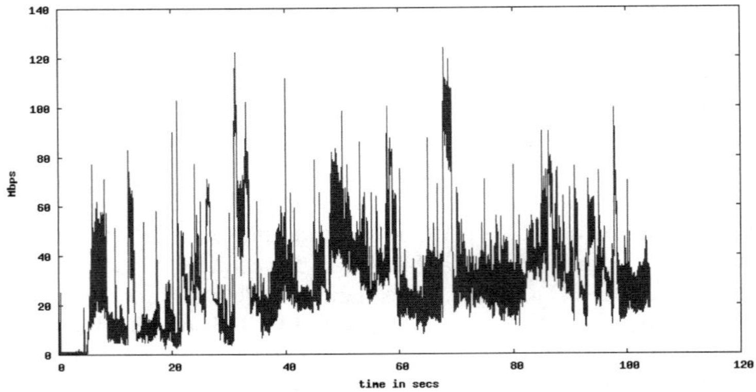

Figure 9.37: Instantaneous rates for five multiplexed 3D HD VC-1 videos (average = 26.15 Mbps).

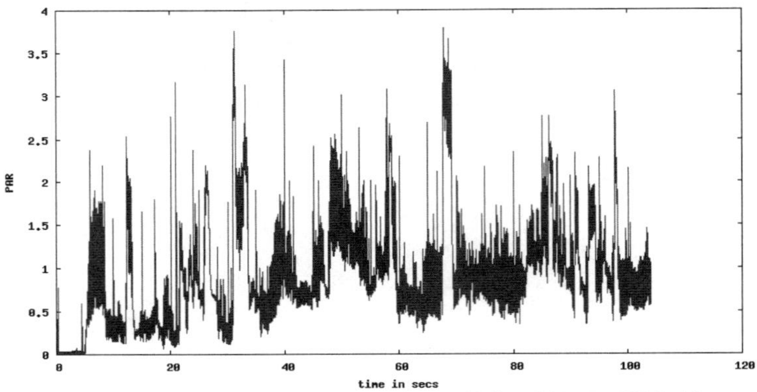

Figure 9.38: Long-term average PAR for five multiplexed 3D HD VC-1 videos.

9.9 Conclusions

In this chapter, we have evaluated the performance of compressed video traffic smoothing and streaming algorithms. It is shown that smoothing reduces the bit rate variability of the video stream and improves the overall STB buffer utilization. For a fixed buffer size, the rate limit, CoV, pause time, and buffer utilization are all connected. A low value for the CoV typically results in higher pause time (and hence, a lower rate limit) and higher buffer utilization. When streaming is initiated from another source, then the rate limit is related to the link capacity. The pause time of the smoothing algorithm (at the video source) should track this rate limit (i.e., the link rate) accordingly. We have also evaluated the performance of progressive streaming and frame smoothed streaming. Frame smoothed streaming shows great potential and can be employed more efficiently with optimizations applied to it. These streaming techniques can be enhanced with compound TCP. Finally, we illustrated the benefits of statistical multiplexing to conserve bandwidth for composite VBR video streams, including 3D and 2D HD streams. We have shown that the improvement in multiplexed bandwidth efficiency for bonded channels may not always scale linearly when compared to the single channel case. This is due to the long-range dependent characteristic of VBR videos, which leads to an increased variability of the unshaped instantaneous rates when a higher number of streams are multiplexed. To achieve high efficiency for composite streams, very aggressive shaping may be needed to control the higher variability. Good estimates for the shaping threshold may be obtained by performing a more in-depth analysis of the LRD of VBR videos.

References

[1] J. Rexford, et al., "Online Smoothing of Live, Variable-Bit-Rate Video," *Proceedings of the Workshop on Network and Operating System Support for Digital Audio and Video*, pp. 249–257, May 1997.

[2] J. Salehi, et al., "Supporting Stored Video: Reducing Rate Variability and End-to-End Resource Requirements through Optimal Smoothing," *IEEE Transactions on Networking*, Vol. 6, No. 4, August 1998, pp. 397–410.

[3] O. Hadar and S. Greenberg, "Statistical Multiplexing and Admission Control Policy for Smoothed Video Streams using e-PCRTT Algorithm," *Proceedings of the International Conference on Information Technology: Coding and Computing*, pp. 272–277, March 2000.

[4] H. C. Chao, C. L. Hung, Y. C. Chang, and J. L. Chen, "Efficient Changes and Variability Bandwidth Allocation for VBR Media Streams," *International Conference on Distributed Computing Systems Workshop*, pp. 289–294, April 2001.

[5] M. Sridharan, et al., "Compound TCP: A New TCP Congestion Control for High-Speed and Long Distance Networks," http://tools.ietf.org/html/draft-sridharan-tcpm-ctcp-00.

Exercises

9.1. In general, commercials and ads present less scene changes and more text. How can this be exploited in online (Internet) video streaming?

9.2. The simplest way to perform smoothing at the transmitting source and achieve a PAR of 1 is to employ a fixed threshold below the average rate of the coded video. Describe the consequences for adopting such a method.

9.3. Unlike UDP, HTTP allows traffic shaping to be performed at the server or player or both by controlling the sending and receive window sizes. List the advantages and disadvantages of performing traffic shaping at the player (client-pull) versus traffic shaping performed at the server (server-push). Design methods to mitigate the initial peak rate at the receiver for HTTP streaming.

9.4. Given that the Internet is never going to provide a constant bandwidth pipe, is traffic shaping necessary? Contrast the ABR approach described in Chapter 1 where the client device can request for the next video segment via a URL in a playlist. Is the automatic adaptation of the video resolution to the highest available bandwidth a good solution for networks with high bit rate variability? Or should the user be allowed to select the video resolution appropriate for the network and device, at the expense of occasional under-utilization of the network bandwidth? Will the ABR method of sending chunks of 2 to 10s video segments enhance or degrade the performance of trick modes (i.e., pause, rewind, fast forward, and so forth)? How will this impact network bandwidth consumption?

9.5. When it comes to dimensioning the buffer resources at the STB or player with HTTP streaming, one can allocate the buffer resources dynamically, based on the duration computed from the number of packets and the data rate of the link or on the number of video frames and the frame rate. Which approach is better? Suppose ABR streaming is employed. For a 1080p video at 30 Hz, how much storage (in Mbyte or Gbyte) will a 2s video segment require, assuming a coding efficiency of 0.25 bit/sample and a 4:4:4 color format?

9.6. Explain whether the values for the PAR computed using the instantaneous rates will be the same as the PAR computed using the frame sizes (Section 3.13.6).

9.7. Explain why the moving average PAR does not result in a PAR that correlates with the long-term average PAR. Will the running average PAR improve the correlation? Why are the 100-ms, 1s, and 10s PARs always greater than 1? Consider a trailer with several consecutive dark scenes at the start and at the end. How will this impact the correlation of the static average PAR with the long-term average PAR? Besides the masking effect of a higher overall average, give another reason to explain why the PAR metric does not work well for a composite set of video streams. Suggest methods to overcome this limitation. Will the coefficient of variability (CoV) defined in (5.1) provide a better metric? Will it work well for both single and multiple video streams? Consider the mixing of an SD video and an HD video. How will the PAR and CoV metrics compare in this case?

9.8. Movie trailers tend to exhibit high bit rate variability. Explain why this is the case. Will the multiplexing of movie trailers generally show a higher bit rate variation compared to the multiplexing of high-action sports videos? Consider the case when different codecs are used (e.g., MPEG-2 versus H.264).

9.9. Some 3D video delivery systems employ two videos stacked side by side. Will this imply that the data rate of a 3D video is roughly twice the data rate of a 2D video of the same content?

9.10. Explain whether the Hurst parameter is still valid for 3D videos. If not, suggest a metric where LRD, if present, can be measured.

9.11. Explain why many rate adaptation methods (e.g., transraters or rate shapers that examine individual video frames and requantize each frame to cap the output at the desired bit rate) cannot be used to manage the rates of multiplexed VBR video streams. Can the video smoothing algorithm described in Section 9.3 be applied to multiplexed VBR video streams? Justify your answer. List the advantages of per stream shaping versus composite stream shaping. Note that rate shaping can be performed in two ways. If adaptive quantization is employed by the coder, then depending on the desired VQ (typically controlled by the rate distortion and quantization algorithms), it may lead to a higher average bit rate because the coding becomes less efficient. For example, a large frame at a scene change that is quantized to a smaller size may cause succeeding frame sizes to be larger. The second method involves rate shaping performed at the coder's VBR output (fixed quantization). This may lead to increased packet buffering delay but faster encoding (why?), lower cost, and no change in the average bit rate or VQ.

9.12. As we have seen, the efficient transport of multiple VBR streams requires the use of shaping but this comes at the expense of increased delay. While the CBR approach is less efficient than VBR, shaping is usually not required. What is another key disadvantage of using CBR?

9.13. In Figure 9.30, verify that only seven 720p H.264 VBR streams can fit into a single 40 Mbps DOCSIS channel (no channel bonding) whereas eight streams can be supported by a 40 Mbps DOCSIS bonded channel. As shown in Figures 9.33 and 9.34, these streams are shaped to a delay of less than 1s. Assuming that a MPEG-2 480p CBR stream is capped to 3.75 Mbps, verify that an MPEG-2 720p CBR stream will require a cap of 10 Mbps. If H.264 is used, roughly how many 720p streams can be accommodated in a 40 Mbps DOCSIS channel? In this case, what is the key drawback of using CBR instead of VBR? Note that a CBR stream may yet generate some bit rate variability due to frames containing small amounts of information, even if these frames are coded with the lowest QP value.

9.14. Will the Hurst parameter value be similar for two VBR videos with the same content but different resolutions? It can be shown that grouping coded videos with similar Hurst parameter values and resolutions but uncorrelated content leads to an aggregated value that is similar to an individual video. Will a composite set of videos with similar Hurst parameter values but different resolutions and content change the overall value? How about grouping videos with data traffic?

9.15. Is there a relationship between the Hurst parameter and PAR? Although both metrics work well for a single video stream, give an example where a relative comparison of the PAR of two videos may not yield consistent results. Between CBR and VBR coding, which method will produce a higher Hurst parameter value? Hint: Think about the impact of the QP variation on the frame sizes in CBR. If a video is coded using MPEG-2 CBR and H.264 VBR, which video will produce a higher Hurst parameter value? For a video coded with one and two reference frames (same QP value), which video will show stronger LRD?

9.16. Figure 9.29 suggests that mixing SD and HD videos may result in multiple peaks in frame size variation. Explain how this can be exploited in bandwidth-efficient multiplexing. Take into account the content type (e.g., sports, movies, newscasting). If a SD video and a HD video both give the same PAR or CoV, will the location of the peaks coincide? What can you conclude about using the PAR or CoV to perform a relative comparison of different videos?

9.17. An adaptive shaping algorithm is needed to manage the instantaneous rate variation for single and multiplexed videos. A possible shaping threshold (s) is computed using the average (m) and standard deviation (σ) of the rates as follows:

$$s = m + w\sigma$$

where w = normalized weighting factor for desired delay

A value of $w = 1$ implies no shaping is performed and the overall rate is allowed to vary according to its natural rate. A value of $w = 0$ implies very aggressive shaping (and buffering) and this may be suitable for two-way network protocols such as HTTP that allow buffer resources at the server and receiver to be controlled dynamically. For $0 < w < 1$, devise a method to compute the desired delay using the packet transmission time, shaping threshold, and instantaneous rates. A sample plot for adaptively shaping 17 30-minute 1080i 60 Hz H.264 videos is shown below. m and σ are calculated for intervals of 60s and w was chosen to be 0.5. As can be seen, the delay rarely exceeds the desired threshold of 100 ms.

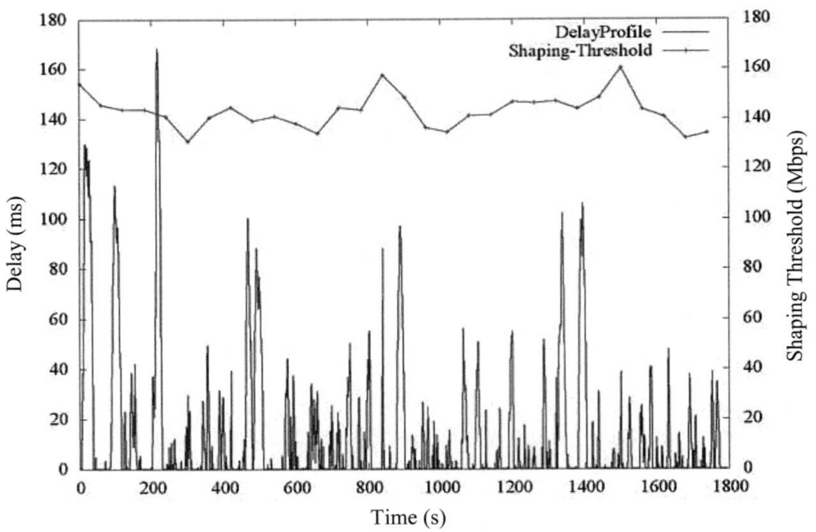

Chapter 10

Intelligent Policy Resource Management

One key challenge facing current broadband access service providers is the need to handle bandwidth intensive applications such as video streaming and peer-to-peer (p2p) applications. However, the star topology that is common in many broadband access deployments creates a local bottleneck at the central headend when all traffic from the end-users is directed to the hub or central office. The problem is aggravated if the bandwidth for the upstream or downstream links is limited. Thus, cable/telco operators may filter p2p traffic if there is evidence of oversubscription. In this chapter, we describe the use of resource allocation and bandwidth management techniques when applied to broadband access networks, focusing on how bandwidth resources can be intelligently scheduled and managed for different traffic types.

10.1 Introduction

Subscribers may occasionally and unintentionally consume bandwidth beyond its subscribed rate when running certain Internet applications (e.g., Skype, p2p file sharing, video streaming). Most of the time, these subscribers are unaware of the oversubscription. However, the star topology that is common in many broadband access deployments creates a local bottleneck at the central headend (e.g., a base station, a CMTS) when handling such applications. One solution is to employ a distributed mesh topology to spread the traffic load [1]. It will also be useful to examine other solutions that do not change the network topology.

Many service providers adopt a metered or usage-based bandwidth approach to counter this problem. This approach is not scalable and has become unpopular among the subscribers [2]. There are several ways to implement metered bandwidth management and several important criteria are taken into account:

- Net neutrality regulations dictate that Internet applications should not be subjected to discrimination (e.g., bias for payTV service over Internet TV);
- Level of network congestion and time of day usage may be taken into account;
- User prioritization may be activated to determine appropriate bandwidth

consumption limits;

- Connections may not be routed based on content but on destinations;
- Invasive methods such as resetting packets (e.g., via TCP) or deep packet inspection (DPI) may not be required.

Based on these criteria, we propose an intelligent resource management (IRM) framework to provide different service tiers and prevent serious security attacks. We employ OPNET simulation and an experimental implementation to evaluate the merits of the proposed scheme. It is shown that the IRM scheme performs better than conventional scheduling schemes. Specifically, the scheme can manage and control bandwidth-intensive applications, and provide fair bandwidth allocation. The broader impacts of the IRM scheme are the ability to mitigate serious security problems (e.g., worm viruses and denial of service attacks), and the ability to deliver quadruple play services while meeting required performance metrics. This capability will become increasingly important with the emergence of mobile broadband access networks such as 4G wireless.

10.2 Policy-Based Approach to Bandwidth Management

A policy-based approach typically employs the lightweight directory access protocol (LDAP) or IETF's Common Open Policy Service (COPS) protocol [3]. We focus on the COPS protocol, which is also part of the IP Multimedia Subsystem (IMS) architectural framework for delivering IP multimedia services. In a cable network, COPS can allow policy enforcement between the application manager and gate controller (policy server), as well as between the PacketCable Multimedia (PCMM) [4, 5] policy server and the cable modem termination system (CMTS) (see Figure 10.1). PCMM defines a system architecture that enables the deployment of real-time multimedia services over a high-speed cable access network. PCMM partitions resource management into distinct segments, namely the access and backbone network. Diffserv and Intserv are two IP quality of service (QoS) methods supported by PCMM. Access network QoS is based on IntServ where QoS is reserved and scheduled for individual traffic flows whereas the DiffServ model is employed to support QoS for the backbone network.

The COPS protocol is a simple query and response protocol and is used to exchange policy information between a policy server (Policy Decision Point or PDP) and its clients (Policy Enforcement Points or PEPs). At least one policy server exists in each controlled administrative, or resource, domain. A client can add advanced services easily using COPS, which specifies a service in unequivocal terms and instructs the edge device (e.g., CMTS) to allocate the resources required to deliver that service. The requirements and rules for resource allocation (known as policies) are often decided in advance (i.e., preallocated). Once an administrator has put such a policy in place, network directives can be issued to meet customer-initiated QoS requirements.

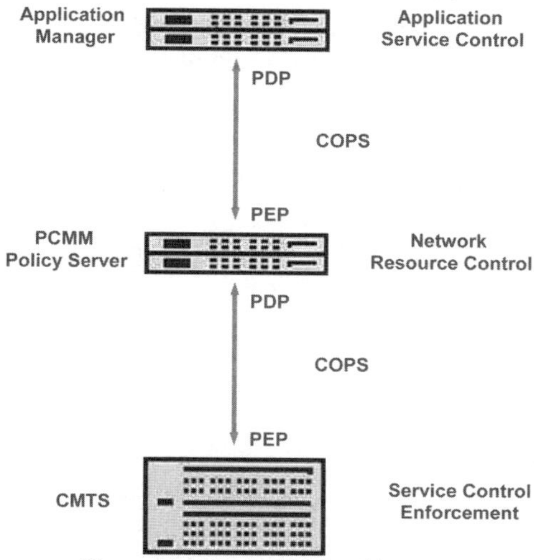

Figure 10.1: Basic COPS architecture.

10.2.1 Scheduling Methods

A typical broadband access traffic scheduling architecture is shown in Figure 10.2. The architecture can be used to support four primary traffic services specified in the Data Over Cable Service Interface Specification (DOCSIS) or the 802.16 standard [6] (which adopts many DOCSIS features and can be viewed as a wireless extension to DOCSIS):

- Unsolicited Grant Service (UGS) (e.g., Voice over IP);
- Real-Time Polling Service (rtPS) (e.g. MPEG video);
- Non-Real-Time Polling Service (nrtPS) (e.g. high-bandwidth Web download);
- Best Effort (BE).

Possible performance metrics include bandwidth utilization, packet drop, or delay rate for each traffic type, number of simultaneous users, and costs (which can be referenced to a fixed-price model).

10.2.2 Surplus Bandwidth

In most scheduling mechanisms, the potential for dropping or delaying packets exists, although ata varying degrees. In order to limit the rate at which packets are dropped, Wone can employ a practical concept of surplus bandwidth allowance that refers to the minimum allocation of excess transmission time by a resource scheduler [7] such that dropped packet rates are bounded. This takes into account the expected packet error probability in addition to the packet drop probability ofinherent with the transmission link. We consider three different cases:

- Single-packet transmission (e.g., control packet transmission);
- Multiple-packet transmission (e.g., Web download);
- Packet streaming (e.g., voice and video streaming).

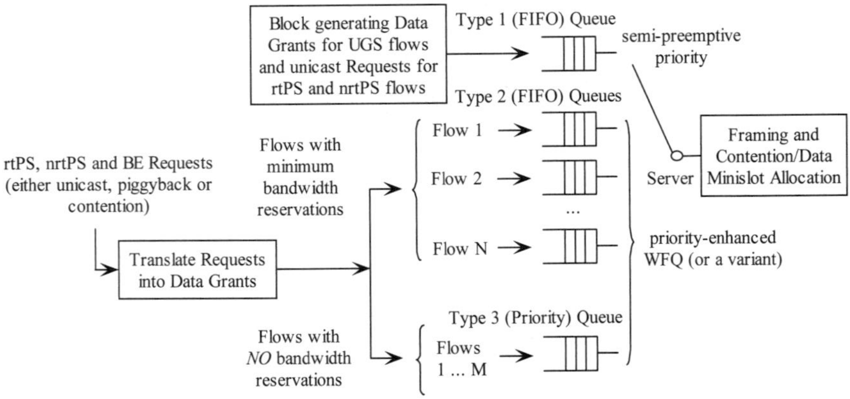

Figure 10.2: Traffic scheduling architecture.

In a single-packet transmission, if the cable channel causes independent errors from packet to packet and the error probability is constant for all packets of the same length, then the probability of dropped packet after n retransmissions is $p_{\text{drop}} = (p_{\text{e}})^{n+1}$ where p_{e} is packet error probability in a single transmission attempt. If $p_{\text{e}} = 0.1$ and $p_{\text{drop}} = 10^{-8}$, then $n = 7$. In other words, the scheduler has to accommodate seven transmit opportunities.

In a multiple packet transmission scenario, suppose N packets need to be transmitted correctly and the packet error probabilities are Bernoulli distributed. Then it can be shown that p_{drop} is given as:

$$p_{drop} = \sum_{k=N_{ex}}^{N+N_{ex}} \binom{N+N_{ex}}{k} p_e^{N_{ex}} (1-p_e)^{N+N_{ex}-k} = B(N_{ex}, N+1) \tag{10.1}$$

where $B(\ )$ is the incomplete beta function, assuming N is large. If $N = 100$, $p_{\text{e}} = 0.1$, $p_{\text{drop}} = 10^{-8}$, then $N_{ex} = 38$. The surplus bandwidth allowance, $S = N_{\text{allocated}}/N_{\text{payload}} = (N + N_{ex})/N = 1.38$. If $N = 100{,}000$, $p_{\text{e}} = 0.1$, $p_{\text{drop}} = 10^{-8}$, then $N_{ex} \sim 12{,}000$, $S \sim 1.12$. The surplus bandwidth decreases as N increases. For an infinite packet stream with no delay constraints, S approaches the lower bound of 1.111.

10.3 Intelligent Resource Management (IRM)

A flowchart of the intelligent algorithm is shown in Figure 10.3. Although there are several choices for the prediction method, in the simulation and implementation sections that follow, we employ a prediction method based on an artificial intelligence algorithm proposed in [8]. This genetic-based admission Control (GAC) algorithm has proved to be adaptive and efficient in predicting future traffic requirements. GAC incorporates multiple admission criteria with the overall goal of finding the minimum bandwidth cost for each QoS request. Parameters are fed into a cost function, which is derived based on the Markov decision process. This cost function is optimized using genetic algorithms (GAs) to minimize latency and maximize network utilization. The specific steps are

- Encoding a set of parameters (e.g., bandwidth cost function, QoS requirements) into a string;
- Evaluating the fitness function;
- Evaluating the final cost function;
- Employing fairness to provide consistent services.

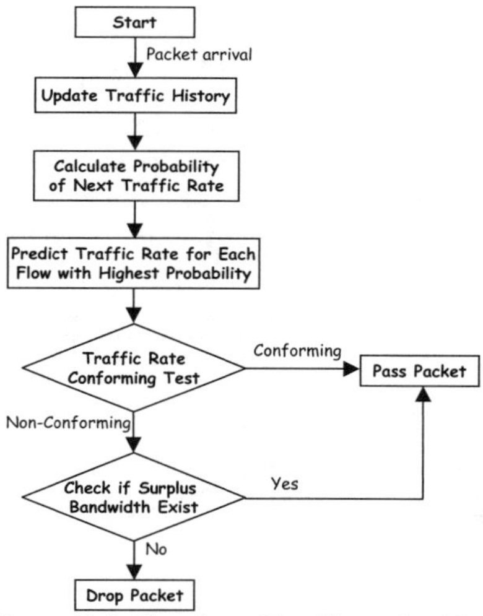

Figure 10.3: Flowchart of intelligent algorithm.

Note that these steps can be implemented dynamically to deal with real-time traffic flows or operated in the background to predict future traffic requirements. A typical exchange of messages using the IRM architecture is illustrated as follows:

- End user requests application service via CMTS;
- Service request received by application manager;
- Policy server receives resource requests (traffic type, protocol, required resources) from application manager;
- Policy server queries resource manager;
- Resource manager processes request using IRM algorithm (GAC-based) and informs policy server;
- Based on the response from the resource manager, the policy server issues a service request to the CMTS;
- CMTS schedules uplink/downlink slot using scheduling algorithm;
- End user runs application using allocated bandwidth.

10.4 Performance Analysis

OPNET simulations and experimental tests (based on a Cisco 7246 CMTS, a CMTS emulator, and a COPS software development kit) have been carried out and they show optimistic results. For instance, one can employ the intelligent algorithm to dynamically detect oversubscribed traffic (such as peer-to-peer traffic) and allocate bandwidth in a fair and efficient manner. A graphical user interface that is integrated with an SQL database has also been built. The following sections provide more details.

10.4.1 OPNET Simulation Setup for a Cable Network

The OPNET simulation setup and the sequence of exchanged messages are shown in Figure 10.4. The configuration parameters are as follows. The CMTS is configured with a 10.24 Mbps upstream and a 42.884 Mbps downstream. There are 32 contention (request) minislots for the MAP (i.e., the bandwidth allocation frame). The propagation delay is assumed to be negligible. We employ the DOCSIS cable model available in OPNET. The DOCSIS parameters for optimizing the MAP throughput are covered in the Appendix.

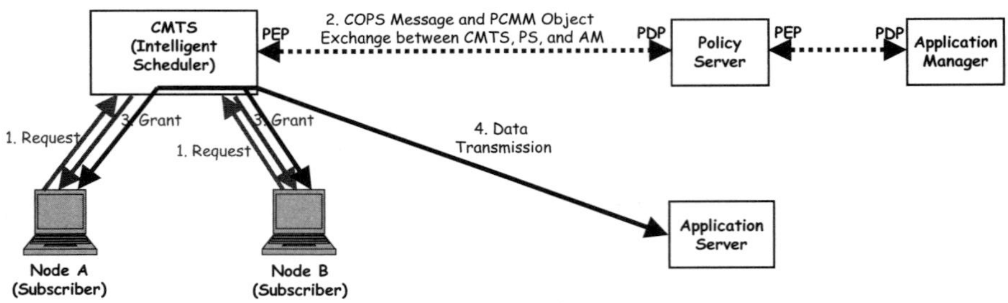

Figure 10.4: Simulation setup.

10.4.2 Dynamic Bandwidth Limitation

For illustration purposes, we employ two equal-priority nodes that generate non-real-time polling (NRTP) traffic with a polling interval of 0.01s. An upstream bandwidth of 690 Kbps is the average transmission rate specified in the volume-limit profile. Each node is expected to transmit traffic at or below this average rate. The combined average upstream transmission rate is therefore 1.38 Mbps. The temporal traffic generation characteristics from the two nodes are shown in Figure 10.5. From time 0 to 60s, nodes A and B each generate a 690 Kbps traffic stream. From 60 to 120s, node A's traffic rate is doubled. Clearly, node A is misbehaving. In addition, the combined transmission rate is above 1.38 Mbps. From 120 to 180s, node B's transmission rate is decreased by half. In this case, the combined transmission rate of the two nodes is about 1.38 Mbps. The time window for computing the average rate is 1s.

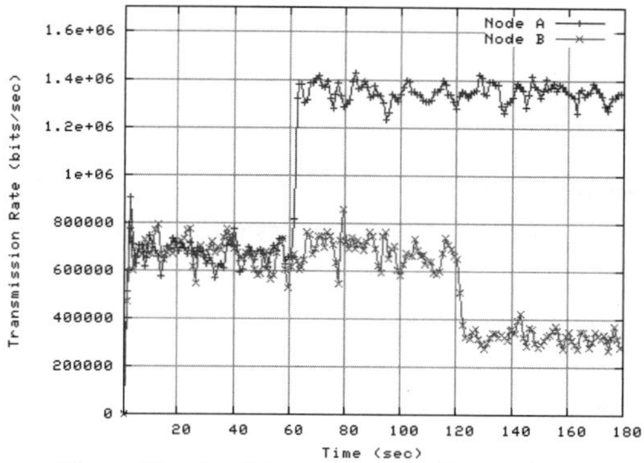

Figure 10.5: Traffic transmission characteristics.

In Figure 10.6, we illustrate the use of first-come-first-serve scheduling. From 60 to 120s, the combined data rate is now within the 1.38 Mbps limit and the bandwidth usage for each node is also lesser. However, during this period, more bandwidth is still allocated to the misbehaving node, which is unfair. Figure 10.7 illustrates the use of hard-bandwidth-limit scheduling to control the transmission rate of the misbehaving node and provide fairness in bandwidth allocation. Each node now has a 690 Kbps hard bandwidth limit. From 60 to 120s, since node A can only transmit up to 690 Kbps both node A and node B are allocated the same bandwidth, which improves the fairness in bandwidth allocation. From 120 to 180s, node B decreases its traffic rate, but node A can only transmit up to 690 Kbps. Since the maximum available combined bandwidth limit is 1.38 Mbps, some bandwidth is not utilized. In the IRM scheme, full bandwidth utilization is possible for the period of 120 to 180s (Figure 10.8). In this case, node B decreases its traffic rate but the bandwidth limit for Node A is increased to utilize the maximum bandwidth. Note that the prediction algorithm provides both volume and time limits to the bandwidth allocation. The frequency of update (e.g., prediction and changing limits) is 0.01s while the traffic buffer history is 10 sec. In addition, there is less bandwidth variation compared to Figures 10.6 and 10.7.

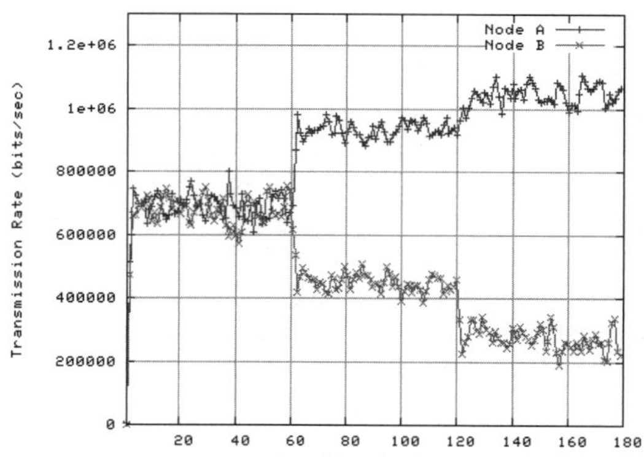

Figure 10.6: Transmission rate using first-come-first-serve scheduling.

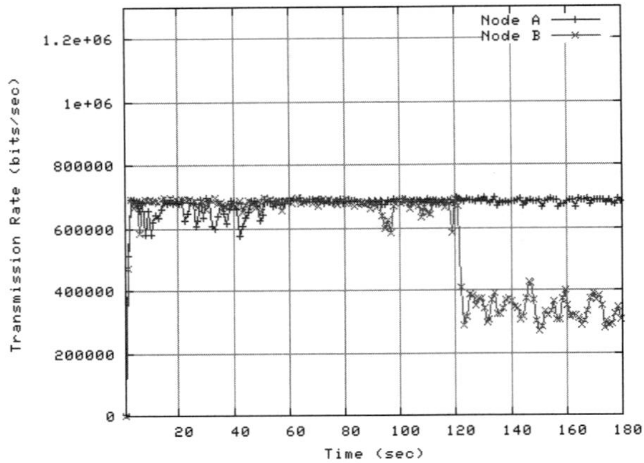

Figure 10.7: Transmission rate using hard-bandwidth-limit scheduling.

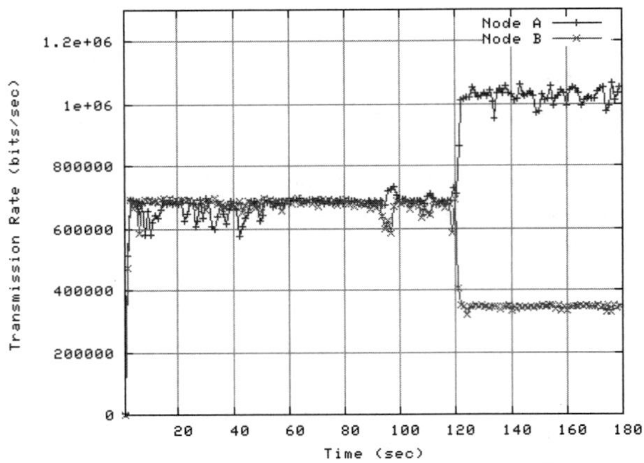

Figure 10.8: Transmission rate using IRM scheduling.

10.4.3 Implementation and Measured Results

The simulation results were validated with an experimental setup using a Cisco 7246 CMTS. A unified interface has been created for the subscriber device (in this case, the cable modem) management, gate management, usage limit control, gate state transition, and COPS management (Figures 10.9 to 10.11).

The traffic generation characteristics of the experimental setup (Figure 10.12) are similar to that of the OPNET simulation (Figure 10.5). Figures 10.13 to 10.15 show the measured results obtained using an experimental setup that is identical to the simulation setup of Figure 10.4. As can be seen, the measured results correlate well with the OPNET simulations.

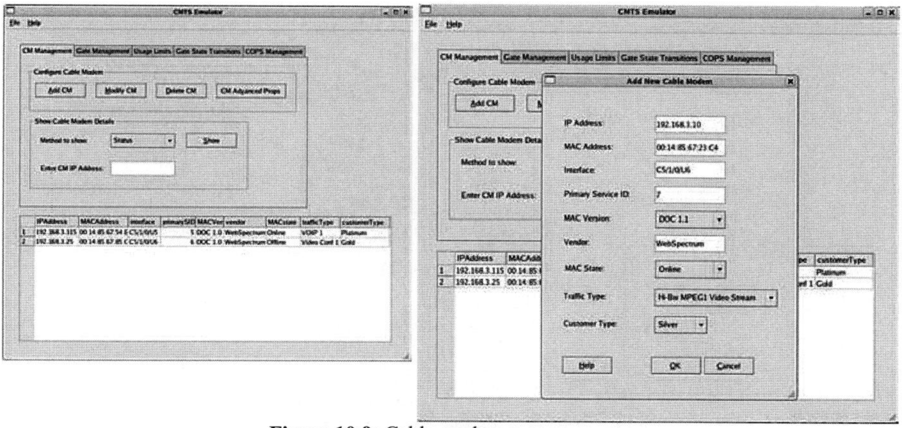

Figure 10.9: Cable modem management.

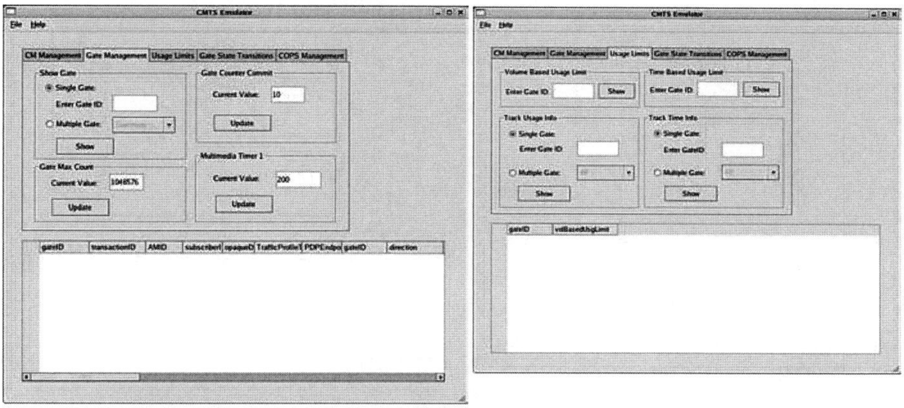

Figure 10.10: Gate management and usage limit (time and volume-based).

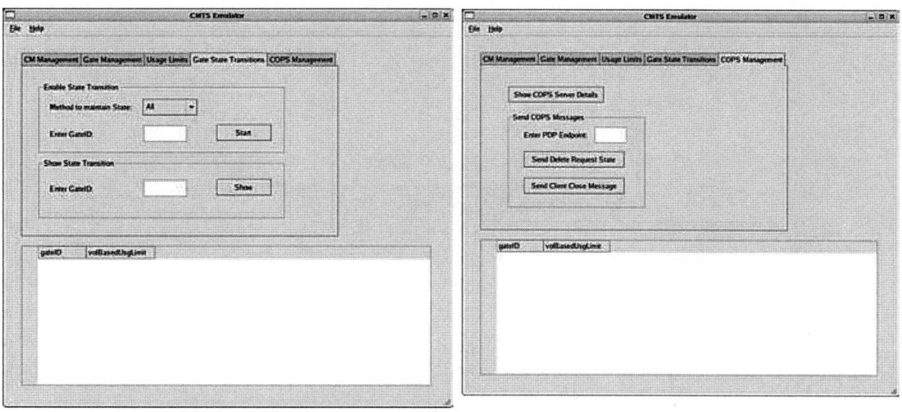

Figure 10.11: Gate state transition and COPS management.

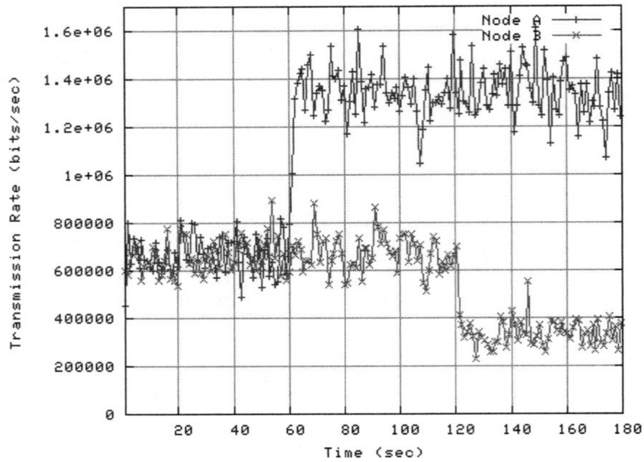

Figure 10.12: Traffic transmission characteristics.

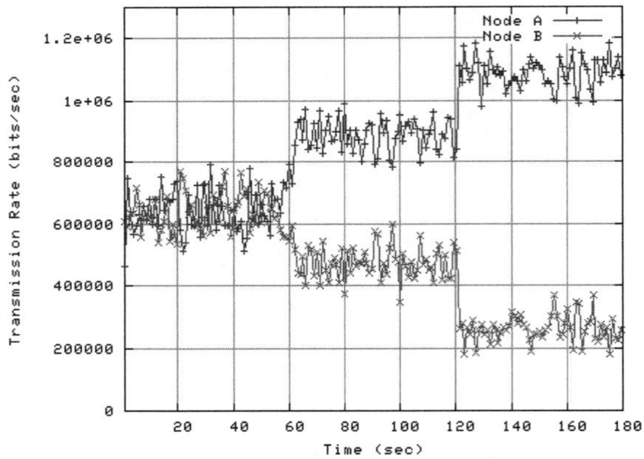

Figure 10.13: Transmission rate using first-come-first-serve scheduling.

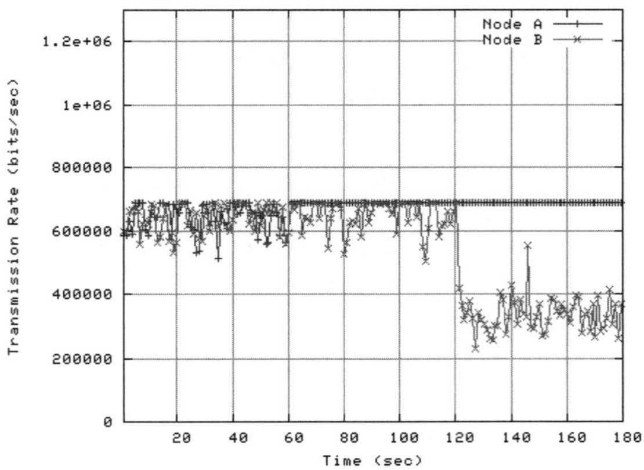

Figure 10.14: Transmission rate using hard-bandwidth-limit scheduling.

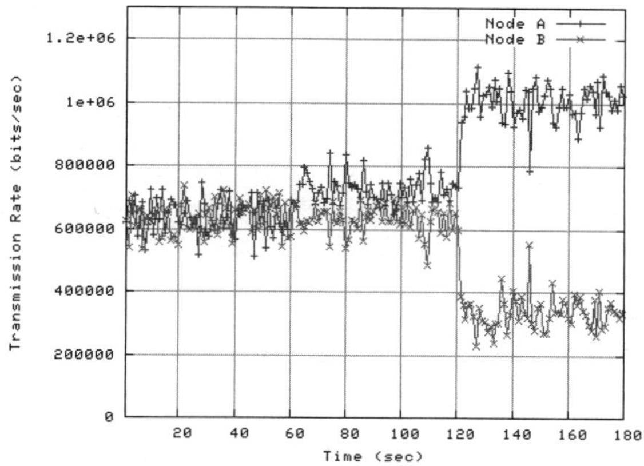

Figure 10.15: Transmission rate using intelligent scheduling.

10.5 Conclusions

In this chapter, we have described and evaluated the performance of an intelligent bandwidth management algorithm suitable for broadband access deployments. Although the simulation and experimental setups are limited to a cable network, the research can be extended to other broadband access settings. The simulation and experimental results have been validated and they show optimistic results. For instance, one can employ the intelligent algorithm to dynamically detect oversubscribed traffic (such as peer-to-peer traffic) and allocate bandwidth in a fair and efficient manner. There is also a possibility of extending the use of the IRM algorithm to manage location-based policies for mobile wireless networks.

References

[1] B. Bing, "Ubiquitous Broadband Access Networks with Peer-to-Peer Application Support," *IEC Annual Review of Communications: Volume 59*, 2006.

[2] D. O'Shea, "TV and the Usage-based Pricing Evolution," *FierceIPTV*, December 15, 2009, http://www.fierceiptv.com/story/tv-and-usage-based-evolution/2009-12-15?utm_medium=nl&utm_source=internal.

[3] K. Haddadou, et. al., "Designing Scalable On-Demand Policy-Based Resource Allocation in IP Networks," *IEEE Communications Magazine*, Vol. 44, No. 3, March 2006, pp. 142–149.

[4] E. Miller, F. Andreasen, and G. Russell, "The PacketCable Architecture," *IEEE Communications Magazine*, Vol. 39, No. 6, June 2001, pp. 90–96.

[5] *PacketCable Multimedia Specification*, PKT-SP-MM-I02-040930, April 30, 2004, Cable Television Laboratories.

[6] IEEE 802.16e: Air Interface for Fixed and Mobile Broadband Wireless Access Systems. Amendment 2: Physical and Medium Access Control Layers for Combined Fixed and Mobile Operation in Licensed Bands, February 2006.

[7] IEEE P802.11e standard, Wireless LAN Medium Access Control (MAC) and Physical Layer (PHY) specifications: MAC Enhancements for Quality of Service, December 2005.
[8] D. Karabudak, C. Hung, and B. Bing, "An Intelligent Admission Control Scheme for Next Generation Wireless Systems using Genetic Algorithms," *IEEE WCNC*, Atlanta, GA, USA, 21–25 March 2004.
[9] B. Bekele, B. Bing, and R. Jackson, "Genetic-Based Policy Resource Management for Advanced DOCSIS and PacketCable Multimedia Networks," CableLabs Whitepaper, 2006.

Appendix: Optimized MAP Throughput

This appendix analyzes the DOCSIS parameters needed for optimizing the throughput. An OPNET simulation setup is shown in Figure 10.A.1. The configuration parameters are as follows. The CMTS is configured with 16 QAM for the upstream (10.24 Mbps) and 256 QAM for the downstream (42.884 Mbps). There are 32 contention (request) minislots for the MAP. The propagation delay is negligible.

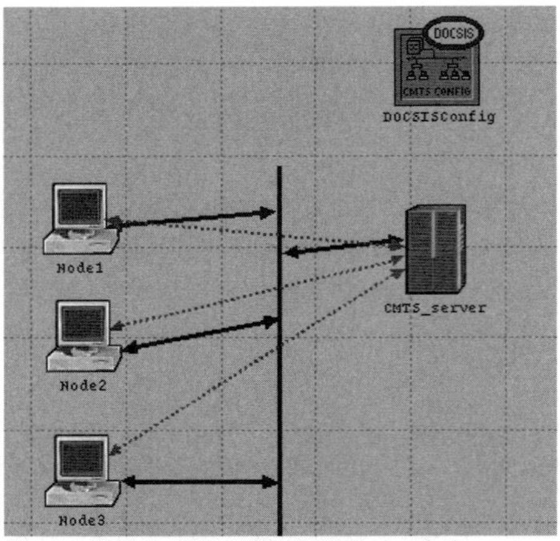

Figure 10.A.1: OPNET DOCSIS simulation setup.

The upstream performance is shown in Figure 10.A.2. Since the upstream throughput is maximized for a MAP interarrival time of 0.008s (i.e., 125 MAP frames are sent periodically per second), this time is used throughout the performance evaluation described in the main paper. There are 160 minislots (32 request and 128 data minislots) in each MAP frame with each minislot occupying 64 bytes (50 microseconds at 10.24 Mbps upstream data rate). The packet length is exponentially distributed with an average length of 1,000 bytes and the packet interarrival time is exponentially distributed. Packet interarrival time is 0.1333s and applies to both non-real-time polling (NRTP) traffic and best effort (BE) traffic nodes. Clearly, UGS has the best throughput and delay performance (Figures 10.A.3 and 10.A.4), followed by NRTP and then BE traffic.

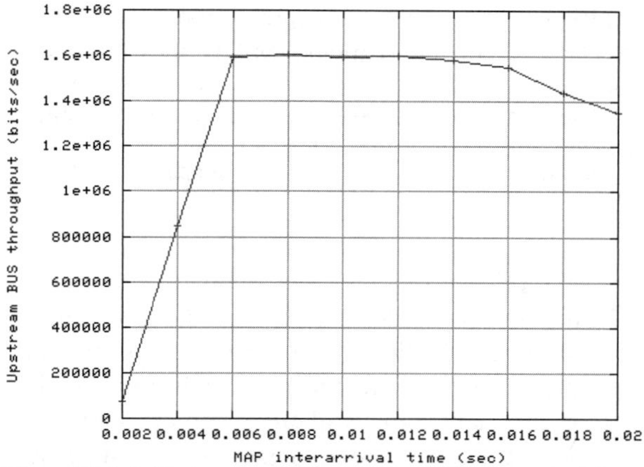

Figure 10.A.2: Upstream throughput versus MAP interarrival time.

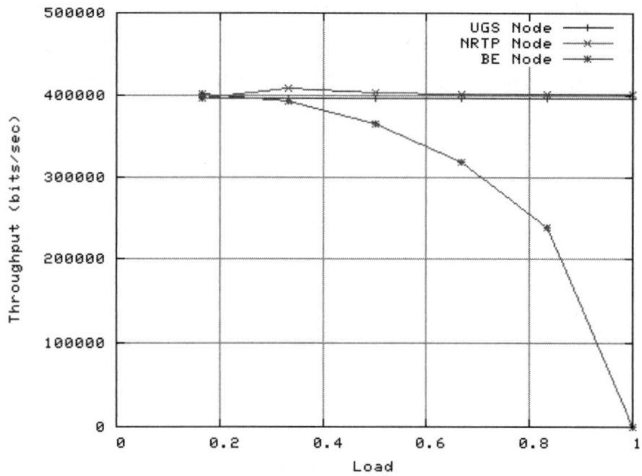

Figure 10.A.3: Average throughput versus traffic load.

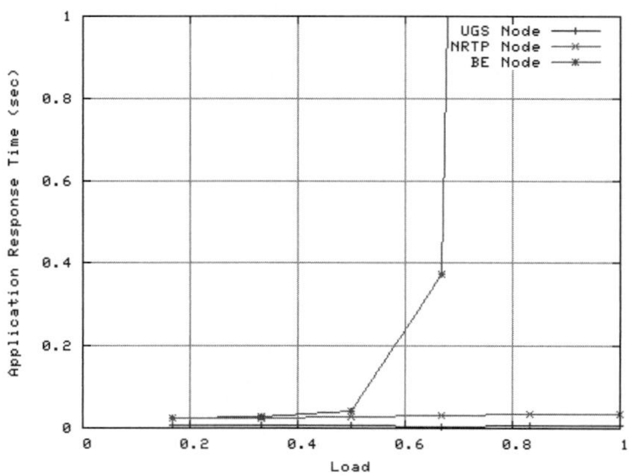

Figure 10.A.4: Average delay versus traffic load.

RTP and UGS traffic cannot employ the contention minislots after they have been activated. For NRTP and BE traffic, contention is allowed to cater for changing bandwidth requirements. This implies that it is easier to allocate bandwidth to NRTP and BE traffic dynamically. NRTP is employed in the simulation described in the main text.

Exercises

10.1. With wireless service providers now removing unlimited data plans for smartphones and enforcing bandwidth caps (primarily due to the increase in video traffic), how will this impact online video streaming and video calls on mobile devices? Design a traffic policing method that will allow both data and video traffic to coexist in the same pipe without compromising QoS. Modify your method to accommodate traffic offload to higher-capacity Wi-Fi networks.

10.2. Are UGS and RTP services necessary given that the DOCSIS MAC protocol is reservation-based? A reservation MAC protocol is essentially a request-and-grant mechanism where the subscriber unit transmits a request to the CMTS which then grants bandwidth resources.

Chapter 11

Supporting Compressed Video Applications over DOCSIS Cable Networks

This chapter presents the key issues and challenges associated with the optimization of switched digital video (SDV) and video-on-demand (VoD) services for the DOCSIS 3.0 Hybrid-Fiber Coax (HFC) cable access network. In our optimization framework, we will consider the video program access pattern, rate-limited video smoothing, and scheduling policies. These three items, when considered together with efficient QAM allocation at the physical layer, will ultimately determine the cost for deploying the broadband cable access network.

11.1 Introduction

While SDV typically deals with real-time video, a VoD service focuses on prerecorded videos. VoD subscribers can view selected movies or TV programs on demand. VoD traffic is currently small compared to broadcast TV but may use substantial amounts of bandwidth. SDV/VoD traffic primarily consists of large MPEG file transfers, which can be classified under high definition (HD) or standard definition (SD). MPEG traffic is characterized by high peak rates and drastic short-term rate changes, which can lead to undesirable losses within the STBs' buffers. The video traffic (in particular HD video and VoD traffic) will normally occupy more downstream (DS) than upstream bandwidth (US) (i.e., asymmetric bandwidth demands). The need to support packetized video breaks traditional Internet bandwidth utilization models since the video traffic is real-time constant bit rate (CBR) or variable bit rate (VBR) with peak viewing hours. MPEG-2 video is the most prevalent video compression standard used in SDV and VoD installations today. The real bandwidth savings of H.264 are obtained from coding HD video. Since HD video will grow in importance in the future, we will see a corresponding demand for H.264 coded video.

Layer 2 traffic engineering techniques can be used to model these video traffic types. This is ideal for Ethernet-based networks such as the DOCSIS 3.0 cable architecture, where the central component is a gigabit Ethernet switch.

Quality metrics should not be discounted. These include subscriber-perceived video quality of experience (QoE) such as audio/video (A/V) synchronization, latency in channel change time, channel service access delays due to popular broadcast events, and so forth.

Both SDV and VoD cable architectures must currently receive CBR video streams to operate seamlessly. CBR traffic simplifies and maximizes the utilization of each available QAM signal at the PHY layer. SD VoD streams are typically compressed to 3.75 Mbps CBR, allowing 10 VoD streams to fit into a 38.88 Mbps 256-QAM modulator. In SDV, a session resource manager (SRM) tracks and computes the available per-channel capacity. Many operators tend to fix different bit rates for different streams based on the video complexity of those streams. The main drawback for doing this is that it presets the total throughput of the video to a maximum bit rate. The SRM tracks the number of provisioned video streams and the edge QAM modulators to deliver the streams, as well as the total throughput. For CBR videos, statistical multiplexing offers little benefit. For VBR videos, we have studied the gains of multiplexing these streams in Chapter 9. However, the complexity and overheads of real-time VBR multiplexing may outweigh the gains.

Our focus in this chapter is on optimizing the video transmission by judicious management of bandwidth at the cable headend. We also describe the evolution of DOCSIS 1.1, 2.0, and 3.0 standards. We first cover the measured QoS performance of the DOCSIS 1.1 cable network. The intent of the study is to characterize the behavior of the network in terms of throughput, packet loss rate, and average delay under different degrees of network loading and prioritization profiles. Based on the measurement results obtained, a QoS model using the weighted fair queuing algorithm was developed for the CMTS scheduler. Simulation results from OPNET demonstrate a close match between this model and the measured results. The empirical results and QoS model presented will be useful for network designers who wish to understand the practical characteristics of the DOCSIS network. We then present the measurements and performance improvement of peer-to-peer applications when operated over a symmetrical DOCSIS 2.0 network under different network conditions. Finally, we present a new video-aware DOCSIS 3.0 architecture and describe some video program scheduling challenges, including a new scheduling method.

11.2 Measured Performance of a DOCSIS Cable Network

QoS metrics such as throughput, average delay, and packet loss rate play important roles in providing various levels of service quality. Past research conducted on QoS for the DOCSIS standard focuses exclusively on algorithm analysis and software simulations [1–4] with no practical measurements. Based on some measured results, we propose a QoS model for the CMTS scheduler using the weighted fair queuing algorithm. Simulation results from OPNET demonstrate a close match between this model and the measured results.

11.2.1 Experimental Setup

The experimental setup is shown in Figure 11.1. Eight Arris Touchstone™ CM300A cable modems (CMs) are connected to an Arris Cornerstone™ 1500 C3 CMTS using a tap and a power splitter. A computer running the dynamic host configuration protocol (DHCP) and trivial file transfer protocol (TFTP) server programs is connected to the CMTS through an Ethernet switch to provide IP addresses and configuration profiles for the CMs in the initialization phase. The Spirent SmartBits 6000B is a network performance analysis tool that can test, simulate, analyze, troubleshoot, and certify the performance of network equipment. A six-port 10/100 Mbps Ethernet card of the SmartBits 6000B is used in this experimental setup. Four ports from the card are used to generate IP traffic for the CMs. Another port is used to collect and analyze the received data packets, as well as to generate measurement reports. QPSK modulation is used for the US and 64 QAM is used for the DS.

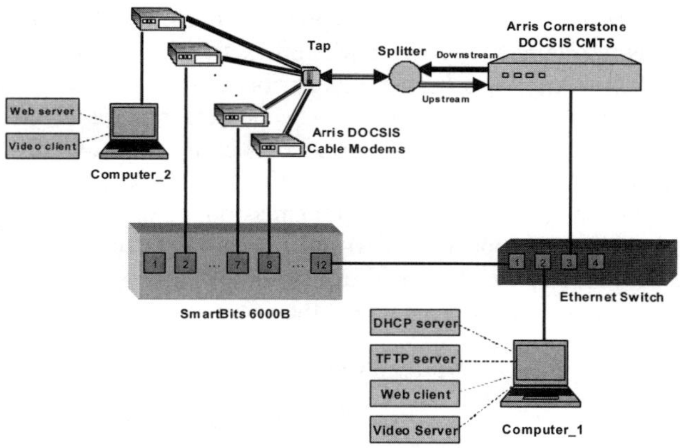

Figure 11.1: Experimental setup.

11.2.2 Measured Results

By keeping only one CM active, contention among CMs is removed and the maximum US and DS throughput of the CM can be measured. The SmartBits 6000B generates a CBR traffic flow with varying transmission rates and packet lengths. As shown in Figure 11.2, both US and DS throughputs increase with increasing packet length. The throughput refers to the bit rate associated with the transmission of the data packet, which is an Ethernet frame. A maximum throughput of about 2 Mbps can be obtained for the US link (from CM to CMTS) while the maximum DS throughput (from CMTS to CM) is about 25.7 Mbps. In both cases, the maximum throughput occurs when packets with the longest length (i.e., 1,500 bytes) are transmitted. For all packet lengths, the saturation throughput occurs when the US or DS links reaches capacity and packet discard occurs.

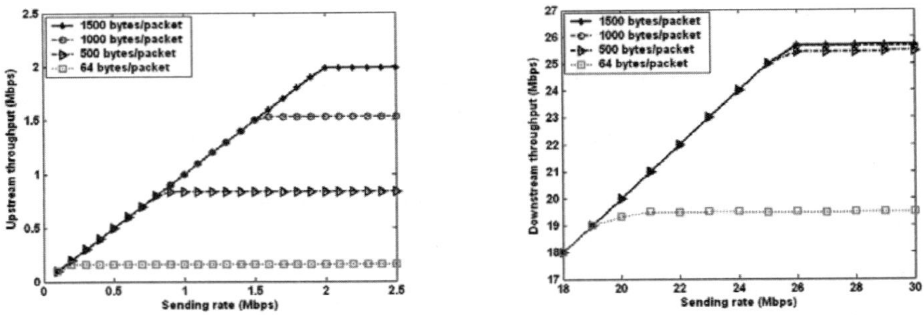

Figure 11.2: Upstream and downstream throughputs of one CM with different packet lengths.

In the experiments that follow, we employ two CMs with two different priorities. Because the maximum US throughput of one CM is about 2 Mbps when the packet length is 1,500 bytes (as shown in Figure 11.2) and the QPSK US channel capacity is about 4.5 Mbps, having two CMs with different priorities will not generate sufficient traffic load to stress the network. To confirm this, we perform some experiments using two CMs, one set to priority 4 while the other is set to priority 3 (priority 4 is higher than priority 3). The results are shown in Figure 11.3. When the sending rate per CM is less than 2 Mbps, there is no congestion and each CM can transfer one packet in each MAC frame period. Hence, the loss rate is zero and the delay of priority 4 packets is only smaller than priority 3 packets by about one packet transmission time. When the sending rate per CM is higher than 2 Mbps, each CM reaches its upper throughput limit, which is not related to the prioritization profiles. This means the overall performance of two CMs with different priorities is almost the same.

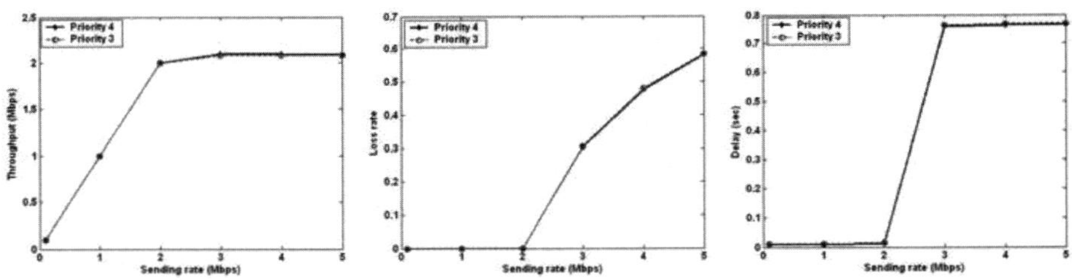

Figure 11.3: Upstream performance of two CMs with different priorities and fixed packet length.

We now measure the prioritized US performance by keeping four CMs active with two CMs assigned priority 3 and 2 CMs assigned priority 4. The SmartBits 6000B generates CBR traffic flows with varying rates but the packet length is fixed at 1,500 bytes. The results are shown in Figure 11.4. The sending rate refers to the data rate of each individual CM while the delay is computed from the time a packet arrives at the CM to the time the packet is received by the CMTS. The loss rate is normalized to the ratio of the number of discarded packets against the total number of packets presented to the CM. It can be observed that when the sending rate is low, the throughput and loss rate of each prioritized packet stream are virtually the same. The distinction between different prioritized streams arises when the sending rate increases to a breakout point of 1 Mbps. Beyond 1 Mbps, the higher priority 4 traffic streams obtain higher throughput and experience smaller delay. The throughput and delay also start to saturate when the CM reaches a capacity of 1 Mbps and packet discard occurs (as evident in the graph showing the loss rate).

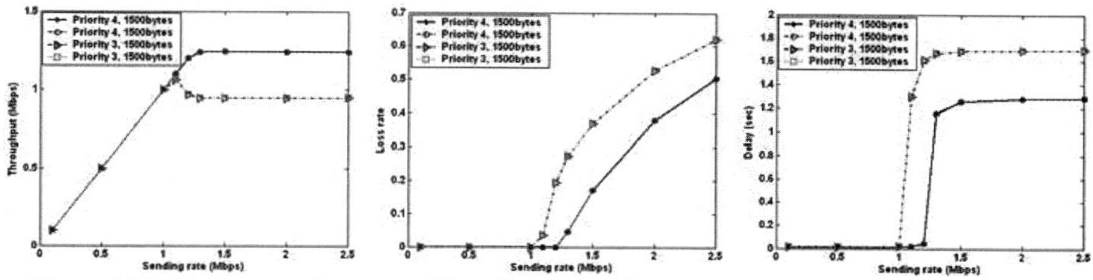

Figure 11.4: Upstream performance of four CMs with different priorities and fixed packet length.

In the next experiment, the SmartBits 6000B generates four CBR traffic flows with varying rate and priorities. However, unlike the previous experiments, the packet lengths are now varied. As shown in Figure 11.5, the DOCSIS 1.1 network achieves higher throughput and lower delay for higher priority and larger packets. Clearly, the packet length affects the performance of the DOCSIS 1.1 packet streams, as evident in the different saturation rates.

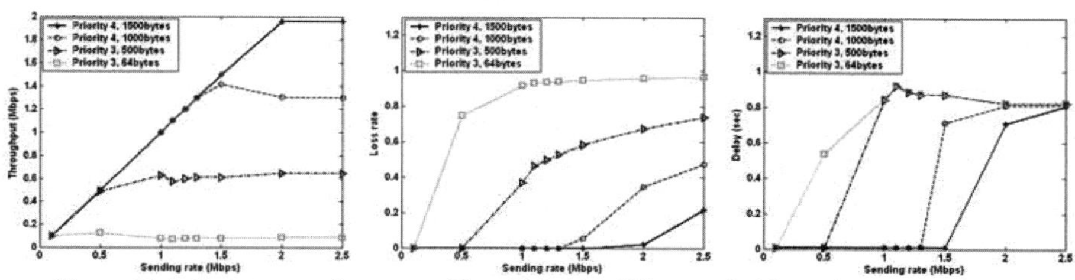

Figure 11.5: Upstream performance of four CMs with different priorities and packet lengths.

In the final experiment, the SmartBits 6000B generates eight CBR traffic flows with varying rate and priorities. Priority 0 corresponds to the highest priority. The packet size is fixed at 1,500 bytes. Once again, it can be observed that when the sending rate is low, the throughput and loss rate of each prioritized packet stream are virtually the same. The distinction between different prioritized streams arises when the sending rate increases and three saturation points appear. Beyond the saturation points, the higher priority traffic streams achieve higher throughput and experience smaller delay. When the throughput and delay start to saturate, the CM reaches capacity and packet discard occurs (as evident in the graph showing the loss rate). As shown in Figure 11.6, although eight priorities can be allocated to CMs, only three QoS levels are implemented in the CMTS under test. These QoS levels are high (priority 7), medium (priority 6, 5, 4), and low (priority 3, 2, 1, 0).

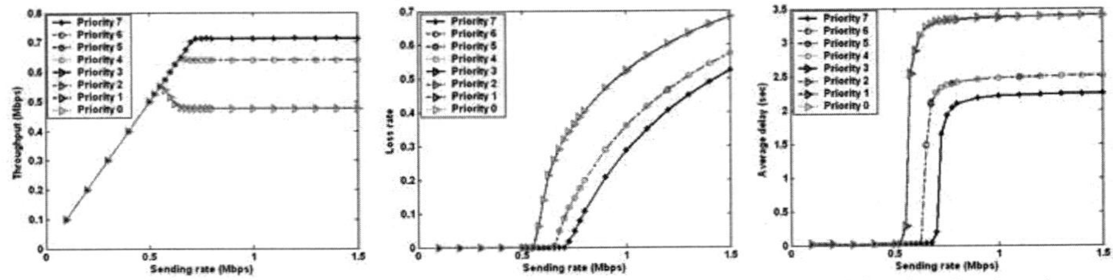

Figure 11.6: Upstream performance of eight CMs with different priorities and fixed packet length.

11.3 A QoS Model for the CMTS Scheduler

The QoS performance of the DOCSIS 1.1 standard is dependent on the CMTS scheduling algorithm. To investigate the scheduling scheme, we employ a weighted round robin queuing algorithm to compute and monitor the delay requirement of every packet, and allow fairness for on-demand bandwidth assignments to bursty traffic. Here, the CMTS acts as an arbiter serving multiple prioritized queues (Figure 11.7). Each CM sends a data packet to the queue of its priority. The four parameters for this model are arrival rate λ (packets/s), departure rate μ (packets/s), queue size L in packets, and scheduling weights representing the different priorities. An OPNET simulation for this model is carried out.

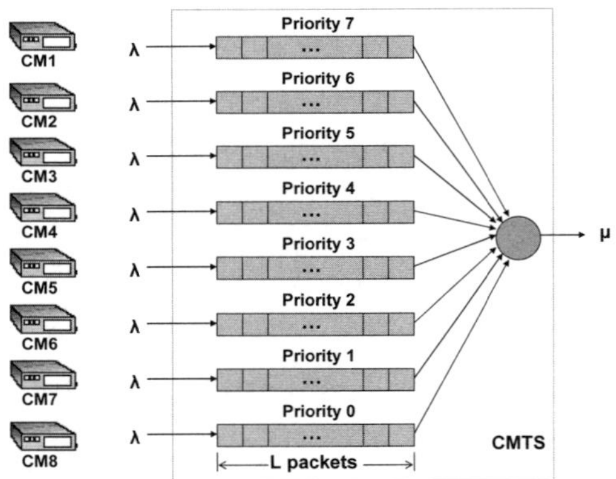

Figure 11.7: Weighted round robin queuing model for the CMTS scheduler.

We employ simulation parameters similar to the measurements. The CMTS bandwidth is set to 4.5364 Mbps. The arrival rate (λ) varies with the sending rate, the departure rate (μ) is set to 378 packet/s with packet length fixed at 1,500 bytes, the queue size (L) is 134 packets per CM, and the scheduling weights are 0.157 (high priority), 0.141 (medium priority), and 0.105 (low priority). As shown in Figure 11.8, for eight CMs with priority 0 to 7, the throughput, loss rate, and delay simulation results are very similar to the measured results from Figure 11.6. This implies that the weighted fair queuing algorithm with appropriate parameters provides a convenient means to accurately model the scheduling algorithm of the CMTS without having to perform physical measurements.

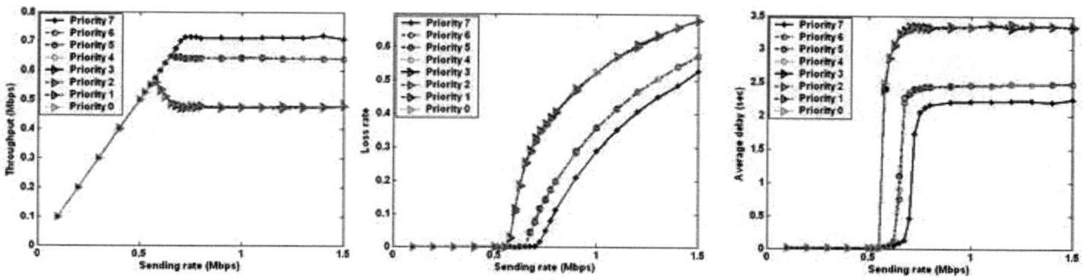

Figure 11.8: OPNET simulation of the weighted fair queuing CMTS scheduler.

11.4 Peer-to-Peer File Sharing

P2p communications is a scalable way of delivering broadcast content. It decreases the load on congested servers, minimizes distribution cost, and reduces downloading time for users. There are essentially two kinds of p2p network nodes. Peers are computers that download and upload data. Trackers are servers assisting peers to find other peers. Peers typically interact with the tracker using a simple protocol that resides on top of HTTP. To demonstrate the potential of p2p performance gains, we employ BitTorrent (BT), a leading open source p2p distribution protocol, for p2p file sharing. BT file sharing may involve the transfer of compressed video files. In addition, applets such as Bitlet (http://www.bitlet.org) may launch BT video streaming.

BT requires peers to organize themselves into an overlay network called a torrent. There are two types of peers that are part of the torrent: leecher and seed. A seed is a client that has a complete copy of the file and remains in the torrent to serve other peers. A leecher is a client that is downloading the file. Unlike other p2p technologies, BT does not provide a search function and it uses a file-level sharing policy as opposed to a directory-level sharing policy. BT employs mechanisms such as tit-for-tat (TFT) and rarest first for cooperative file distribution among participating peers. The TFT mechanism operates as follows. Unless a peer contributes to the ongoing downloading of file pieces, it will not be able to download the file. BT uses two kinds of files, namely a content data file (called seed) and a small file (called .torrent metainfo) that provides tracker location, currently active peers within a torrent, and data file description (e.g., data file length, piece identification, hashing information). The basic mechanisms are:

- Publish: Generate BitTorrent file and run a centralized tracker;
- Join: Contact tracker, obtain list of active peers;
- Piece: Break data file into smaller pieces of fixed size, each downloaded piece is reported by all participating peers;
- Piece selection: Rarest first, if not available, then random first;
- Fetch: Download file pieces from peers, upload file pieces to peers.

To initiate a BT session, a Web server, a centralized tracker, and a seed node are activated. A .torrent file is generated at the server. A new peer joins the session by downloading the .torrent file and then contacting the tracker for a list of other participating peers. The peer then selects a set of neighboring peers from the list of peers provided by the tracker and then establishes connections with them to exchange file pieces. BT uses pipelining to keep the HTTP connections at full capacity. To facilitate this, a fixed-size piece (typically 256 Kbyte) is broken down into several sub-pieces (typically 16 Kbyte). Five requests for sub-pieces are normally kept pending in the pipeline. Because the data file is broken down into smaller pieces, each peer reports to all other peers the pieces it is holding. This enables one peer to obtain several different pieces from a number of different peers (usually 4) simultaneously, in addition to the seed node. As a result, the download performance improvement for each peer is proportional to the number of peers in a session and the load on the seed is reduced substantially. In contrast, in conventional client/server applications like the File Transfer Protocol (FTP), the server can easily become a bottleneck as the number of clients increases.

We tested the performance of BT over a DOCSIS 2.0 cable network. DOCSIS 2.0 provides symmetrical US/DS data rates, which suits p2p applications better than asymmetrical data links since each peer can act as both a client and a server. We tested two scenarios:

- When there is congestion on the backbone network (Figure 11.9);
- When there is congestion on the US (Figure 11.10).

The background traffic on the backbone network and on the US link is generated using the SmartBits traffic generator. The average throughput per client is measured by averaging the overall traffic load with the number of clients. The individual traffic load variance is negligible while the overall traffic load for BT and FTP are the same. The measured results of the average throughput of one client are shown in Figure 11.11. As the number of clients (peers) increases, the throughput of one FTP client decreases while the throughput of one BT peer increases. In addition, as the background traffic increases, the FTP throughput decreases much faster than BT. In both experimental setups, the measured results show that p2p file distribution achieves higher throughput than FTP as the background traffic or as the number of peers increases. This is because information can be downloaded from multiple peers as opposed to one server in FTP. As the number of peers increase in BT, the downloading rate for all peers increases. BT outperforms FTP when at least two clients download at the same time, except when the background traffic is very light.

11.5 Real-Time Peer-to-Peer Streaming

In addition to p2p file sharing, video p2p streaming is another popular application. It is used in video Skype for video conferencing applications and for broadcasting live sports (http://www.myp2p.eu) and movies (http://www.vuze.com, http://www.babelgum.com). By using the p2p model, the video sender only needs to deliver the video stream directly to a limited number of peers, which will in turn, redirect the video content to other peers. Although each peer still keeps a control connection with the sender, the video traffic is delivered in a distributed manner.

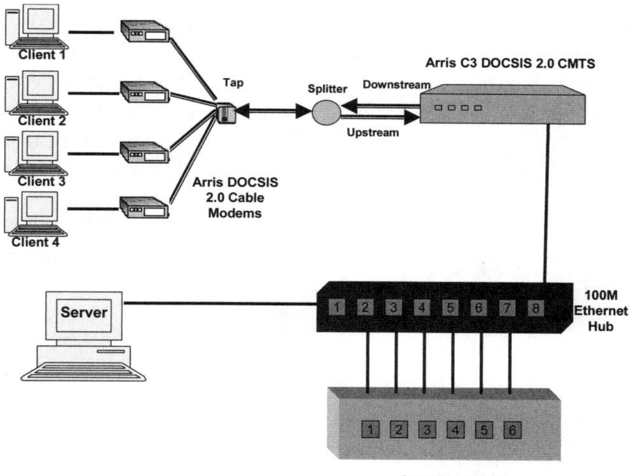

Figure 11.9: Background traffic on the backbone (Setup 1).

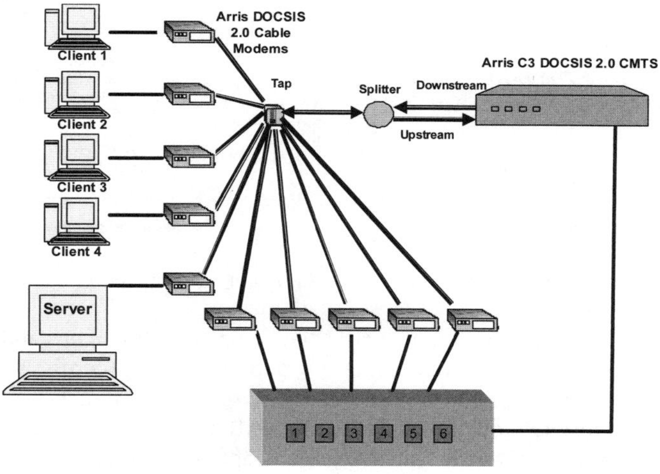

Figure 11.10: Background traffic on the upstream (Setup 2).

We tested real-time p2p video streaming experiment using the p2p radio software (http://p2p-radio.sourceforge.net, see also http://www.theora.org) over a DOCSIS 2.0 cable network with five peers (Table 11.1). P2p radio provides audio/video p2p Internet "broadcast" similar to a radio station broadcast and can deliver audio or video live or on-demand. Supported formats include MPEG Layer 3 (MP3), Ogg Vorbis, and Nullsoft Streaming Video (NSV). Figure 11.12 shows the experimental setup, which clearly reduces the bottleneck in unicast transmission (Figure 11.13). At the video sender, the NSVCAP software is used to capture and compress video from a camcorder in real time, and deliver the compressed video to a local SHOUTcast server (http://www.shoutcast.com). Each peer runs the p2p radio software, including the video sender, to distribute video content from SHOUTcast server to all other peers. The p2p stream relay using p2p radio is more scalable but may introduce longer time delay. However, noninteractive applications like video streaming are more sensitive to jitter and loss than a fixed delay. Each peer is configured to relay to no more than two other peers. In p2p radio, peers are dynamically connected in a tree architecture.

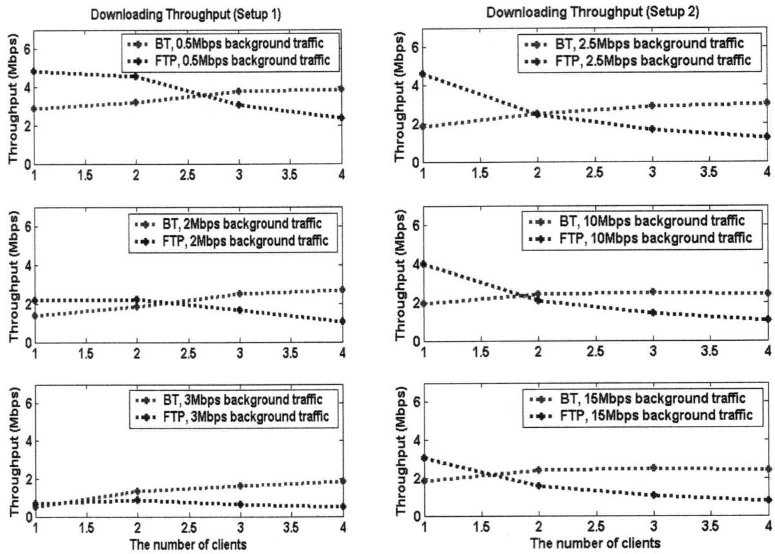

Figure 11.11: Peer-to-peer file sharing performance.

Good quality p2p streaming was achieved using 400 Kbps nullsoft video (NSV) that is captured real-time using a video camcorder (configured with CIF resolution). Voice information was also captured real-time using a microphone and synchronized with the video transmission. A Sniffer analyzer is used by each peer to monitor the real-time traffic flow. A screen capture of video sender is shown in Figure 11.14. The traffic map indicates the different peers exchanging information while the monitor shows the data pieces kept by a specific peer. All peers can view the video streaming smoothly. However, the video quality can be degraded during the transition period when a peer withdraws from the session and re-establish the tree topology.

Table 11.1: IP Addresses of Peers.

Computer	IP Address
Peer 1 (sender)	10.1.1.101
Peer 2	10.1.1.102
Peer 3	10.1.1.103
Peer 4	10.1.1.104
Peer 5	10.1.1.105

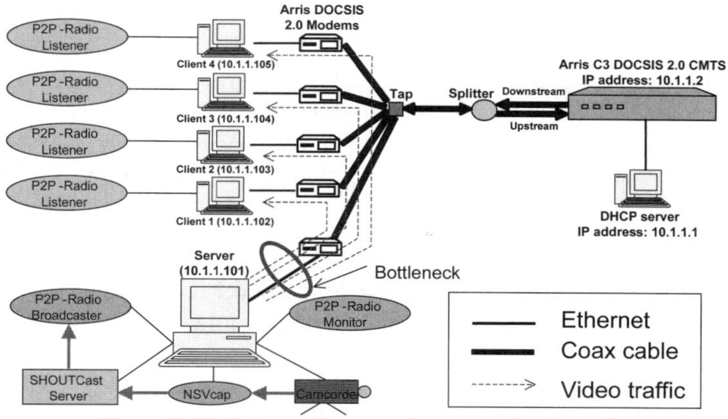

Figure 11.12: Unicast video streaming communication paths.

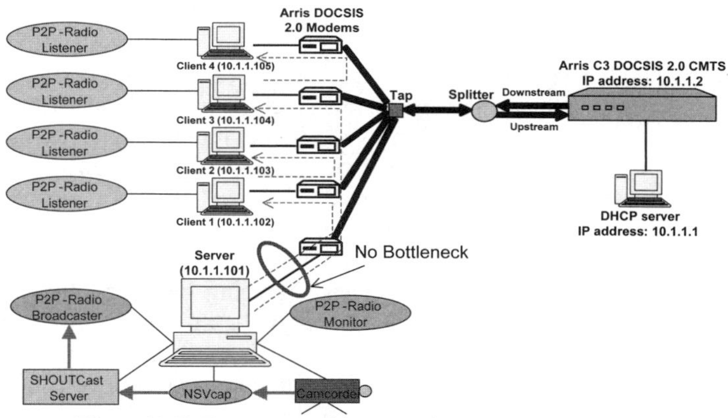

Figure 11.13: Peer-to-peer video streaming communication paths.

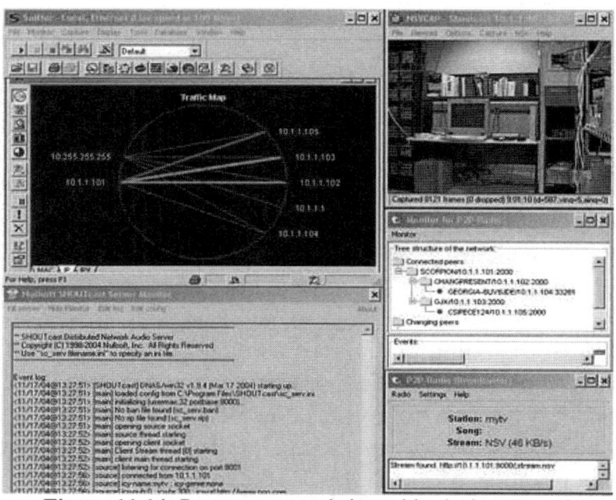

Figure 11.14: Peer-to-peer real-time video/voice streaming.

11.6 Video-Aware DOCSIS 3.0 Architecture

In a video aware DOCSIS 3.0 architecture, the forwarder and edge QAM modules are enhanced. At the forwarder, video traffic characterization and smoothing, policy and QoS management, and video-aware QAM resource scheduling and optimization can be performed. The efficient multiplexing of multiple streams of video traffic can be done at the edge QAM module. In this module, video transmission can be optimized at various levels: at the individual stream level (video stream smoothing), at the multiplexed stream level (all streams that fit a 256-QAM channel), and at the channel-bonding level (comprising multiple 256-QAM channels). Each optimization level can be performed individually or jointly.

11.7 Program Scheduling Challenges

The type of video service will determine the effectiveness of the scheduling policies. There are essentially two broad categories of video service:

- Interactive video service: the selected video program is transmitted to the subscriber via a dedicated channel at the exact moment the subscriber requests for it, enabling the video program to be started at anytime with controls of a digital video decorder (DVR): pause, rewind, fast-forward, and so forth.
- Quasi-video-on-demand service: transmission of a video program on a broadcast/multicast channel is based on the number of viewer's requests.

For cable systems that perform request batching based on the number of requests, it is difficult to determine the optimal "threshold" for each video program. The waiting times can be unpredictable if the requests are buffered for scheduling at later intervals. Moreover, popular programs need to be scheduled regularly while unpopular programs must not be prejudiced. The long-tail distribution of the popularity of the video program access pattern (see Figure 11.18) worsens this problem. One solution is to batch requests at regular intervals (e.g., every 5 minutes). There are two batching methods:

- Batching without queueing policies: requests batched during the scheduling interval are either granted or rejected so no requests are reconsidered in the next interval (fairness problem: unpopular video programs may never get granted due to insufficient bandwidth).
- Batching with queueing policies: requests that cannot be scheduled in the current interval are buffered for the next interval.

One method of batching without queueing policies is the batching with scheduling (BS) policy. Here, video programs are scheduled according to the number of requests. If the number of free channels is n, then the CMTS will select up to n video programs for broadcasting, and the video programs are ranked according to the number of requests. If the total number of requested video programs is more than n, the requests for video programs with fewer requests are rejected. The key benefit of this approach is the ability to provide DVR-like functions for popular video programs (since these programs are likely to be scheduled for broadcasting at each time interval). The main drawback is that it suffers from the "memory effect": during prime time, all channels are occupied at around the same time and the system remains saturated for a substantial period of time—new requests are rejected after the system becomes saturated. In addition, it can potentially suffer from unstable behavior: if the durations of all video programs are about the same and the request rate remains high, all channels will be released at about the same time and then occupied again quickly—the system sometimes accepts most of the requests while at other times, it rejects most of the requests. This reduces the request rejection rate of popular programs and can therefore be unfair to unpopular programs. Batching with preallocated channels (BP) solves the memory effect of the BS policy. Here, the number of free channels for scheduling at each interval has an upper bound that is computed beforehand. Unfortunately, this method also suffers from the same fairness problem as the BS policy.

One method of batching with queueing policies is the batching with accumulated waiting time (BA) policy. This takes into consideration both waiting time and the number of requests queued. It uses the sum of waiting times of all requests as a priority for scheduling a video program. The BA policy is biased towards popular video programs. Yet another method is batching with average waiting time (BM) policy. As the name implies, video programs are scheduled according to their average waiting times. Unlike the BA policy, the BM policy is biased towards unpopular video programs.

We note that the average waiting times of popular video programs will increase slowly (since there are many requests and the requests tend to accumulate). In an adaptive system such as the one shown in Figure 11.15, video programs are scheduled according to the average waiting time, the viewing pattern associated with the time dependence of the request rate and the popularity of video programs, and the overall bandwidth utilization. Unlike backbone networks where congestion in packet routing can increase the packet loss rates, in an access network, the time-dependent request rate is a key parameter that affects traffic variability and peak demands. When bandwidth is available and link conditions are good, the highest quality video will be transmitted. When bandwidth is available but if the channel is poor and/or if bandwidth utilization is heavy (e.g., during peak periods), error concealment and bandwidth conservation methods can be exploited to reduce the overall transmission bandwidth of the video. In doing so, the number of rejected requests is reduced (since more requests can be accepted if bandwidth is available) while fairness to video programs with different degrees of popularity can be maintained.

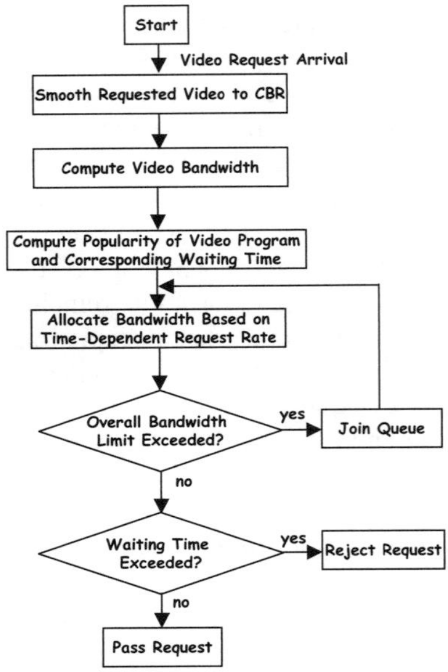

Figure 11.15: Proposed adaptive scheduling algorithm.

11.8 Simulation Model and Results

The simulation model and parameters are described next. We assume viewer requests arrive at the CMTS according to a random Poisson process. This implies that the aggregate arrival rate at time t as seen by the CMTS is normal distributed:

$$\lambda_{agg}(t) = \frac{N\frac{\lambda}{7}}{\sigma\sqrt{2\pi}} e^{-(t-\mu)^2/2\sigma^2} \tag{11.1}$$

where (in all simulations):
λ (user request rate per week) = 2,
σ = 45 minutes,
μ = 20 hours (or 8 p.m.),
N (number of subscribers) = 5,000.

Suppose the DOCSIS cable network reserves 500 channels for VoD/SDV services and the number of selectable video programs (L) is 200. The duration of the video program is uniformly distributed between 90 and 120 minutes.

The typical time-dependent request rate of a cable system is shown in Figure 11.16. Clearly, the normal distribution of (11.1) (plotted on Figure 11.17) can fit the peaks of the actual profile (that are shifted in time) although this may not be necessary—working solely on the highest peak (corresponding to the highest demand) may be sufficient since this will determine the maximum bandwidth requirements.

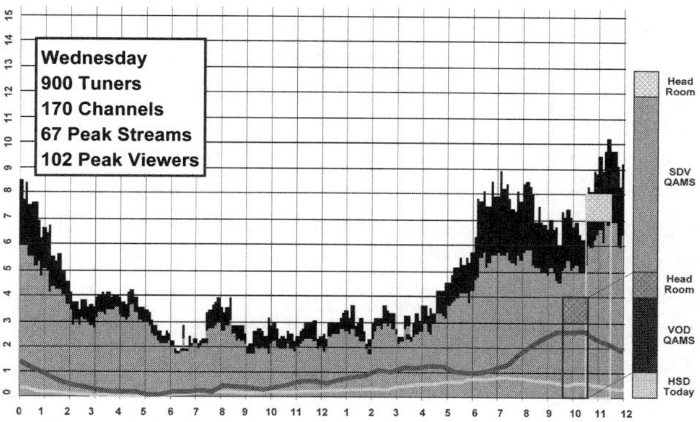

Figure 11.16: Typical time-dependent request rate distribution.

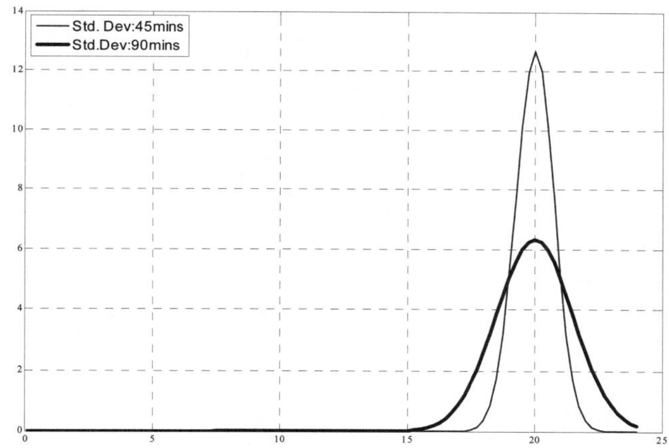

Figure 11.17: Plot of (11.1) with $\sigma = 45$ and 90.

The popularity of certain video programs [5, 6] can be exploited to conserve bandwidth. For instance, popular programs or channels can be broadcast continuously whereas less popular programs are broadcast only on-demand. The popularity of video programs can be modeled using the Zipf distribution where the probability that the *i*th video program is selected is given by:

$$p_i = \frac{1}{i^{1-q} \sum_{j=1}^{L} \frac{1}{j^{1-q}}} \tag{11.2}$$

where:
q is the skew factor;
L is the number of selectable video programs.

Values close to 0 yield a highly skewed distribution but results in better batching performance. Values close to 1 yield a uniform distribution.

The Zipf function with 200 video programs and a skew factor of $q = 0.3$ is shown in Figure 11.19. By comparing the actual video program access pattern (Figure 11.18), it is clear that there is a close match and that values of q close to 1 (i.e., uniform distribution) are not practical. The results are shown in Figure 11.20. Due to channel preallocation, the BP policy starts rejecting requests at an earlier time than the BS policy but is able to reduce the rejection rate at around the peak hour. The BA policy accepts more requests per day than the other policies. The BS policy without queueing suffers the worst performance. As shown, our proposed scheme performs the best when compared to the other policies. More specifically, the scheme is able to reduce the number of rejected requests during the peak period while maintaining the lowest overall rejection rate.

Service ID	Aggregate Viewing Minutes	Min per Day
1	74,391,837	34.96
2	65,636,334	30.84
3	57,167,771	26.86
4	51,591,374	24.24
5	47,214,135	22.19
6	45,827,624	21.54
7	44,971,682	21.13
8	42,719,368	20.07
9	41,623,165	19.56
10	40,795,084	19.17
11	35,242,590	16.56
12	33,486,559	15.74
13	29,028,231	13.64
14	28,409,579	13.35
15	28,202,088	13.25
16	27,103,379	12.74
17	25,969,624	12.20
18	24,437,261	11.48
19	23,540,362	11.06
20	21,740,764	10.22
21	20,129,669	9.46
22	19,448,868	9.14
23	19,263,275	9.05
24	19,199,877	9.02
25	18,963,005	8.91
26	18,850,166	8.86
27	18,783,985	8.83

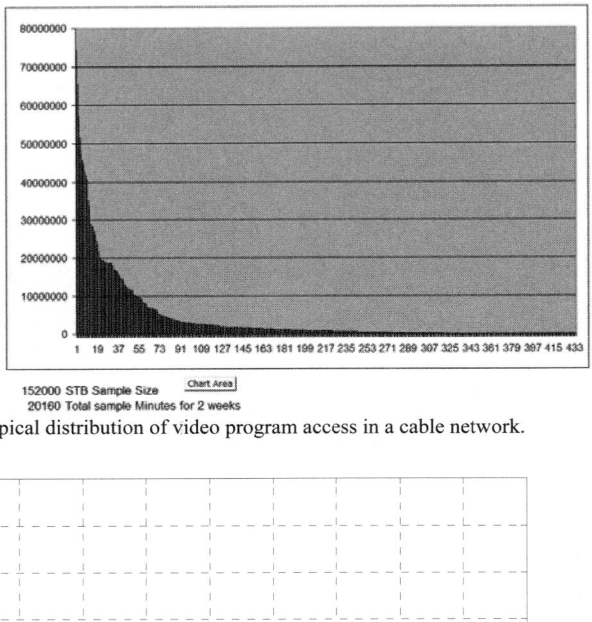

Figure 11.18: Typical distribution of video program access in a cable network.

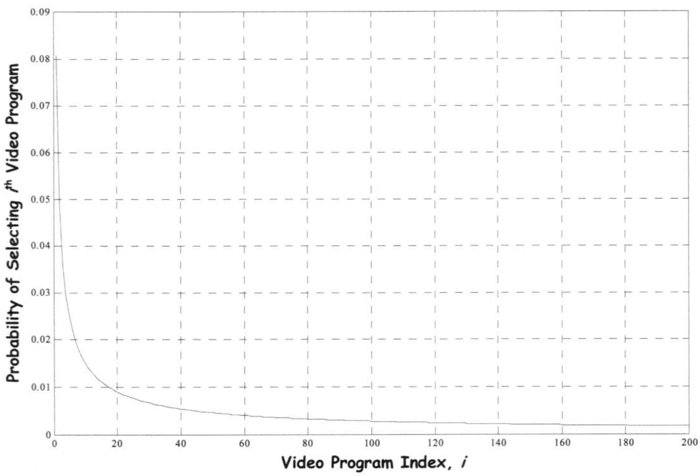

Figure 11.19: Zipf function with 200 video programs and skew factor $q = 0.3$.

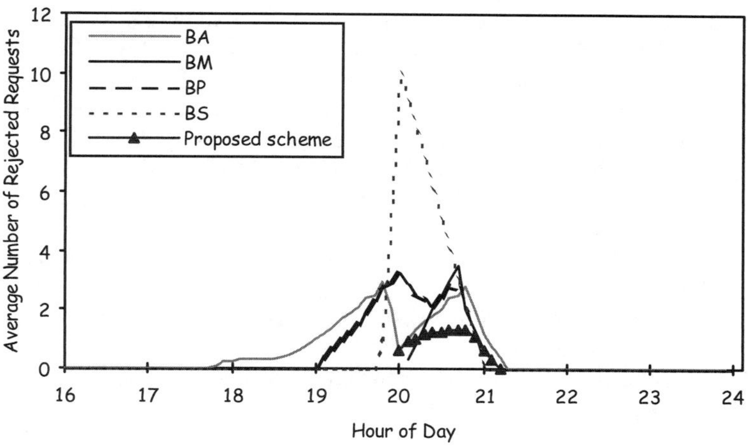

Figure 11.20: Performance of proposed scheme.

11.9 Conclusions

In this chapter, we have described a new video-aware DOCSIS 3.0 architecture, and evaluated the performance of an adaptive video program scheduling suitable for broadband cable access deployments. The primary video bandwidth savings are determined by the scheduling policies. The predictability of the video program access pattern and time-dependent request rate can be exploited to match the scheduling policies. These techniques can be used to enhance the forwarder and edge QAM modules of the DOCSIS 3.0 architecture.

In a commercial broadband access network, video programs are scheduled according to the average waiting time, the time dependence of the request rate, the overall bandwidth utilization, and the popularity of video programs. Unlike backbone networks where congestion in packet routing can increase the packet loss rates, in an access network, the time-dependent request rate is a key parameter that affects traffic variability and peak demands. When bandwidth is available and link conditions are good, the highest quality video will be transmitted. If the channel is poor and/or if bandwidth utilization is heavy (e.g., during peak periods), error concealment methods and bandwidth conservation methods can be exploited to reduce the overall transmission bandwidth of the video. In doing so, we can maintain the quality of the transmitted video while reducing the bit rate variability of the encoded video.

References

[1] V. Sdralia, C. Smythe, P. Tzerefos, and S. Cvetkovic, "Performance characterization of the MCNS DOCSIS 1.0 CATV protocol with Prioritized First Come First Serve Scheduling," *IEEE Transactions on Broadcasting*, Vol. 45, pp 196–205, June 1999.

[2] M. Droubi, N. Idirene, and C. Chen, "Dynamic-Bandwidth Allocation for the HFC DOCSIS MAC Protocol," *Proceedings of the 9th IEEE Conference on Computer Communications and Networks*, pp 54–60, October 2000.
[3] C. Heyaime-Duverge and V. K. Prabhu, "Traffic-Based Bandwidth Allocation for DOCSIS Cable Networks," *Proceedings of the 11th IEEE Conference on Computer Communications and Networks*, pp. 586–590, October 2002.
[4] J. Fijoleck, et al., *Cable Modems: Current Technologies and Applications*, IEC-IEEE, Press, 1999.
[5] J. Gong and Z. Zhang, "VoD Traffic Analysis," *CableLabs Technical Report*, July 2007.
[6] Top 25 Cable Program Networks, NCTA, 2009, http://www.ncta.com/Stats/TopNetworks.aspx.
[7] M. Davis, "Optimizing SDV Bandwidth Gains and Applications," *CableLabs Winter Conference*, March 2007.

Exercises

11.1. While p2p technologies have been used for various kinds of traffic, including file transfer (data), voice communications, and low-resolution video streaming, explain how p2p can be adapted for HD video streaming. Can 16 Kbyte sub-pieces be used? Justify your answer with some sample figures using a 720p video and a HTTP overhead of 40 bytes per IP packet.

11.2. Is ABR with progressive streaming different from p2p streaming? Since both technologies employ HTTP, can ABR be applied to p2p systems? Compare the impact of network congestion on these two streaming approaches.

11.3. Supporting QoS provisioning may not necessarily lead to better service for all users of the access network. For instance, if a headend starts prioritizing the packets, then users with no prioritization will suffer in performance, even with a broadband connection. Will this constitute discrimination for unprioritized users? Note that service pricing is critical. For example, QoS cannot be enforced using Skype (why?) but many people choose to use the service because it is free.

11.4. Skype addresses network address translation traversal issues by dynamically assigning peers to function as proxies and distributed file-sharing servers. How will this impact video streaming?

Chapter 12

Intelligent Activity Detection Techniques for Advanced Video Surveillance Systems

Video surveillance systems are becoming increasingly popular due to the emergence of high-speed wireless Internet (enabled by WiMax and LTE), bandwidth-efficient video compression schemes (such as H.264), and low-cost (and high-resolution) IP video cameras. This chapter presents two applications of an advanced surveillance system, specifically in suspicious activity detection and human fall detection, for both indoor and outdoor environments. The implemented prototype captures and analyzes live 1080p high-definition (HD) video that is streamed from a remote camera. We will show that by combining the strengths of ellipse modeling and shadow removal and other novel algorithms, the false alarms in the detection can be significantly reduced.

12.1 Introduction

With 4G wireless systems such as WiMax currently being deployed worldwide (over 500 deployments in 145 countries), network operators, including major U.S. WiMax operator Clearwire, recognize digital video surveillance as a core service and application [1]. Such surveillance systems already complement many existing WiFi mesh network installations [2] and can be set up quickly to monitor events of interest, including natural disasters. These systems will become even more popular when alternative 4G wireless systems such as LTE are fully deployed. In addition, because video surveillance systems can improve public safety and transportation significantly, it has become a national initiative in several countries and a key strategic plan for many municipal governments to deploy these systems.

The demand for video surveillance systems is not only driven by high-bandwidth wireless systems. Highly efficient video codec such as H.264 are now readily available. H.264 typically provides a two fold improvement in compression efficiency over legacy MPEG-2 codec. Coupled with inexpensive and high-resolution IP-enabled cameras (a 1080p webcam now costs less than $60), crystal-clear live videos can be captured, compressed, and relayed to a remote monitoring office.

Many video surveillance systems aim to minimize human dependency by notifying the operator only of salient information as it occurs, and engaging the operator minimally to alter platform operations. Some advanced surveillance tasks involve detection of newly added objects in a rapidly changing scene or a complex background (e.g., a bag left unattended at a train station for several minutes). These tasks enable only the pertinent information of interest to be captured, and can filter unwanted background movement (e.g., snow, rain, falling leaves).

In this chapter, we present an intelligent video surveillance system that includes advanced detection methods. The real-time system is set up using a 1080p webcam, a loss-free streaming protocol, a campus-wide WiFi network that is connected to the Internet, and a real-time player running on a laptop (Figure 12.1). This is the first time a 1080p HD video surveillance scheme has been reported.

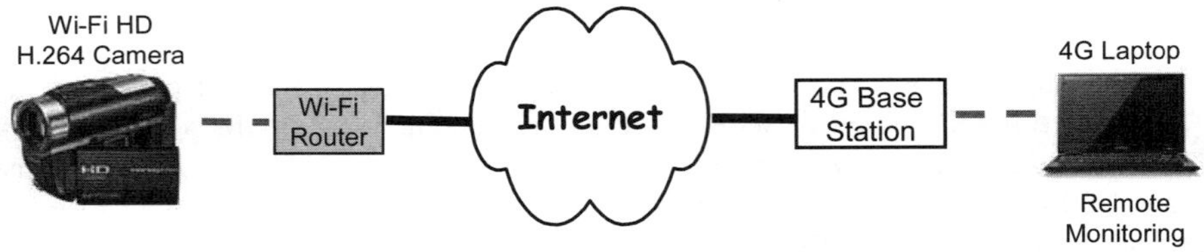

Figure 12.1: Real-time 1080p HD video surveillance system setup.

12.2 System Overview

12.2.1 Suspicious Activity Detection

The first task is to model the background. Although existing techniques such as frame differencing, median filtering, Kalman filtering [3], mixture of Gaussian [4], and Baye's algorithm [5] generally perform well, they assume a simple background. Static objects added in the modeled background are compared with a reference background and the newly added object can thus be detected. Another approach is to use midground object detection, which classifies objects into multiple realms [6]. A midground realm is a temporal window that not only follows the object's appearance but also employs adaptive background modeling to detect scene variations (e.g., occlusion, small brightness changes). Although such an algorithm can model a more complex background, it suffers from high complexity and may not be suited for real-time surveillance applications.

Our system uses the background modeling algorithm described in [4] with some important modifications. For instance, we classify the background and foreground information based on Baye's decision theory, so that background from a real-time video stream containing a complex background can be modeled. In an outdoor environment, the background can be quite complex with moving objects (e.g., humans, vehicles) that can potentially interfere with the activity detection. The block diagram of our background modeling algorithm is shown in Figure 12.2. It consists of four parts:

- Change detection;
- Change classification;
- Foreground object segmentation;
- Background learning and maintenance.

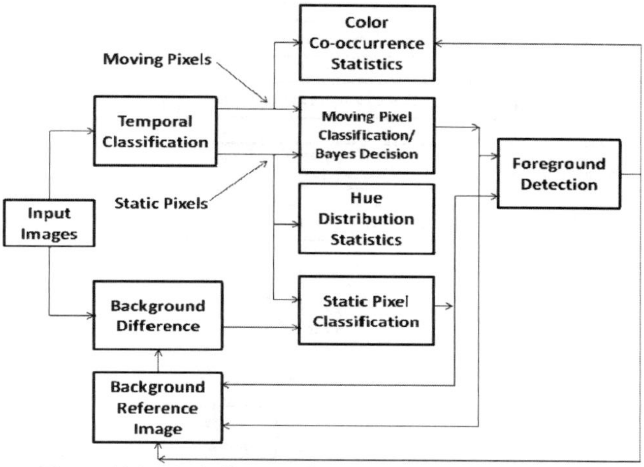

Figure 12.2: Block diagram of background modeling algorithm.

The first step involves removing static pixels using simple background and temporal change detection techniques. These detected pixels are labeled as pixels belonging to stationary and moving objects according to interframe pixel modification. In the next step, by using the statistics of color and color co-occurrences acquired from the Baye's decision rule, the pixels are further classified as foreground or background pixels. In the third step, segmentation of foreground objects are carried out by combining the classification results from both stationary and moving parts. In the fourth step, background models are updated. Meanwhile, a reference background image is maintained to ensure that the background difference is accurate, up-to-date, and adaptive to the changing background. The thresholding algorithm used for change detection in the first step adapts itself to changing scenes, and varying illumination. This is necessary since a predetermined threshold parameter may fail when the scene illumination deviates from the ideal value. For example, if parameters are prechosen to be daylight-centric, the system may fail at night due owing to reduced illumination.

Having modeled the background, the next step involves detecting the suspicious activity. Our algorithm continuously updates a buffer of images that contains the background models generated from the streamed images. Median filtering is then applied on the obtained buffer. The advantage of using median filtering over averaging is that it can eliminate the effect of large noise values that may have occurred in the background modeling step. The filtered image is then compared with a previously stored reference of the background. Static changes in the background can thus be detected by simply subtracting the two images. If the change detected is bigger than a threshold, the operator is notified. Subsequent new frames are stored in the buffer and the reference background frame is updated. The flowchart of our detection algorithm is shown in Figure 12.3.

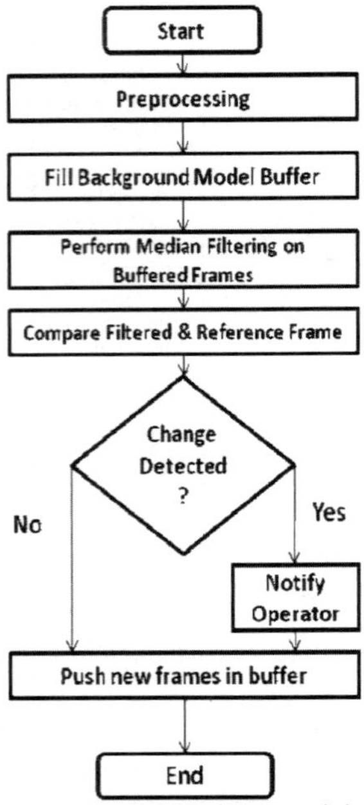

Figure 12.3: Flowchart of suspicious activity detection.

12.2.2 Human Fall Detection

In teleheath applications, the majority of falls for old people can prove to be fatal if no medical assistance is provided within a few minutes. The typical solution to detect falls is to employ wearable sensors such as accelerometers or help buttons. However, the key drawback of such detectors is that older people often forget to wear them. Moreover, a help button can be useless if the person is unconscious or immobilized. Our system uses video image processing to detect falls and notify a centralized operator of an impending emergency. We implemented a modified version of the fall detection mechanism described in [7]. The authors in [7] propose ellipse fitting on motion mask obtained by computing the motion history images (MHI) of the scene. We enhanced the system using shadow removal to reduce the number of false detections.

There are many techniques where motion can be accurately detected. Examples include optical flow and MHI. The drawback with optical flow is that it does not provide a very high density of motion vectors and hence, it cannot be used for abrupt human falls. MHI directly encode the system time into the color information of the scene. Pixels representing recent motion are allotted brighter color intensities.

We first obtain a silhouette of the region of interest using Fourier descriptors (FD). The MHI is then computed using a two-step procedure [8]:

- Obtain the silhouette, $D(x, y, t)$ of the motion by a simple frame differencing (FD) technique.
- Compute the motion history of a pixel $H(x, y, t)$ as

$$H(x,y,t)=\begin{cases} k, & \text{if } D(x,y,t)=1 \\ \max(0, H(x,y,t-1)-1), & \text{otherwise} \end{cases} \tag{12.1}$$

Once the MHI is obtained, an ellipse is fitted around the pixels with intensity of more than 250. This value is chosen because it represents the most recent set of pixels. The reason an ellipse is chosen instead of a rectangle or a circle is because the ellipse approximates, with minimum area, the upright shape of a human. Once the ellipse is approximated, its parameters like eccentricity, center of ellipse, and angle of rotation are analyzed to detect falls. Standard deviations are calculated for the eccentricity (σ_ρ), angle of orientation (σ_θ) of the ellipse, and the coordinates of center of the ellipse. There are large variations in these three parameters whenever a sudden motion is observed in the frame. These variations can be quantified and analyzed to deduce a probable fall in the video frame. The basic assumption is that the person becomes immobilized on the ground after a possible fall. Based on this assumption, the ultimate step is to check for zero movement in the frame. If there is no movement for more than say 5 seconds, then a fall is detected. The flowchart of the fall detection implementation is depicted in Figure 12.4.

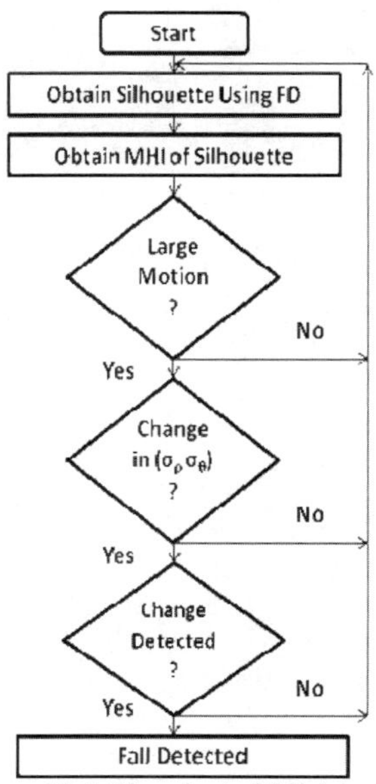

Figure 12.4: Flowchart of fall detection.

12.3 Experimental Results

Internet streaming imposes additional constraints on the coded video size, delivery times, and may suffer information loss due to packet losses. A variety of IP-enabled cameras can be used for streaming. Coding standards such as H.264 allows for efficient usage of bandwidth. Our system was implemented using a standard PC at frontend equipped with a HD camera. The video was streamed using the HTTP protocol. Dynamic rate control was implemented to match the video processing rate and streaming rate. Processing of the video stream was performed on Microsoft Visual C++ using the OpenCV library [9].

12.3.1 Suspicious Activity Detection

Figure 12.5 shows a snapshot of the original scene and its corresponding modeled background. In Figure 12.6, the person drops a bag and leaves the background. This bag is detected and a bounding box is drawn around the object of interest, as shown in Figure 12.7. Figures 12.8 and 12.9 show similar detection capabilities in an outdoor setting and the corresponding modeled backgrounds.

Figure 12.5: (a) Original scene and (b) modeled background.

Figure 12.6: (a) Person leaves bag behind and (b) modeled background.

Figure 12.7: (a) Suspicious object detected and (b) modeled background.

Figure 12.8: (a) Suspicious car detected and (b) modeled background.

Figure 12.9: (a) Car breakdown and (b) modeled background.

12.3.2 Human Fall Detection

We tested 15 test sequences containing normal day-to-day activities interspersed with different types of fall. After intensive testing, the threshold parameters chosen for mean motion, standard deviation of eccentricity (σ_ρ), and standard deviation of angle of orientation (σ_θ) are 50, 0.15, and 30, respectively. These parameters were obtained after multiple trial-and-error processes, and appear reasonable in representing a fall. A motion is labeled as a fall when the values of the parameters are more than their thresholds. Snapshots of the implementation on a few frames with its corresponding MHI are shown in Figures 12.10 to 12.13. The fall detection algorithm was able to distinguish between a person sitting down or walking normally versus someone falling.

The fall detection algorithm can be extended to detect multiple persons in the scene by modeling ellipses on individual MHI blobs (Figure 12.14). However, occlusion may pose an issue and this will be investigated in future work. Nevertheless, such a situation is less common in telehealth applications since single old people are usually targeted.

$$\text{Motion} = \frac{\sum\limits_{\text{moving pixels}} H(x,y,t)}{\text{Number of Moving Pixels}} \tag{12.2}$$

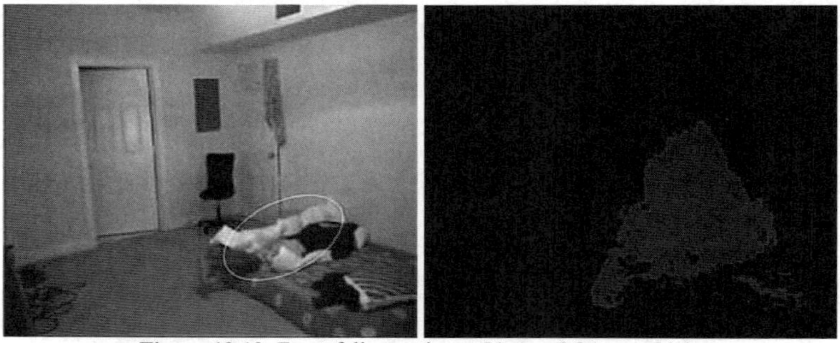

Figure 12.10: Front fall – motion = 54, $\sigma_\rho = 0.24$, $\sigma_\theta = 30.3$.

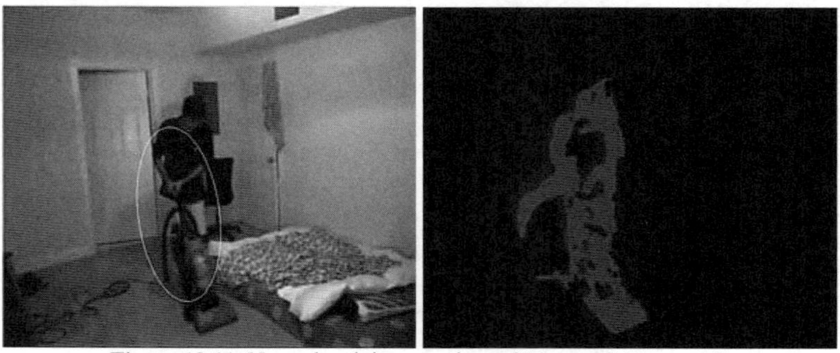

Figure 12.11: Normal activity – motion = 80.1, $\sigma_\rho = 0.11$, $\sigma_\theta = 15$.

Figure 12.12: Sitting down – motion = 81, $\sigma_\rho = 0.04$, $\sigma_\theta = 4$.

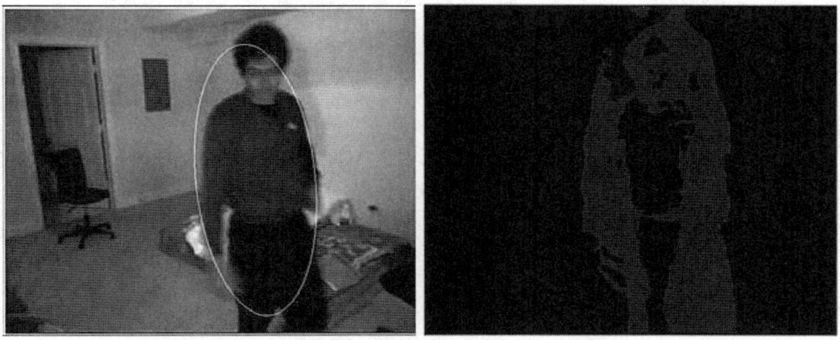

Figure 12.13: Walking – motion = 75, $\sigma_\rho = 0.089$, $\sigma_\theta = 8$.

Figure 12.14: Fall detection for multiple persons – $\text{motion}_{\text{standing}} = 70$, $\sigma_{\rho\text{-standing}} = 0.076$, $\sigma_{\theta\text{-standing}} = 7$, $\text{Motion}_{\text{falling}} = 70$, $\sigma_{\rho\text{-falling}} = 0.16$, $\sigma_{\theta\text{-falling}} = 35$.

12.3.3 Shadow Removal Enhancement

It is often necessary to separate shadows from the moving objects in applications such as fall detection. The effect of a shadow is pronounced in fall detection because it can cause an abnormal inflation of the modeled ellipse size [11] as shown in Figure 12.15. Under usual indoor lighting conditions, shadows tend to have a weak edge and hence it will not be detected by relying entirely on edge detection schemes such as a Canny Filter [10]. The block diagram of our implementation is shown in Figure 12.16. Figure 12.17(c) shows the elimination of the shadow that was captured in the frame differencing step in Figure 12.17(b). As can be seen from Table 12.1, shadow removal improves the detection of a fall.

Table 12.1: Fall Detection Enhancement Using Shadow Removal

	Positives	False Positives
Fall detection without shadow removal	10	4
Fall detection with shadow removal	12	1

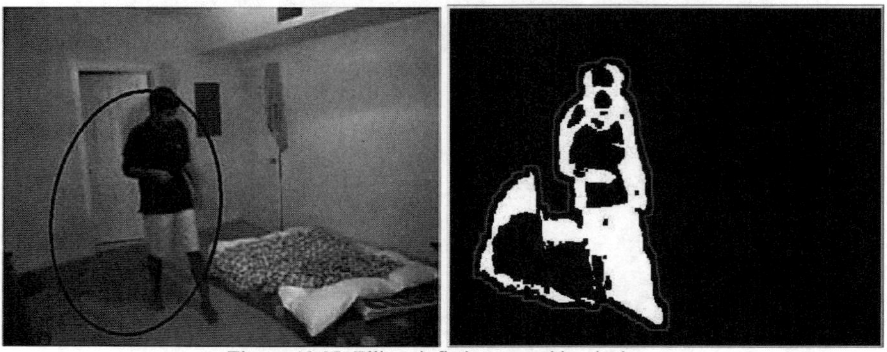

Figure 12.15: Ellipse inflation caused by shadow.

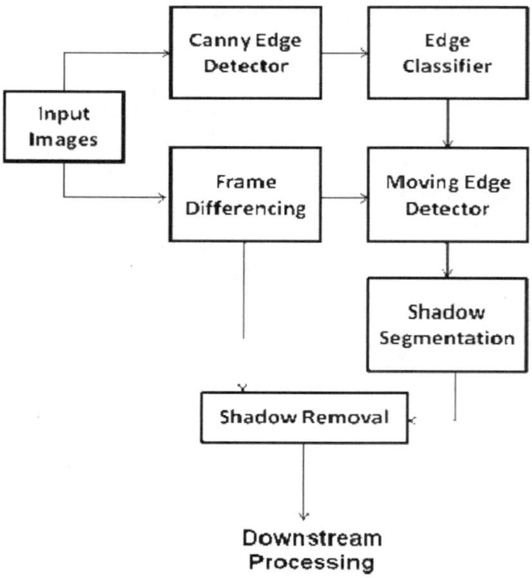

Figure 12.16: Block diagram of shadow removal technique.

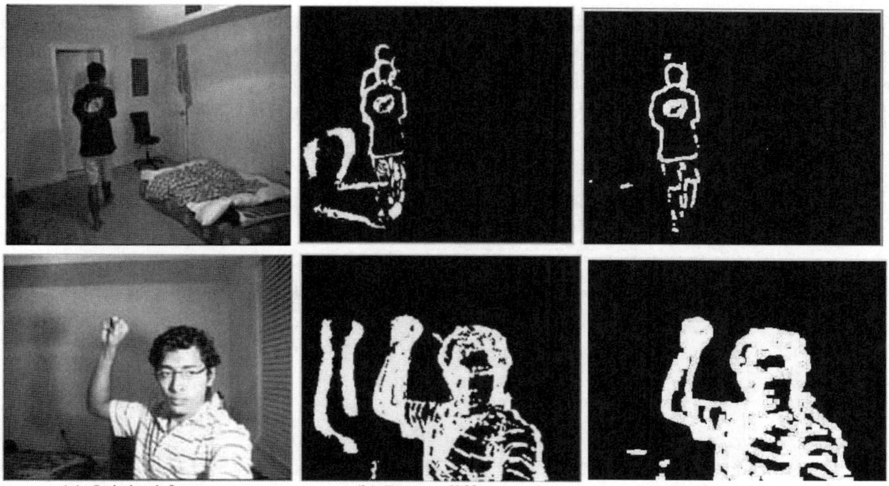

(a) Original frame (b) Frame differenced (c) Shadow eliminated

Figure 12.17: Implemented shadow removal prototype.

12.4 Conclusions

We have presented two key activity detection methods in this chapter. The suspicious activity detection method has widespread applications including detecting vehicles that have stopped for an unusually long amount of time on interstate highways as well as detecting suspicious objects in a busy scene. A robust shadow-elimination implementation removes shadows from the downstream processing steps and improves the accuracy of human fall detection. Our system currently streams H.264 HD videos, which is converted to the raw RGB video by the OpenCV system. This increases the processing time. We plan to investigate if the H.264 video can be directly dissected to obtain MVs corresponding to 16 × 16 MBs and these vectors can then be used to generate a motion mask for use by the downstream processing modules. Because the 16 × 16 MBs are much smaller than a video frame, this may speed up the processing of some of the motion information considerably. The false detections may be further reduced by defining normal zones of inactivity as described in [12, 13].

References

[1] B. Bowman, "Sprint Clearwire Comcast Time Warner–Their hold on 2.5 GHz Educational Broadband Spectrum (EBS)," 2008, http://www.accessdelray.org/pdf/Municipal%20Broadband%20Future%20%28print%29.pdf.

[2] K. Russell, "Instant Deployment of Video Surveillance for the Boston Marathon Enabled by Strix Systems Wireless Mesh Technology," April 2008, http://www.strixsystems.com/press/boston-marathon.asp.

[3] S. Cheung, et al., "Robust Techniques for Background Subtraction in Urban Traffic Video," *Video Communication and Image Processing*, Vol. 5308, pp 881–892, SPIE Electronic Imaging, San Jose, January 2005.

[4] Liyuan Li, et al., "Foreground Object Detection from Videos Containing Complex Background," *ACM Multimedia*, November 2–8, 2003, UC Berkeley.

[5] N. Friedman, et al., "Image Segmentation in Video Sequences: A Probabilistic Approach," *Proceedings of the Thirteenth Annual Conference on Uncertainty in Artificial Intelligence*, San Francisco, CA, May 1997.

[6] L. Wills, et al., "Midground Object Detection in Real World Video Scenes," *Proceedings of the IEEE International Conference on Advanced Video and Signal-based Surveillance*, London, England, September 2007.

[7] R. Caroline, et al., "Fall Detection from Human Shape and Motion History using Video Surveillance," *Advanced Information Networking and Applications Workshop*, Ontario, Canada, May 2007.

[8] G. Bradski, et al., "Motion Segmentation and Pose Recognition with Motion History Gradients," *Machine Vision and Applications*, 2002.

[9] OpenCV, http://opencv.willowgarage.com/documentation/index.html.

[10] I. Canny, "A Computational Approach to Edge Detection," *IEEE Transactions on Pattern Analysis and Machine Intelligence*, Vol. 8, pp. 679–698, 1986.

[11] Dong Xu, et al., "Indoor Shadow Detection for Video Segmentation," *International Conference on Multimedia and Expo*, 2004.

[12] T. Lee, et al., "An Intelligent Emergency Response System: Preliminary Development and Testing of Automated Fall Detection," *Journal of Telemedicine and Telecare*, pp. 194–198, 2005.

[13] H. Nait-Charif, et al., "Activity Summarization and Fall Detection in a Supportive Home Environment," *Proceedings of the 17th International Conference on Pattern Recognition (ICPR)*, Vol. 4, pp. 323–326, 2004.

Chapter 13

Hand Gesture Control for Broadband-Enabled HDTVs and Multimedia PCs

This chapter presents the development of a gesture-based user interface that gives the consumer unrivaled living-room experience while enjoying premium-quality videos. The human-TV interface eliminates the need to lean forward and makes channel flipping easy, futuristic, and fun. To this end, we describe several methods for hand gesture detection and recognition. These methods aim at recognizing hand signs and positions using a single webcam, which may be used to control a broadband-enabled HDTV. The hand gesture can be trained to suit the user preference. We first provide an analysis of pattern matching, histogram back projection, and the use of Fourier's descriptors. These methods achieve good reliability and acceptable resource consumption. We compare these methods with a new method based on H.264 motion vectors that directly analyzes the captured video in the compressed domain. It will be shown that this technique provides a faster and accurate way to recognize motion trajectories. The extracted gesture or trajectory information can then be used for various multimedia applications, including improving human-TV and human-PC interaction. Our prototype recognizes gestures for reasonably long distances and is suitable for use with even the largest HDTV screens. We also present a method to control the movement of a mouse cursor using simple hand gestures and a webcam. A real-time tracking algorithm is implemented based on adaptive skin detection and motion analysis. Using the history of motion, the trajectory of the hand's movement is drawn and used to identify a gesture, without a hand reference. A region of interest algorithm is employed to scale the motion when the user is located far away from the point of capture and the mouse cursor's motion is scaled accordingly. Finally, we illustrate the use of a 3D webcam to automatically extract a hand reference. This removes the need to prerecord and store hand references for gesture recognition.

13.1 Introduction

Hand gestures enable humans to communicate naturally. The recognition of these gestures leads to many applications. For example, hand gestures can be used to characterize human activity such as interpreting sign language and music conducting action [1, 2]. Hand gestures can also be used to control the behavior of robots to help humans [3]. An active research area is the use of hand gesture recognition for human computer interaction. Some work has been done to allow gestures to recognize letters and simple signs [4].

Gesture control and movement tracking are popular in new game consoles and are largely based on proprietary accelerometer and optical sensor hardware. For example, the Wii remote has the ability to sense acceleration along three axes using an accelerometer sensor. Project Natal is a software technology that enables advanced gesture recognition, facial recognition, and voice recognition for computer games (http://www.youtube.com/xboxprojectnatal). Project Natal does not require any game remotes but employs infrared LED sensors that are housed in a sensor bar. The depth sensor consists of an infrared projector combined with a monochrome sensor, and allows the sensor to see in 3D under any ambient light conditions. The sensing range of the depth sensor is adjustable. The skeletal mapping technology is capable of simultaneously tracking up to four users for motion analysis, with a feature extraction of 48 skeletal points on a human body at a frame rate of 30 Hz. Depending on the subject's distance from the sensor, Project Natal is capable of tracking and identifying individual fingers.

With the emergence broadband-enabled HDTVs, the Web and TV are converging. Instead of using the remote, we can employ a webcam and gesture recognition to improve human-TV interaction. Such a touch-free and wireless approach removes the need to lean forward when browsing, selecting, or replaying videos, and can be extended to cursor control for multimedia PCs as well as next-generation game consoles.

13.1.1 Related Work

Different techniques have been used to detect and track hand gestures. One of the most common methods is to wear colored gloves or colored tags on the fingers to identify the hands [5]. In this case, color detection is used. However, adding colored gloves will be an inconvenience to simple human-TV interaction. In addition, the system will fail if the background has a similar colored object. Another common way is to track hand gestures using two cameras. The cameras can be placed in front or on the side [5] to enable triangulation to identify the exact 3D position and location of the hand. Similarly, a stereoscopic 3D camera [2], which consists of two cameras positioned side by side at a distance corresponding to the distance between the eyes, can provide depth information to simplify hand gesture recognition. Although these methods achieve good results, the need for multiple cameras may not be practical in some situations. For example, some desktop monitors and laptops only come with a single embedded webcam. A third method is to use infrared sensors, which requires special hardware. Using infrared sensors may also limit the number of gestures that can be recognized. With our system, there are virtually unlimited gestures that we can train, including numbers and letters. The ultimate goal is to be able to write in the air and be able to capture the words on the screen. Technically, this is very challenging, especially when using a single webcam. Commercially, it has many potential applications and setup can be achieved easily using software.

A simple way of describing a complex scene is to use background modeling and extraction. Static or slowly adaptive background modeling [7] requires the scene to be static at the beginning of the tracking, which may not be the case for a living room environment. For example, someone may stand in front of the TV when it is turned on. Adaptive background models [8] perform better and can be used to select regions of interest in the image to speed up processing, but still suffer from the same problem as described previously.

Skin color detection is a common method to detect the hand [4, 6, 10]. This method works in most situations but requires either extracting the skin color of the hand or making assumptions about it [6]. Making loose assumptions on the skin color may result in false positives, hence other techniques must be included to improve the reliability of the hand recognition. In addition, extraction of the skin color must be performed regularly as it is sensitive to indoor lighting variation.

Another approach is to detect edges using either a Sobel filter or a Canny filter, and then extract a hand shape from these edges [10]. This method appears to be efficient on a clear background. Nonetheless, these filters are sensitive to detected edges and in particular, to edges located in the background. This is why these filters are often combined with background modeling.

Yet another method to remove background information is to use some pattern matching methods that search for a pattern as defined by a set of references. As these methods take up much processing time, a decision tree using small Haar-like features and Adaboost method have been reported [9]. These pattern-matching methods require a very large database and a long training time. In order not to rely on a large hand gesture database, we prefer to use a simpler pattern matching method based on the user's hand reference.

Trajectory analysis can be performed by many ways, either by simple matching such as features points or direction [3]. More complex modeling theories have emerged to recognize complex gesture trajectories using hidden Markov models [2] or neural networks. These methods require a lot of training and therefore cannot adapt to user-specific gestures and cannot be used for real-time applications such as video browsing.

We wish to recognize gestures on a complex background such as inside the home, which can be potentially surrounded by moving persons and objects in the background. The webcam is either embedded in the screen of the computer or placed on top with the user facing it. As the gesture and motion we aim to recognize are simpler, we intend to first use a very basic method called "motion pattern matching." Then we design a way to handle complex backgrounds using skin color detection and the use of Fourier's descriptors to handle the matching. These methods will be designated as "skin color matching" and "Fourier's descriptors" respectively. We will show a new method using the H.264 motion vector (MV) to perform real time gesture recognition. Although the H.264 compressed video has been used to classify videos according to the MVs [11], this work is simplified in the sense that it does not rely on any assumptions about the camera position or motion type. We will show that the hand size and a fix camera can be used to perform specific analysis of trajectory gestures captured with the H.264 video codec. All methods are experimented on 640 × 480 videos acquired with different webcams. The goal is to process at 25 Hz and perform real-time processing using a standard PC. We employ the OpenCV library [12].

13.2 Gesture Matching Methods

13.2.1 Motion Pattern Matching

This method computes the difference between two successive video frames, which is usually stable enough to enable hand recognition using a reference. Figure 13.1 presents the flowchart of this method.

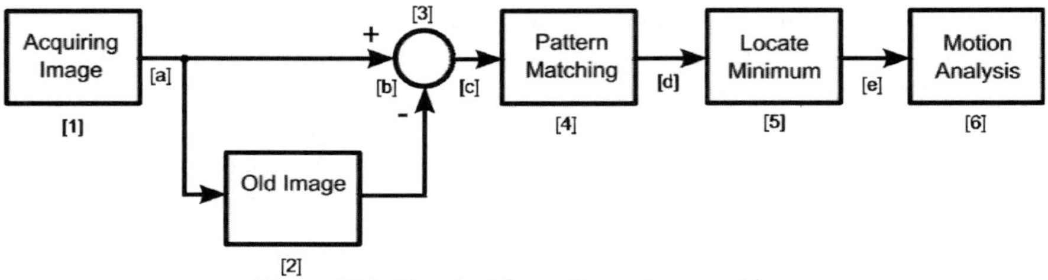

Figure 13.1: Flowchart for motion pattern matching.

We first compute the difference between the current and previous images, and convert it to grayscale. Then, we establish a distance map between a reference and the motion observed with the webcam. The map is obtained by computing the difference between the grayscale image and the reference throughout the grayscale image. Equation (13.1) is used to create the correlation map:

$$\mathrm{Dist}(x,y)=\sum_{x',y'}\left(\mathrm{Img}(x'+x,y'+y)-\mathrm{Ref}(x',y')\right)^2 \tag{13.1}$$

where Img is the residual image resulting from the difference of two consecutive images, Ref is the reference image, (x, y) are the coordinates of the destination, and (x', y') the coordinates of the reference.

We select the maximum correlation (minimum residual) and compare it to a threshold. If the threshold is reached, we can accurately predict that the hand is at the desired position. The reference from the user must be acquired offline. Figure 13.2 presents two examples of the patterns to be matched and a difference (residual) frame. The major drawback of this method is inherent in all pattern matching methods: the processing time is very long for one pattern but increases proportionally to the number of patterns to match, including scaled references. This means that we can only process either a few references with a constant scale or use a wider scale with only one reference. Both reference and motion maps obtained from the webcam are downsampled to reduce the processing time. Table 13.1 presents the impact of downsampling and the amount of scaling applied to the reference. We choose to use a range of 4 for scaling and a downsampling factor of 4, which is a reasonable choice in terms of minimizing processing time. According to Table 13.1, in this case, the processing time is 13.1 ms, the total time to match two different gestures becomes 26.2 ms. Note that a larger downsampling factor affects our method's reliability whereas a lower one will not allow the processing of several references.

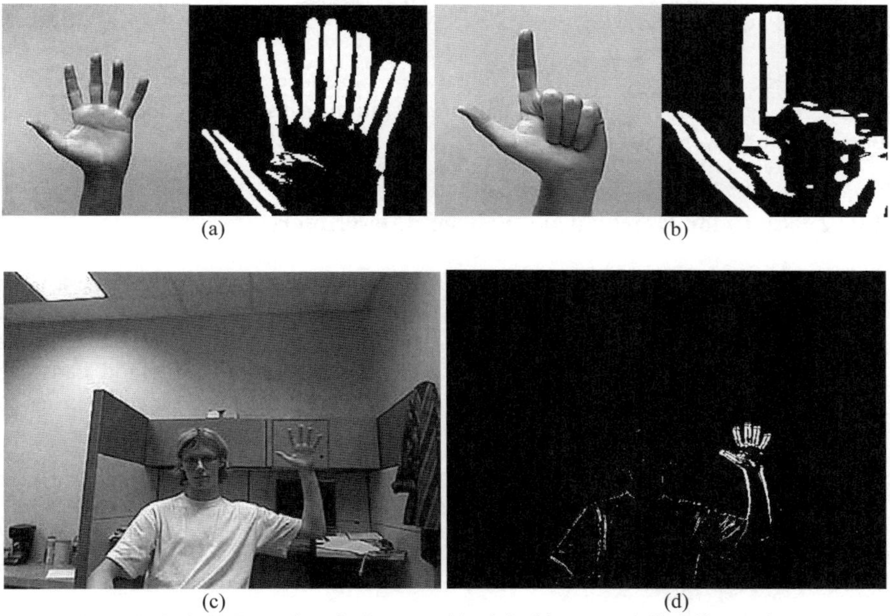

Figure 13.2: (a, b) Examples of references, (c) original image, and (d) difference image.

Figure 13.3 presents the variation of processing time and memory usage according to the steps of the flowchart using the previous parameters. The total processing time, including image acquisition and display, is approximately 40 ms, which gives the desired frame rate of 25 Hz. We can see that step 4 incurs the most processing time because it requires much computation for the scaling of each reference. This algorithm can be applied to simple hand recognition but cannot accommodate many different hand gestures or a large set of distances between the user and the display.

Table 13.1: Processing Time using Different Downsampling Range and Amount of Scaling

Scaling Factor	**Downsampling Range (ms)**					
	1	**2**	**3**	**4**	**5**	**6**
1	47.1	11.6	5.41	2.89	2.02	1.31
2	95.3	22.9	11.5	6.56	3.53	2.61
4	192	46.6	22.2	13.1	8.30	5.83
6	289	72.6	32.2	20.0	12.6	8.80
8	394	99.0	44.1	23.5	16.8	12.6

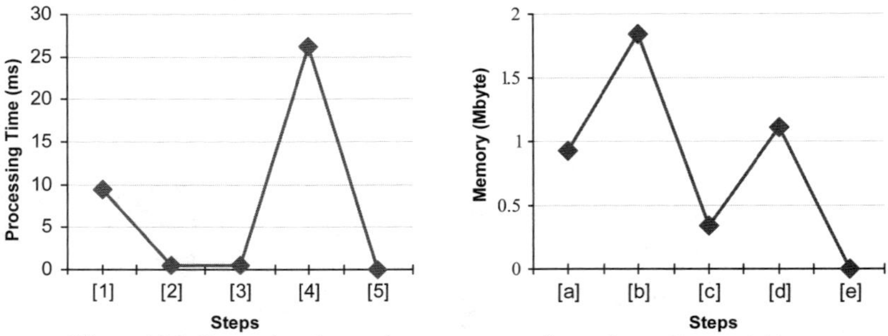

Figure 13.3: Processing time and memory usage for motion pattern matching.

13.2.2 Skin Color Matching and Fourier's Descriptors

This method extracts pixels with the skin color and attempts to recognize a hand gesture from the contours. Figure 13.4 presents the flowchart of this method. During calibration time, the skin color is detected by using the color of the cheeks retrieved from face detection. The previous pattern matching method can also provide a way to initially detect the hand and record the skin color in order to initialize this algorithm.

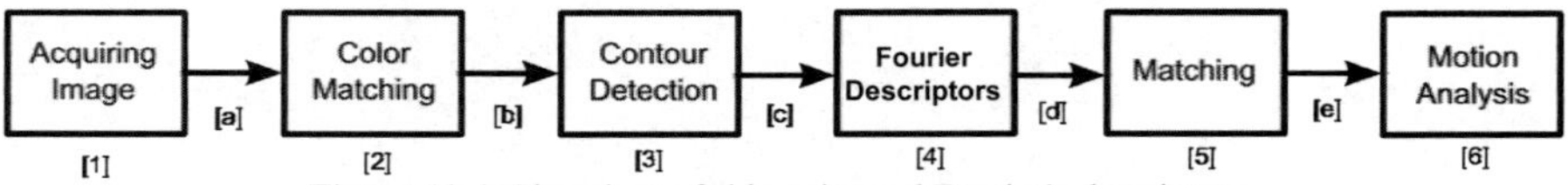

Figure 13.4: Flowchart of skin color and Fourier's descriptors.

The skin match is performed using the hue channel from hue saturation value (HSV) color space. We first match the pixels with similar hue as in the sample. Then we retrieve contours from the binary mask obtained and discard very small contours to avoid false positives. We only take the upper part of the binary object so that any exposed part of the arm will not interfere. We set a maximum ratio for the height of the object on the width of the object to 1.1, which corresponds to the normal ratio for the palm of a hand. Figure 13.5 presents the result of the skin color extraction.

We first distribute the contour's points at regular intervals along the contour, and then we transform the resulting vector of points in the frequency domain using Fourier's descriptors. Equation (13.2) shows the computation of the Fourier's descriptors.

$$FD(\omega) = \sum_{k=0}^{N-1} \left(Pt_k \cdot x + j \times Pt_k \cdot y \right) e^{2\pi j \left(\frac{\omega k}{N} \right)} \tag{13.2}$$

where Pt_k is the kth point of the contour, N is the number of points in the contour, and ω is the frequency.

We set the first coefficient of Fourier's descriptors to zero to have a translation invariant pattern and we divide each coefficient by the second coefficient so that our pattern becomes scale invariant. We apply a low-pass filter and calculate the distance between the contour and a reference and then perform the inverse transform. We use (13.3) to compute the distance between the reference and a contour retrieved from the image.

$$\text{Dist} = \sum_{k=0}^{N=1} \sqrt{\left(Pt_k \cdot x - \text{Ref}_k \cdot x\right)^2 + \left(Pt_k \cdot y - \text{Ref}_k \cdot y\right)^2} \tag{13.3}$$

where Ref_k is the kth point of the reference contour.

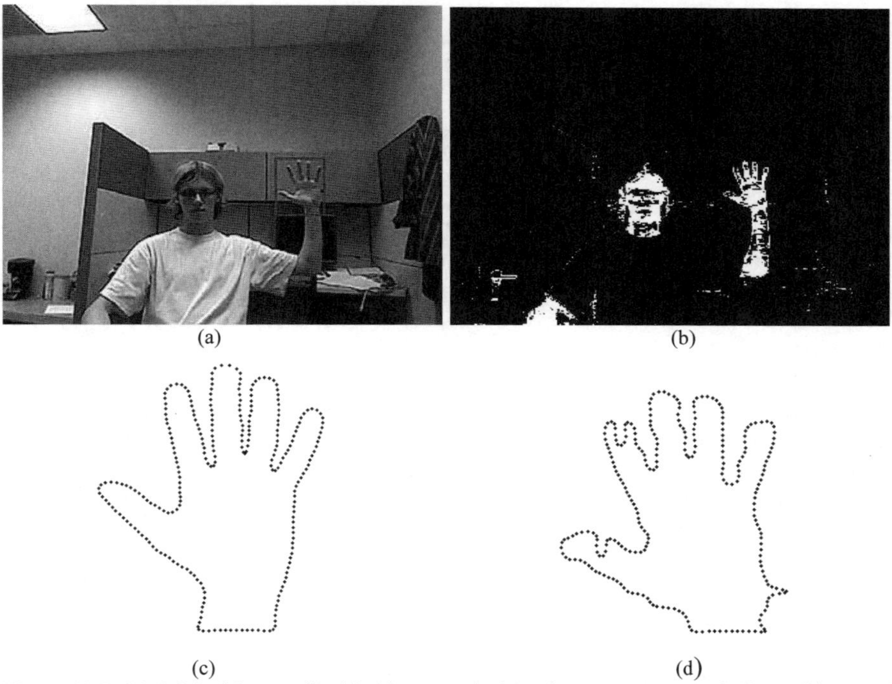

Figure 13.5: (a) Original image, (b) skin binary mask, (c) reference contour, and (d) resulting hand contour from binary mask.

Figure 13.6 shows the processing time and memory usage for each step. We observe that most of the processing time and memory are used at the beginning of the image processing to create the binary mask. Therefore, during gesture matching, we need very little time to process each pattern. This is why we are able process as many references as the user defines. Unlike the pattern matching method, we do not need to do scaling thanks to the Fourier's descriptors properties.

The drawback of color matching is the poor accuracy of the hue coefficient in very bright or dark scenes. In our experiment, we employ electrical lights to light the area but in the presence of strong daylight, we observe some deterioration of our results. There are potential problems if the background has exactly the same color as the skin, but this is not likely to happen in typical home settings.

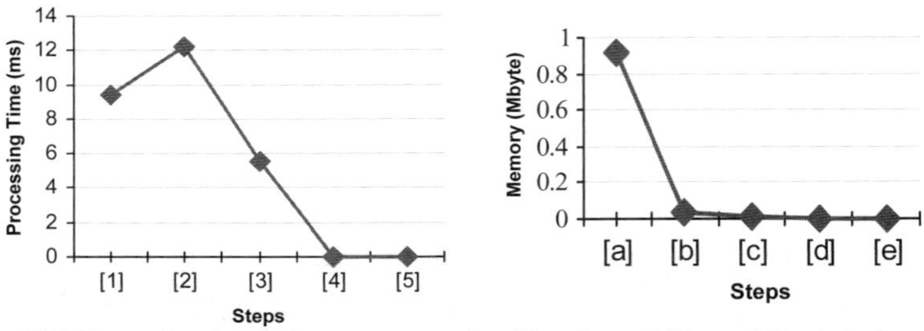

Figure 13.6: Processing time and memory usage for skin color matching and Fourier's descriptors.

13.3 Using H.264 Motion Vectors for Motion Tracking

In this section, we employ an H.264 webcam so that the hardware device can perform the computing intensive part of the video compression. The rest of the processing (e.g., movement recognition) can be performed by a broadband-enabled HDTV, a PC, or STB. As described in Chapter 3, H.264 creates MVs of macroblocks (MBs) for video coding purposes. A related method for object detection using H.264 videos is reported in [11]. The authors in [11] estimate the global MV to discard local motion generated by a moving camera. Then, they use motion history of images (MHI) to identify and track objects, and object history of images (OHI) to estimate object trajectories contained in the video sequence. This method was designed to be very generic. It can handle several moving objects but no information is extracted to characterize the nature of the object. In this section, we show how this can be done.

We wish to detect the hand motion trajectory with a static webcam. Global motion estimation is not needed in this case but we need to make assumptions on the hand characteristics such as the size and speed of movement. We also wish to distinguish the hand from the rest of the body, which can also move. Figure 13.7 presents the flowchart of this method.

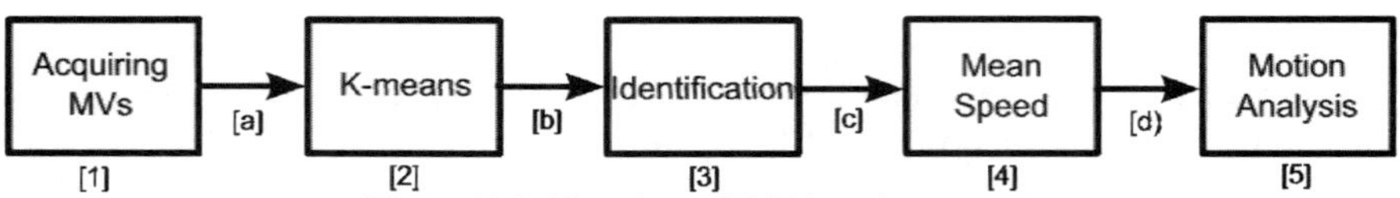

Figure 13.7: Flowchart of H.264 motion vectors.

In this method, we first extract the MVs in the coded video. Then, we isolate the different moving objects. We cluster the MVs by range and direction using the k-means method. This algorithm aims to partition a data set by minimizing the magnitude of the vector projected from the center of the cluster in (13.4).

$$\sum_{i=0}^{N}\sum_{j=0}^{M_i}\left|\overrightarrow{Pt_jC_i}\right| \quad (13.4)$$

where N is the number of cluster, M_i the number of MVs in cluster i, C_i the center of cluster i, and Pt_j the jth point from cluster i.

We divide the MVs into five clusters and proceed with five iterations. Figure 13.8 presents the result of k-means in the speed and space domains. This gives five MV clusters, which represent five objects moving in different directions at different speeds. We use a median filter on these objects to suppress isolated vectors. As the hand and the arm should move in the same direction, we calculate the spatial dispersion of each cluster of vectors. To identify the hand, we set a threshold for the size and the dispersion of the clusters. The hand should have a minimum size to be detected and big objects or dispersed objects can be reasonably rejected as potential hands. Then, if several clusters remain, we choose the one with the highest speed, which is the most likely to be the hand.

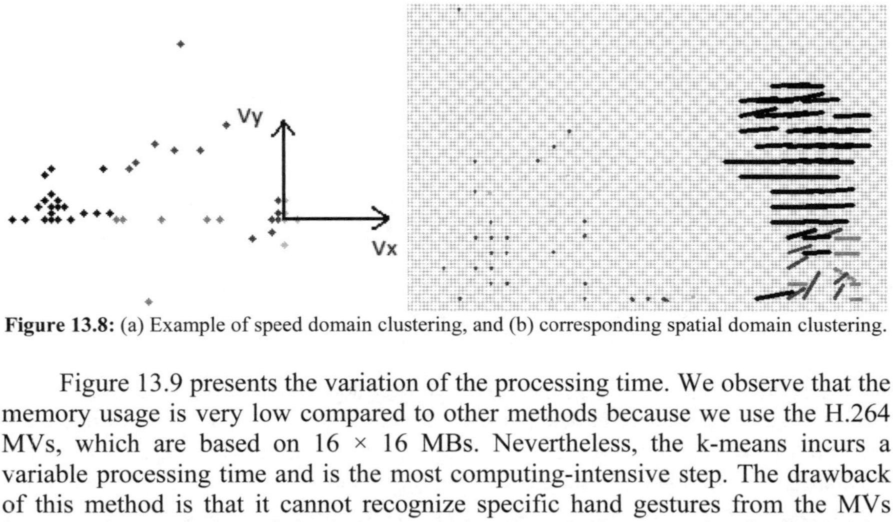

Figure 13.8: (a) Example of speed domain clustering, and (b) corresponding spatial domain clustering.

Figure 13.9 presents the variation of the processing time. We observe that the memory usage is very low compared to other methods because we use the H.264 MVs, which are based on 16 × 16 MBs. Nevertheless, the k-means incurs a variable processing time and is the most computing-intensive step. The drawback of this method is that it cannot recognize specific hand gestures from the MVs because the resolution of the image resulting from MVs extraction is divided by 16, the size of the MBs. Note that a moving background may also interfere with the hand trajectory recognition. If, for example, another person is moving in the same direction as the hand, the k-means will merge the hand and the person and we will lose track of the hand. However, it directly provides the speed of the hand by averaging the value of the MVs, which is useful for gesture recognition.

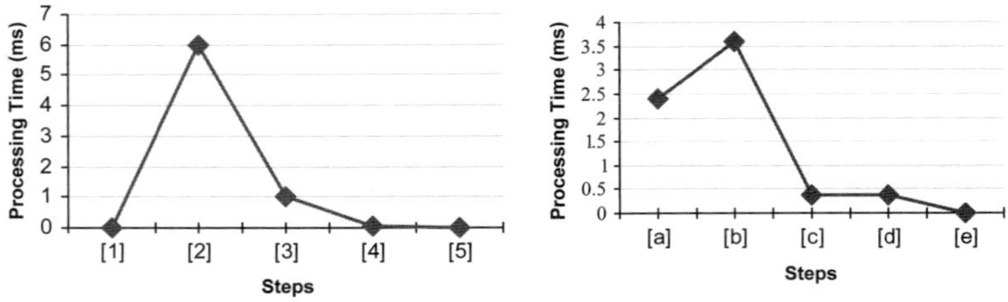

Figure 13.9: Processing time and memory usage for H.264 motion vectors.

13.3.1 Histogram Matching for Trajectory Recognition

To compare our methods, we tested several motion trajectories. For each trajectory, we compute the number of correct detections and false positives. We classify our trajectories by gesture type, including the hand reference, which is either the index finger and the thumb pointed or the palm with five fingers spread out, as shown in Figures 13.2(a) and (b). We also distinguish whether the gesture is horizontal or vertical. We compute the number of false positives as the total number of mistakes made using these sequences. Tables 13.2 and 13.3 present the results of these tests.

Table 13.2: Detection Accuracy for Each Method

Gesture	Motion	Color	H.264
Palm horizontal	41%	87%	87%
Finger horizontal	43%	79%	62%
Palm vertical	6%	93%	61%
Finger vertical	2%	65%	61%

Table 13.3: Total Number of False Positives

	Motion	Color	H.264
False positives	0	3	NA

The index finger reference is generally harder to detect because it has a less concrete shape and less edges than the hand. We observe that motion pattern matching provides very poor results with vertical motion. There are two reasons for that. First, our hand references have stronger vertical edges than the horizontal edges. As frame differencing emphasizes edges that are perpendicular to the motion, horizontal motion makes hands easier to detect. In addition, we use horizontal motion to create our references that are more likely to match a similar motion. The skin color matching achieves good results on all hand sequences but a few false positives are caused by variation of the skin color that can sometimes delete a finger in the contour. The H.264 method performs better for horizontal trajectories because the arm moves less for such gestures and therefore creates less interference. Despite the lower resolution of the MVs, this method performs almost as well as skin color for horizontal gestures.

We use two trajectories to test for false positives and the results are shown in Table 13.4. The first row corresponds to some random activity that does not include hand gestures. In the second row, hand gestures with postures different from the "palm" and "finger" are used. Thus, there should not be any matching gestures. The skin color method presents a higher number of false positives in both cases because of the variation of the hue for skin pixels. The H.264 method lacks precision because no shape matching is done and it detects many small objects moving. However, as we will see in the next section, it can compensate for this using histogram matching.

Table 13.4: Number of False Positives for Each Method, Sequence, and Posture

Palm			Finger		
Motion	Color	H.264	Motion	Color	H.264
9	10	81	12	14	NA
0	6	NA	4	15	NA

We now test our trajectory recognition algorithm using the histogram matching method. In [2], time segmentation is done by tracking the hand gesture at the start of the motion and at the end. However, because we are now using H.264 MVs, segmentation is performed manually as follows. We first compute the speed and allocate the value to one of the eight directions presented in Figure 13.10. We store this value in a histogram. We then compare our histogram to the reference histograms in Figure 13.11.

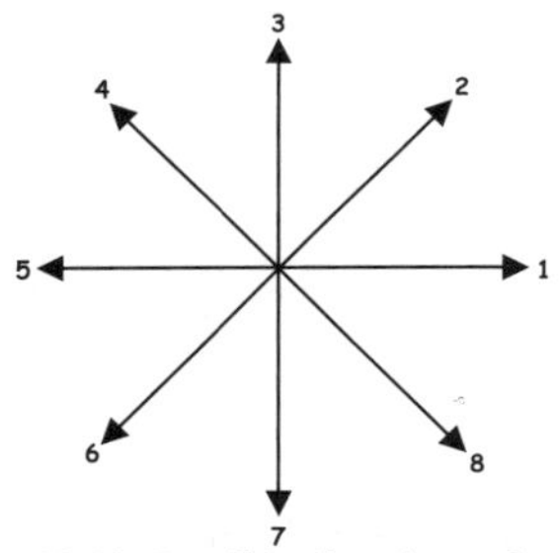

Figure 13.10: Possible directions of vectors.

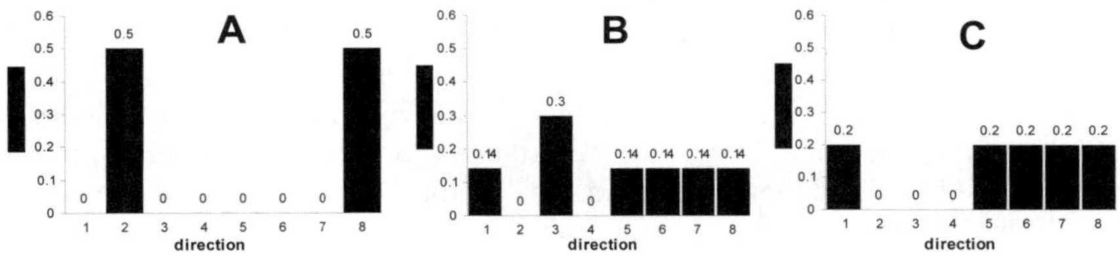

Figure 13.11: Histogram examples for A, B, and C.

Table 13.5 shows the results for the gesture recognition of five trajectories, with some trajectories representing letters A, B, and C. We observe that the motion method is not very efficient because it misses most of the vertical motion. The color method and the H.264 MV method show good results. The excellent results of the H.264 method can be attributed to the accuracy of the speed calculation as described in the previous paragraph.

Table 13.5: Gesture Recognition Accuracy for Each Gesture

	Motion	**Color**	**H.264**
Right	100%	100%	100%
Up	0%	100%	100%
A	27%	100%	100%
B	78%	78%	100%
C	38%	92%	100%

13.4 Hand Tracking for Mouse Cursor Control

The hand-tracking algorithm is designed to be simple, efficient and fast so that it can be applied to real-time mouse cursor control. The algorithm is based on the detection of motion and skin color. No hand reference is needed. Motion is indicated by the change in the pixel values. The frames are first converted into grayscale images. Then, a frame-differencing algorithm is used to analyze the region where the movement has taken place. Equation (13.5) gives the image that has nonzero pixel values in the regions where motion has taken place. A thresholding algorithm based on (13.6) gives a binary image with white pixels indicating the region of motion. The threshold of 30 is chosen by evaluating the frame difference under different lighting conditions and this value gives sufficient white pixels to track the location of the hand.

$$F_{diff}(x,y) = F_{current}(x,y) - F_{previous}(x,y) \qquad (13.5)$$

$$F(x,y) = \begin{cases} 0; & \text{if } F_{diff}(x,y) < 30 \\ 255; & \text{if } F_{diff}(x,y) \geq 30 \end{cases} \qquad (13.6)$$

It is assumed that the only moving object performing the gesture is the hand. Since the video is captured from a regular webcam, random camera noise may cause large variations in the pixel values in successive frames. These variations result in white pixels in the thresholded images. Figure 13.12 illustrates the nonzero pixel values of a stationary user.

These unwanted white pixels can be removed by applying the morphological operator of erosion twice on the captured frame. The erosion operator reduces the area of interest, and many subsequent dilation operations are applied to compensate for this reduced area and also to increase the area of interest. The erosion operator removes the pixels at the borders of objects in the image it is applied to, which shrinks the objects [13]. The dilation operator adds pixels to the borders, which make the objects larger. The flowchart for the above steps is shown in the Figure 13.13. In order to increase the reliability of detecting the area of motion, an adaptive skin color detector based on hue thresholds is used [14]. The detector employs a skin classifier based on the hue histogram of skin color pixels. The limitation of this skin detector is that in the presence of very high-intensity lighting conditions, the background is included in the output of the filter. A flowchart of the adaptive skin color detector is shown in Figure 13.13. Motion detection is used to find the new global thresholds required for filtering the image. The flowchart for obtaining a region representing the location of the hand is shown in Figure 13.14. In Figure 13.15(a), the image after frame differencing contains a lot of irrelevant information that is removed after erosion and dilation, as shown in Figure 13.15(c). The combination of motion detection and the skin detector gives an approximate region of the hand, as shown in Figure 13.15(d).

Figure 13.12: Image of a stationary subject after frame differencing and thresholding.

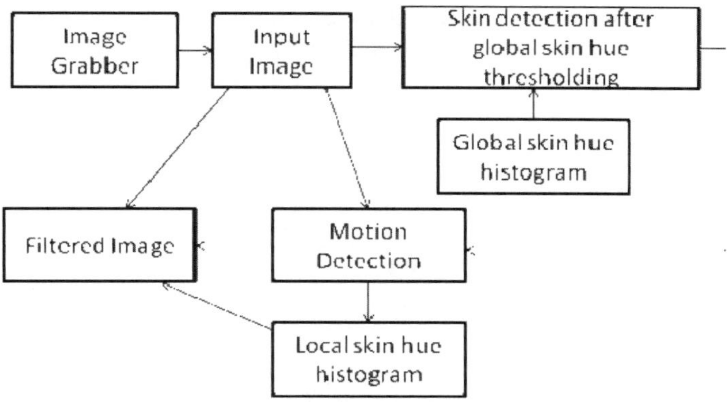

Figure 13.13: Overview of adaptive skin detector.

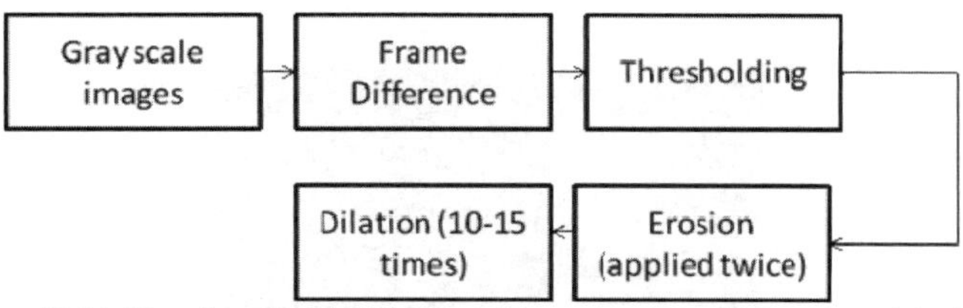

Figure 13.14: Flowchart for obtaining a region representing the location of the hand.

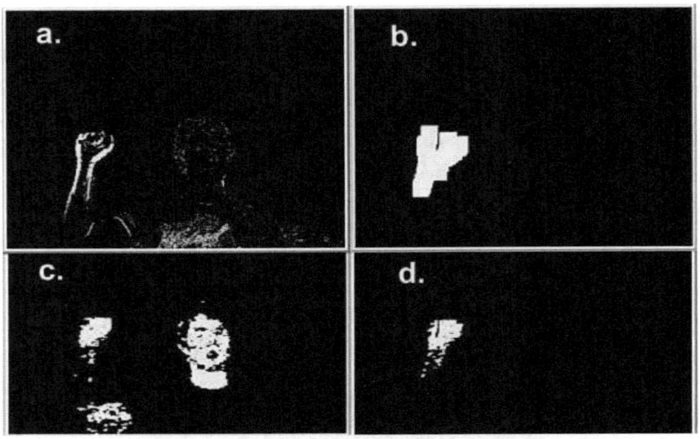

Figure 13.15: (a) Frame difference image, (b) after dilation and erosion, (c) skin detector, and (d) combination of skin detection and motion.

The detected area may be larger than expected. This can be due to the movement of an exposed arm. To overcome this, the following algorithm is used to find the center of the hand region. First, the entire image for each frame is scanned and the boundary of the detected motion region is noted. Let the coordinates of the boundary be (x_1, y_1), (x_1, y_2), (x_2, y_1), (x_2, y_2). The center of gravity is found by (13.7).

$$x_{cog} = \frac{1}{S}\sum_{x=x_1}^{x_2} x$$

$$y_{cog} = \frac{1}{S}\sum_{y=y_1+\frac{3(y_2-y_1)}{4}}^{y_2} y \tag{13.7}$$

where S = total number of white pixels in the marked region shown in Figure 13.16.

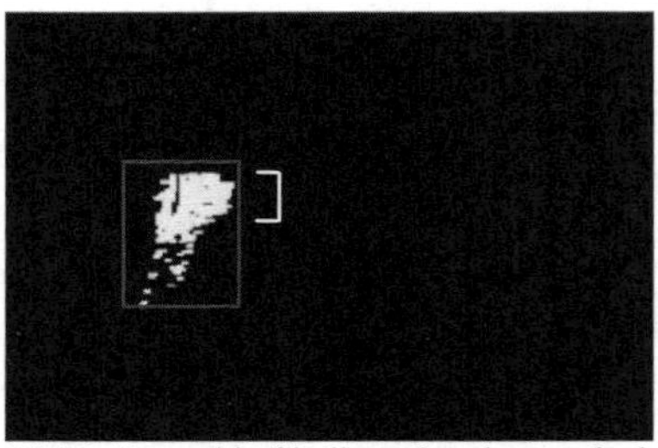

Figure 13.16: Centroid of indicated region tracks location.

Only the centroid of the top portion of this region is calculated to track the location of the hand. In order to avoid detecting small head movements and small random movements, a suitable threshold count is kept for detecting motion. If the number of pixels indicating motion is below this threshold, the detection is simply ignored and the previous tracked location is retained, as in (13.8). By testing the system with slight head movements, the pixels affected by these movements came in the range of 1,000 to 1,500 pixels (for a 640x480 video capture). Thus, an approximate threshold of $0.004S_{\text{total}}$ is chosen to avoid detecting these movements.

$$C = \begin{cases} x_{new}, y_{new}; & \text{if } p > 0.004S_{total} \\ x_{prev}, y_{prev}; & \text{if } p \le 0.004S_{total} \end{cases} \tag{13.8}$$

where

p = Number of detected white pixels after performing procedures in Figure 13.13.

S_{total} = Total number of pixels captured by the camera (28,800 pixels for a 640 x 480 video).

C = Coordinates of current tracked location.

13.4.1 Trajectory Formation Using Motion History

Evaluating the motion history in the video frame leads to the direction of movement. The motion gradient of successive frames over a predefined interval indicates the direction of motion. The trajectory of motion is drawn based on MHI. History of the difference images forms a silhouette. The gradient of this silhouette is updated with every frame. The amount of time for which the previous images stay in the silhouette is 0.3s. The frame capture rate in webcams normally varies from 20 to 30 Hz. Thus, an approximate time of 0.3s will capture between 6 to 9 frames. Any pixel that is older than this timestamp duration is set to 0. The motion gradient is found by applying a 3 × 3 Sobel operator on the MHI silhouette [10]. This procedure gives a mask in which each nonzero value of the image indicates motion in a specific direction [15]. Using these motion gradient values, a global MV is computed, as shown in Figure 13.17.

13.4.2 Using the Global Motion Vector to Track Trajectory

The angle of the global MV is used to track the trajectory of motion. The tracked trajectory of the hand relates to commands and gestures only if the hand motion is relatively stable (slow and consistent). The stability of the tracked position is determined by finding the standard deviation of this tracked position (σ_x and σ_y). The first point of the trajectory is drawn using a stable tracked position. The subsequent points of the trajectory are drawn using the detected angle in each frame and a fixed length $l = 3$ pixels.

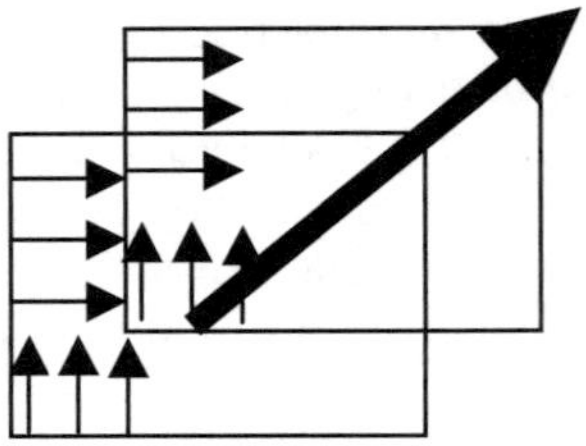

Figure 13.17: Global vector.

The trajectory is updated if both conditions in (13.9) are satisfied.

$$T_{update} \text{ if } \begin{cases} \sigma_x < 0.07W \text{ pixels and } \sigma_y < 0.09H \text{ pixels} \\ \text{Count of } \left(\text{MHI}_{\text{silhoutte}}\right) > 0.002 S_{total} \end{cases} \quad (13.9)$$

where W and H are the width and height of the captured video.

Then the trajectory coordinates are given by

$$\begin{aligned} t_{x,new} &= t_{x,prev} + l\cos\phi \\ t_{y,new} &= t_{y,prev} + l\sin\phi \end{aligned} \quad (13.10)$$

where
$l = 3$ pixels.
S_{total} = Total number of pixels captured by the camera.
ϕ = Angle of global MV.

These conditions are needed to avoid drawing the trajectory when the hand is moving quickly around the screen. Thus, when the tracked position exceeds the standard deviation ($\sigma_x > 0.070W$ and $\sigma_y > 0.093H$), the captured trajectory is purged. In this case, the length of the trajectory is set to zero and all points stored in trajectory array are purged.

Figure 13.18 shows two of the four gestures that were implemented. Fast motion of the hand is used for scrolling around the screen and slow motion in a particular direction causes the formation of a trajectory. The left, right, up, and down gestures are identified when the trajectory length reaches $0.08W$ of capture in the desired direction. This value minimizes the false alarms when the user is just scrolling about the screen. The trajectory formed can also be extended to recognize other gestures such as letters or shapes.

13.4.3 Scrolling When User is Located at Varying Distances

In sufficiently lit environments, the above system can be used when users are at distances of about 3m. In order to use the system in situations when the user is far away from the point of capture, it is necessary to scale the motion accordingly on the screen. This can be done by first locating the region of the motion using the following conditions:

- Motion for at least 150 captured frames.
- No motion for 40 frames after that.

The first condition is needed to ensure there is enough motion (above the threshold of $0.002S_{total}$) to detect a user in front of the camera. The second condition is needed to ensure that the position is stable and an identifiable region of interest can be extracted. As shown in Figure 13.19, since the user is standing far away from the point of capture, he can perform motion within a limited region. Figure 13.19(c) shows the maximum movement of the user's right hand in that region. The maximum movement is recorded in every frame and the boundary of the region of interest is expanded if the user exceeds the previous recorded maximum. Any motion within the extracted box is scaled to the whole screen. This provides effective scaling of the movement throughout the screen even if the user is not standing close to the webcam.

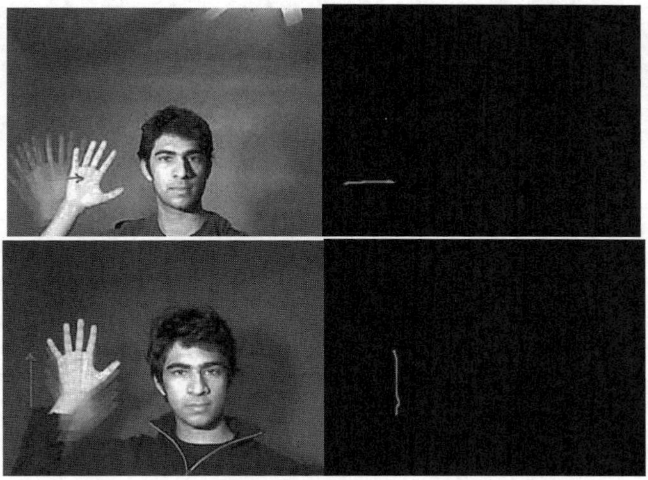

Figure 13.18: Drawn trajectory of left-moving and up-moving gestures.

Figure 13.19: Extracting the region of interest.

13.4.4 Experimental Setup

The system was tested using Visual Studio 2008 and OpenCV 2.0 platform on a Pentium 4 Core 2 Duo 2.1 GHz system. The motion of the hand was used to move the mouse cursor. The two gestures of slow movement to the left and to the right were assigned to left-click and right-click, respectively. The system has an accuracy of about 70% in recognizing these four gestures. Incorrect gestures are recognized when the user does not perform them correctly. For example, when the intended gesture is slowly moving in the up direction but the user unintentionally moves in the top-right direction, this triggers the recognition of up and right gestures instead of the up gesture. Figure 13.20 shows an example of our system where the white box indicates the tracked location. The left selection is activated due to the slow movement of the user's hand in the left direction.

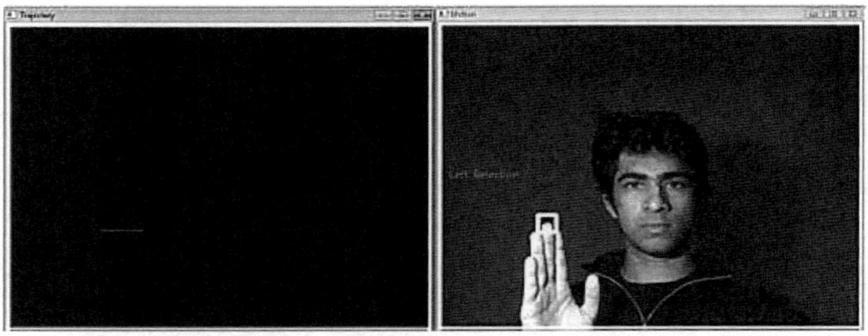

Figure 13.20: Left selection performed by slow hand motion to the left direction.

13.4.5 Comparison of Trajectory Tracking Methods

The Lucas Kanade (LK) optical flow method [16] can be used to find MVs and form the trajectory. The LK method turns out to be slower and less accurate than the MHI based trajectory formation. This is because the performance of the LK method highly depends on stable values of the pixels surrounding the tracked point. With fast movement of the hand and variations in the brightness of the video captured by the webcam, the original tracked position can be easily displaced, giving an incorrect angle of motion estimate.

The speed of the LK algorithm was tested to form the trajectory with each video frame being processed at 110 ms (captured video resolution is 640 × 480). The MHI algorithm took 65 ms in comparison. The slowness of the LK can be attributed to the fact that it is necessary to obtain the good features to track and then estimate the angle according to the tracked features. A comparison of two camera resolutions is shown in Table 13.6.

Table 13.6: Performance of MHI and LK

Resolution	**MHI (ms)**	**LK (ms)**
640 × 480	65	110
320 × 240	45	75

13.5 Hand Reference Extraction Using a Stereo 3D Webcam

A stereo 3D webcam consists of two webcams placed side by side, separated by approximately the distance between two eyes. With the recent progress of 3D display systems, these webcams are likely to become widespread in the future. From the two images generated by the webcam, a stereo matching algorithm can be used to extract a depth map of the viewed scene. From this depth map, we can estimate the relative distance of different objects from the camera. As we want to extract a hand reference, a simple way to segment it is to have the user reach out his hand in front of the camera. In this configuration, the hand is located in the first plane of the webcam image, the user in the second plane, and the background in the next plane. To segment the hand from this configuration, we employ the Otsu's thresholding method [17], which allows the depth map to be divided in two regions: the first plane where the hand is located and the other planes. However, the binary mask obtained from the previous step contains a lot of noise generated by the stereo matching method. To overcome this, we successively erode and dilate the image several times to reduce the noise area relative to the object area representing the hand. To estimate the position of the hand, we use the centroid of the binary mask. To estimate its size, we use a value proportional to the second order moments of the binary mask. From these parameters, the region surrounding the hand can be extracted.

Since the stereo matching algorithm gives a noisy depth map, it cannot be used to segment the hand region. A background model is used instead and this gives a more precise and accurate spatial segmentation. The background model needs to be established before the user starts to show his hand in front of the camera. To do this, the user stands in front of the webcam without showing his hand for a few seconds, and then brings out his hand to be recorded. The chosen background modeling technique is the Mixture of Gaussians [18] because it trains quickly and gives accurate results for this simple use. Once the background model is established, the algorithm attempts to identify any object in the first plane. To ensure that the identified object is the hand, we use two criteria. We verify that the object size is larger than the noise but smaller than half of the frame. The second criterion is that the object is static. We can then check the hand position for 30 consecutive frames. The steps to extract the hand are summarized in Figure 13.21.

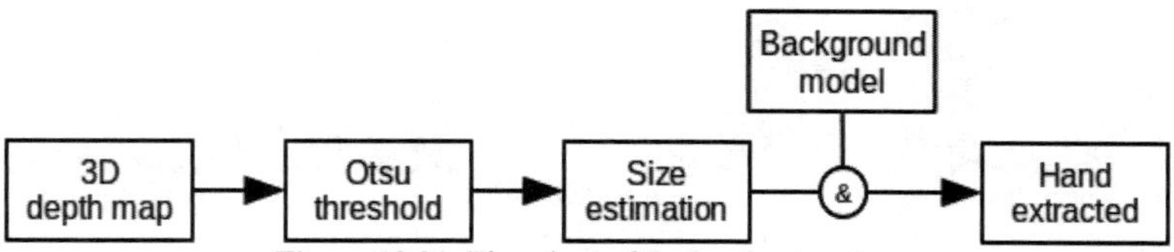

Figure 13.21: Flowchart of the hand extraction.

We have been able to successfully extract hand gestures from various scenes, light environments, and users. Figure 13.22(a) shows an example of a depth map generated by the stereo matching algorithm. The grayscale areas correspond to the relative depth of the object. The light areas are closest to the webcam whereas the dark areas are further back. The black area represents the zone where no reliable depth estimate is found and thus, no depth information is produced. This usually happens with mono colored walls. Figure 13.22(b) shows the image after Otsu thresholding and noise filtering are applied. The user's hand is correctly segmented between the first plane and the user/background planes.

(a) Depth map (b) Filtered image

Figure 13.22: Depth map segmentation

Figure 13.23 presents the results of the hand extraction steps. Figure 13.23(a) shows the region extracted using the centroid and the moments of Figure 13.22(b). The area selected is correctly fitted to the hand size. Figure 13.23(b) shows the binary mask generated by the background model on the region. The white pixels represent the foreground and the black pixels represent the background. It can be noted that the image includes some noise at the bottom of the hand but overall the hand is correctly segmented. Figure 13.23(c) illustrates the superposition of the two previous images (i.e., the extracted and segmented hand). From the segmented hand, several parameters can be extracted such as color distribution, hand shape, hand gesture, or ambient lighting.

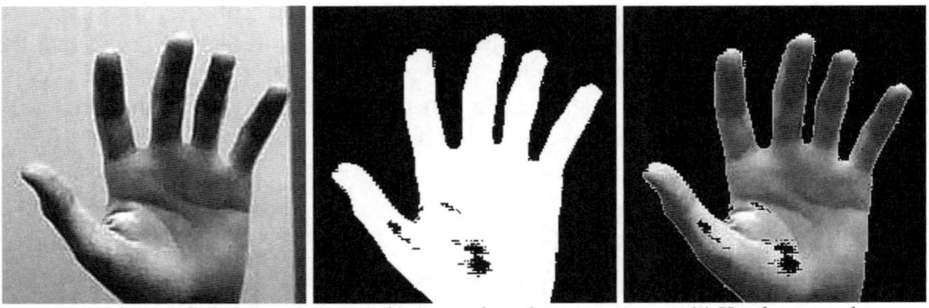

(a) Background model (b) Area selected (c) Hand extracted

Figure 13.23: Hand extraction

13.6 Conclusions

We have presented several methods for hand gesture and trajectory recognition for online video browsing and mouse cursor control. The first is based on a combination of frame differencing and pattern matching, which achieves average detection performance. The second is based on skin color extraction and Fourier's descriptors, which achieves good detection performance but is susceptible to false positives. Another method employs MVs from H.264 videos. This method achieves good performance and can be used to recognize motion trajectories representing numbers and alphabets accurately, but cannot recognize hand specific gestures. The trajectory-based gesture recognition system is shown to be fast and reliable for real-time mouse cursor control applications. By incorporating further trajectory analysis techniques, this system can be effectively used for other multimedia applications. We have also illustrated the use of a 3D webcam to spatially segment and extract a hand reference.

References

[1] G. Saman, A. Sajjad, A. Hameed, and A. Shah, "Visual Sign Language Interpretation Using Spatial Temporal Neural Processing," *IEEE Multitopic Conference (INMIC)*, 2006.

[2] J. Hongmo, K. Jiman, and K. Daijin, "Hand Gesture Recognition to Understand Musical Conducting Action," *16th IEEE International Symposium on Robot and Human interactive Communication*, 2007.

[3] U. Rokade, D. Doye, and M. Kokare, "Hand Gesture Recognition Using Object Based Key Frame Selection," *International Conference on Digital Image Processing*, 2009.

[4] J. Alon, V. Athitsos, Y. Quan, and S. Sclaroff, "A Unified Framework for Gesture Recognition and Spatiotemporal Gesture Segmentation," *IEEE Transactions on Pattern Analysis and Machine Intelligence*, Vol. 31, No. 9, pp. 1685–1699, 2009.

[5] A. Just and S. Marcel, "A Comparative Study of Two State-of-the-art Sequence Processing Techniques for Hand Gesture Recognition," *Computer Vision Image Understanding*, Vol. 113, No. 4, pp. 532–543, 2009.

[6] A. Mitome, and R. Ishii, "A Comparison of Hand Shape Recognition Algorithms," *29th Annual Conference of the IEEE Industrial Electronics Society*, Vol. 3, pp. 2261– 2265, November 2003.

[7] W. Grimson, et al., "Using Adaptive Tracking to Classify and Monitor Activities in a Site," *IEEE Conference on Computer Vision and Pattern Recognition*, 1998.

[8] K. Toyama, J. Krumm, B. Brumitt, and B. Meyers, "Wallflower: Principles and Practice of Background Maintenance," *IEEE International Conference on Computer Vision,* Vol. 1, pp. 255–226, 1999.

[9] P. Viola and M. Jones, "Rapid Object Detection Using a Boosted Cascade of Simple Features," *IEEE Computer Society Conference on Computer Vision and Pattern Recognition*, Vol. 1, pp. I-511–I-518, 2001.

[10] F. Chen, C. Fu, and C. Huang, "Hand Gesture Recognition using a Real-time Tracking Method and Hidden Markov Models," *Image and Vision Computing*, Vol. 21, 2003, pp. 745–758.

[11] C. Kas and H. Nicolas, "An Approach to Trajectory Estimation of Moving Objects in the H.264 Compressed Domain," *3rd Pacific Rim Symposium on Advances in Image and Video Technology*, 2009, pp. 318–329.

[12] G. Bradski and A. Kaehler, *Learning Open CV*, O'Reilly Media, September 2008, pp. 341–345.

[13] M. Nixon and A. Aguado, *Feature Extraction and Image Processing*, Second Edition, Academic Press, 2008, pp. 104–109.

[14] F. Dadgostar and A. Sarrafzadeh, "An Adaptive Real-Time Skin Detector based on Hue Thresholding: A Comparison on Two Motion Tracking Methods," *Pattern Recognition Letters*, Vol. 27, No. 12, September 2006, pp. 1342–1352.

[15] R. Gonzalez and R. Woods, *Digital Image Processing*, Second Edition, Prentice Hall, 2002, pp. 136–137.

[16] B. D. Lucas and T. Kanade "An Iterative Image Registration Technique with an Application to Stereo Vision," *International Joint Conference on Artificial Intelligence*, pp. 674–679, 1981.

[17] Otsu's thresholding method, http://en.wikipedia.org/wiki/Otsu's_method.

[18] Gaussian mixture models, http://www.cs.cmu.edu/afs/cs/Web/People/awm/tutorials/gmm.html.

Glossary

3D	Three Dimensional
3G	Third Generation
3GPP	3G Partnership Project
3G-SDI	3 Gbps SDI
4G	Fourth Generation
AAC	Advanced Audio Coding
ABR	Adaptive Bit Rate
ACK	Acknowledgment
ADTS	Audio Data Transport Stream
AIMD	Additive Increase Multiplicative Decrease
AQ	Adaptive Quantization
ASF	Advanced Systems (Streaming) Format
ASI	Asynchronous Serial Interface
ASM	Any Source Multicast
ASO	Arbitrary Slice Ordering
ASP	Advanced Simple Profile
ATSC	Advanced Television Systems Committee
AU	Access Unit
AVB	Audio Video Bridging
AVC	Advanced Video Coding
AVI	Audio Video Interleave
BDP	Bandwidth-Delay Product
BE	Best Effort
BER	Bit Error Rate
CABAC	Context Adaptive Binary Arithmetic Coding
CAVLC	Context-Adaptive Variable-Length Coding
CATV	Cable TV
CBR	Constant Bit Rate
CDMA	Code Division Multiple Access
CDN	Content Delivery Network
CE	Consumer Electronics
CIF	Common Intermediate Format
CM	Cable Modem
CMTS	Cable Modem Termination System
COPS	Common Open Policy Service
COV	Coefficient of Variability
CPE	Customer Premise Equipment

CRT	Cathode Ray Tube
CTCP	Compound TCP
CTS	Composition Timestamp
DCT	Discrete Cosine Transform
DECE	Digital Entertainment Content Ecosystem
Diffserv	Differentiated Services
DHCP	Dynamic Host Configuration Protocol
DLNA	Digital Living Network Alliance
DOCSIS	Data Over Cable Service Interface Specification
DoS	Denial-of-Service
DP	Data Partitioning
DPI	Deep Packet Inspection
DRM	Digital Rights Management
DS	Downstream
DSG	DOCSIS Set-top Gateway
DSL	Digital Subscriber Line
DSLAM	Digital Subscriber Line Access Multiplexer
DTS	Decoding Timestamp
DTV	Digital TV
DVB	Digital Video Broadcasting
DVD	Digital Video Disk
DVI	Digital Visual Interface
DVR	Digital Video Recorder
EBIF	Enhanced TV Binary Exchange Format
EBU	European Broadcasting Union
EC	Error Concealment
ESG	Electronic Service Guide
FCC	Federal Communications Commission
FD	Fourier Descriptors
FEC	Forward Error Correction
FMO	Flexible Macroblock Ordering
FRExt	Fidelity Range Extension
FTP	File Transfer Protocol
GDR	Gradual Decoding Refresh
GOP	Group of Pictures
HD	High Definition
HDMI	High Definition Multimedia Interface
HD-SDI	High Definition SDI
HE-AAC	High-Efficiency AAC
HFC	Hybrid-Fiber Coax
HMAC	Hash-based Message Authentication Code
HPNA	Home Phoneline Networking Alliance
HRD	Hypothetical Reference Decoder
HSPA	High-Speed Packet Access
HSS	Hypothetical Stream Scheduler
HTTP	Hyper-Text Transfer Protocol

HVS	Human Visual System
Hz	Short for frames per second
IDR	Instantaneous Decoding Refresh
IEC	International Electrotechnical Commission
IEEE	Institute of Electrical and Electronic Engineers
IETF	Internet Engineering Task Force
IGMP	Internet Group Management Protocol
IMS	IP Multimedia Subsystem
Intserv	Integrated Services
IP	Internet Protocol
IPTV	IP Television
IRM	Intelligent Resource Management
ISMA	Internet Streaming and Media Alliance
ISO	International Standardization Organization
ISP	Internet Service Provider
ITU	International Telecommunications Union
ITU-T	ITU Telecommunication Standardization Sector
JM	Joint Model
JPEG	Joint Photographic Experts Group
JVT	Joint Video Team
LAN	Local Area Network
LCD	Liquid Crystal Display
LDAP	Lightweight Directory Access Protocol
LED	Light Emitting Diode
LFE	Low Frequency Effects or Low Frequency Enhancement
LK	Lucas Kanade
LRD	Long-Range Dependency
LTE	Long-Term Evolution
MAP	Bandwidth Allocation Message
MB	Macroblock
MBAFF	MB Adaptive Frame-Field
MBA map	Macroblock Allocation map
MHI	Motion History Images
MIMD	Multiplicative Increase Multiplicative Decrease
MIME	Multipurpose Internet Mail Extension
MKV	Matroska Video
MoCA	Multimedia over Coax
MOS	Mean Opinion Score
MOV	Movie
MP4	MPEG-4
MPEG	Moving Picture Experts Group
MPLS	Multiprotocol Label Switching
MPTS	Multi-Program Transport Stream
MSE	Mean Squared Error
MSO	Multiple Service Operator
MSS	Maximum Segment Size

MTU	Maximum Transmission Unit
MV	Motion Vector
MVC	Multiview Video Coding
MXF	Material eXchange Format
NAL	Network Abstraction Layer
NALU	NAL Unit
NAT	Network Address Translator
NGN	Next Generation Network
NRI	NAL Reference Indicator
NRTPS	Non-Real-Time Polling Service
NTSC	National Television System Committee
OAM	Operational Administration and Maintenance
OC	Optical Carrier
OFDM	Orthogonal Frequency Division Multiplexing
OHI	Object History Images
OLED	Organic LED
OMVC	Open Mobile Video Coalition
OSI	Open System Interconnect
OTN	Optical Transport Network
OTT	Over-the-top
P2P	Peer to Peer
PAR	Peak-to-Average Rate
PBB	Provider Backbone Bridge
PBT	Provider Backbone Transport
PC	Personal Computer
PCMM	PacketCable Multimedia
PDP	Policy Decision Point
Pel	Pixel
PEP	Policy Enforcement Point
PES	Packetized Elementary Stream
PicAFF	Picture Adaptive Frame-Field
PIP	Picture-in-Picture
PLR	Packet Loss Rate
PPS	Picture Parameter Set
PPV	Pay-Per-View
PS	Program Stream
PSNR	Peak Signal to Noise Ratio
QAM	Quadrature Amplitude Modulation
QCIF	Quarter CIF
QoE	Quality of Experience
QoS	Quality of Service
QP	Quantization Parameter
QPSK	Quadrature Phase Shift Keying
QVGA	Quarter VGA
RDO	Rate Distortion Optimization
RGB	Red-Green-Blue

RGC	Request-Grant Cycle
RIA	Rich Internet Application
RM	Real Media
RPE	Relative Percentage Error
RS	Redundant Slices
RSVP	Resource ReSerVation Protocol
RTCP	Real-Time Control Protocol
RTMP	Real-Time Messaging Protocol
RTP	Real-Time Transfer Protocol
RTPS	Real-Time Polling Service
RTSP	Real-Time Transport Streaming Protocol
RTT	Round Trip Time
SACK	Selective Acknowledgement
SAD	Sum of Absolute Differences
SCTP	Stream Control Transmission Protocol
SD	Standard Definition
SDH	Synchronous Digital Hierarchy
SDI	Serial Digital Interface
SDV	Switched Digital Video
SED	Surface-Conduction Electron-Emitter Display
SEI	Supplementary Enhancement Information
SG	Slice Group
SH	Slice Header
SI	Switching Intra
SIM	Security Identity Module
SIP	Session Initiation Protocol
SLA	Service Level Agreement
SMPTE	Society of Motion Picture and Television Engineers
SONET	Synchronous Optical NETwork
SP	Switching Predictive
SPS	Sequence Parameter Set
SPTS	Single Program Transport Stream
SSIM	Structural Similarity
SSM	Source Specific Multicast
STB	Set-Top Box
SUBS	Subscribers
SVG	Scalable Vector Graphics
TCP	Transmission Control Protocol
Telco	Telephone Company
TFT	Tit-for-Tat
TFTP	Trivial File Transfer Protocol
TG	Task Group
TS	Transport Stream
TSP	Transport Stream Packet
TVE	TV Everywhere
UDP	User Datagram Protocol

US	Upstream
UVLC	Universal Variable Length Coding
VBR	Variable Bit Rate
VC-1	Video Codec 1
VCEG	Video Coding Experts Group
VCL	Video Coding Layer
VGA	Video Graphics Array
VHT	Very High Throughput
VLAN	Virtual LAN
VLC	Variable Length Coding
VoD	Video-on-Demand
VoIP	Voice over IP
VPN	Virtual Private Network
VQ	Video Quality
VSB	Vestigial SideBand
VUI	Video Usability Information
WAN	Wide Area Network
Wi-Fi	Wireless Fidelity
WiGig	Wireless Gigabit Alliance
WiMax	Worldwide Interoperability for Microwave Access
WMV	Windows Media Video
Y-PSNR	Luminance PSNR

About the Author

Benny Bing has been a research faculty member with the Georgia Institute of Technology since 2001. He has published over 80 technical papers, 11 books, and 6 pending patents in network and media communications technologies. His publications have appeared in the *IEEE Spectrum* and he has received two best paper awards. In early 2000, his book on wireless LANs was adopted by Cisco Systems to launch Cisco's first wireless product, the Aironet Wi-Fi product. He was subsequently invited by Qualcomm and the Office of Information Technology to conduct customized Wi-Fi courses. Other books were reviewed extensively by *IEEE Communications Magazine* (twice), *IEEE Network* as well as the *ACM Networker*. He is an editor for the *IEEE Wireless Communications Magazine* since 2003, where he also heads the Industry Perspectives section. He has guest edited for the *IEEE Communications Magazine* (two issues) and the *IEEE Journal on Selected Areas on Communications*. All 5 of his online IEEE wireless tutorials have been sponsored by industry with one tutorial sponsored twice. In October 2003, he was invited by the National Science Foundation to participate in a workshop on residential broadband. He also led a team that received the National Association of Broadcasters (NAB) Technology Innovation Award in 2010. The award recognizes outstanding demonstrations of advanced research and development of new media technologies at the NAB Show. He is the founder of a startup focused on enriching and delivering next-generation video entertainment and services. He is a Senior Member of IEEE and an IEEE Communications Society Distinguished Lecturer.

Index

Recent Titles in the Artech House Telecommunications Series

Vinton G. Cerf, Senior Series Editor

Access Networks: Technology and V5 Interfacing, Alex Gillespie

Achieving Global Information Networking, Eve L. Varma et al.

Advanced High-Frequency Radio Communications, Eric E. Johnson et al.

ATM Interworking in Broadband Wireless Applications, M. Sreetharan and S. Subramaniam

ATM Switches, Edwin R. Coover

ATM Switching Systems, Thomas M. Chen and Stephen S. Liu

Broadband Access Technology, Interfaces, and Management, Alex Gillespie

Broadband Local Loops for High-Speed Internet Access, Maurice Gagnaire

Broadband Networking: ATM, SDH, and SONET, Mike Sexton and Andy Reid

Broadband Telecommunications Technology, Second Edition, Byeong Lee, Minho Kang, and Jonghee Lee

The Business Case for Web-Based Training, Tammy Whalen and David Wright

The Business Privacy Law Handbook, Charles H. Kennedy

Centrex or PBX: The Impact of IP, John R. Abrahams and Mauro Lollo

Chinese Telecommunications Policy, Xu Yan and Douglas Pitt

Communication and Computing for Distributed Multimedia Systems, Guojun Lu

Communications Technology Guide for Business, Richard Downey, Seán Boland, and Phillip Walsh

Community Networks: Lessons from Blacksburg, Virginia, Second Edition, Andrew M. Cohill and Andrea Kavanaugh, editors

Component-Based Network System Engineering, Mark Norris, Rob Davis, and Alan Pengelly

Computer Telephony Integration, Second Edition, Rob Walters

Customer-Centered Telecommunications Services Marketing, Karen G. Strouse

Delay- and Disruption-Tolerant Networking, Stephen Farrell and Vinny Cahill

Deploying and Managing IP over WDM Networks, Joan Serrat and Alex Galis, editors

Desktop Encyclopedia of the Internet, Nathan J. Muller

Digital Clocks for Synchronization and Communications, Masami Kihara, Sadayasu Ono, and Pekka Eskelinen

Digital Modulation Techniques, Second Edition, Fuqin Xiong

Disaster Recovery Planning for Communications and Critical Infrastructure, Leo A. Wrobel and Sharon M. Wrobel

E-Commerce Systems Architecture and Applications, Wasim E. Rajput

EMI Protection for Communication Systems, Kresimir Malaric

Engineering Internet QoS, Sanjay Jha and Mahbub Hassan

Error-Control Block Codes for Communications Engineers, L. H. Charles Lee

Essentials of Modern Telecommunications Systems, Nihal Kularatna and Dileeka Dias

FAX: Facsimile Technology and Systems, Third Edition, Kenneth R. McConnell, Dennis Bodson, and Stephen Urban

Fundamentals of Network Security, John E. Canavan

Gigabit Ethernet Technology and Applications, Mark Norris

The Great Telecom Meltdown, Fred R. Goldstein

Guide to ATM Systems and Technology, Mohammad A. Rahman

A Guide to the TCP/IP Protocol Suite, Floyd Wilder

Home Networking Technologies and Standards, Theodore B. Zahariadis

Implementing Value-Added Telecom Services, Johan Zuidweg

Information Superhighways Revisited: The Economics of Multimedia, Bruce Egan

Installation and Maintenance of SDH/SONET, ATM, xDSL, and Synchronization Networks, José M. Caballero et al.

Integrated Broadband Networks: TCP/IP, ATM, SDH/SONET, and WDM/Optics, Byeong Gi Lee and Woojune Kim

Internet E-mail: Protocols, Standards, and Implementation, Lawrence Hughes

Introduction to Telecommunications Network Engineering, Second Edition, Tarmo Anttalainen

Introduction to Telephones and Telephone Systems, Third Edition, A. Michael Noll

An Introduction to U.S. Telecommunications Law, Second Edition, Charles H. Kennedy

IP Convergence: The Next Revolution in Telecommunications, Nathan J. Muller

LANs to WANs: The Complete Management Guide, Nathan J. Muller

The Law and Regulation of Telecommunications Carriers, Henk Brands and Evan T. Leo

Litigating with Electronically Stored Information, Marian K. Riedy, Susman Beros and Kim Sperduto

Managing Internet-Driven Change in International Telecommunications, Rob Frieden

Marketing Telecommunications Services: New Approaches for a Changing Environment, Karen G. Strouse

Mission-Critical Network Planning, Matthew Liotine

Multimedia Communications Networks: Technologies and Services, Mallikarjun Tatipamula and Bhumip Khashnabish, editors

Next Generation Intelligent Networks, Johan Zuidweg

Open Source Software Law, Rod Dixon

Performance Evaluation of Communication Networks, Gary N. Higginbottom

Performance of TCP/IP over ATM Networks, Mahbub Hassan and Mohammed Atiquzzaman

The Physical Layer of Communications Systems, Richard A. Thompson, David Tipper, Prashant Krishnamurthy, and Joseph Kabara

Power Line Communications in Practice, Xavier Carcelle

Practical Guide for Implementing Secure Intranets and Extranets, Kaustubh M. Phaltankar

Practical Internet Law for Business, Kurt M. Saunders

Practical Multiservice LANs: ATM and RF Broadband, Ernest O. Tunmann

Principles of Modern Communications Technology, A. Michael Noll

A Professional's Guide to Data Communication in a TCP/IP World, E. Bryan Carne

Programmable Networks for IP Service Deployment, Alex Galis et al., editors

Protocol Management in Computer Networking, Philippe Byrnes

Pulse Code Modulation Systems Design, William N. Waggener

Reorganizing Data and Voice Networks: Communications Resourcing for Corporate Networks, Thomas R. Koehler

Security, Rights, and Liabilities in E-Commerce, Jeffrey H. Matsuura

Service Assurance for Voice over WiFi and 3G Networks, Richard Lau, Ram Khare, and William Y. Chang

Service Level Management for Enterprise Networks, Lundy Lewis

SIP: Understanding the Session Initiation Protocol, Third Edition, Alan B. Johnston

Smart Card Security and Applications, Second Edition, Mike Hendry

SNMP-Based ATM Network Management, Heng Pan

Spectrum Wars: The Policy and Technology Debate, Jennifer A. Manner

Strategic Management in Telecommunications, James K. Shaw

Strategies for Success in the New Telecommunications Marketplace, Karen G. Strouse

Successful Business Strategies Using Telecommunications Services, Martin F. Bartholomew

Telecommunications Cost Management, S. C. Strother

Telecommunications Department Management, Robert A. Gable

Telecommunications Deregulation and the Information Economy, Second Edition, James K. Shaw

Telecommunications Technology Handbook, Second Edition, Daniel Minoli

Telemetry Systems Engineering, Frank Carden, Russell Jedlicka, and Robert Henry

Telephone Switching Systems, Richard A. Thompson

3D and HD Broadband Video Networking, Benny Bing

Understanding Modern Telecommunications and the Information Superhighway, John G. Nellist and Elliott M. Gilbert

Understanding Networking Technology: Concepts, Terms, and Trends, Second Edition, Mark Norris

For further information on these and other Artech House titles, including previously considered out-of-print books now available through our In-Print-Forever® (IPF®) program, contact:

Artech House
685 Canton Street
Norwood, MA 02062
Phone: 781-769-9750
Fax: 781-769-6334
e-mail: artech@artechhouse.com

Artech House
16 Sussex Street
London SW1V HRW UK
Phone: +44 (0)20 7596-8750
Fax: +44 (0)20 7630-0166
e-mail: artech-uk@artechhouse.com

Find us on the World Wide Web at: www.artechhouse.com